D0993662

Virtual Worlds on the Internet

Virtual Worlds on the Internet

John Vince
Rae Earnshaw

IEEE
COMPUTER
SOCIETY

Los Alamitos, California

Washington • Brussels • Tokyo

Library of Congress Cataloging-in-Publication Data

Virtual worlds on the Internet / [edited by] John Vince, Rae Earnshaw.
    p.   cm.
Includes bibliographical references and index.
ISBN 0-8186-8700-2
1. Internet programming.   2. Virtual reality.   I. Vince, John
(John A.)   II. Earnshaw, Rae A., 1944-
QA76.625.V57   1998
005.2 ' 76—DC21

98-36697
CIP

IEEE Computer Society Press Order Number BP08700
Library of Congress Number 98-36697
ISBN 0-8186-8700-2

Additional copies may be ordered from:

IEEE Computer Society Press	IEEE Service Center	IEEE Computer Society
Customer Service Center	445 Hoes Lane	Watanabe Building
10662 Los Vaqueros Circle	P.O. Box 1331	1-4-2 Minami-Aoyama
P.O. Box 3014	Piscataway, NJ 08855-1331	Minato-ku, Tokyo 107-0062
Los Alamitos, CA 90720-1314	Tel: +1-732-981-0060	JAPAN
Tel: +1-714-821-8380	Fax: +1-732-981-9667	Tel: +81-3-3408-3118
Fax: +1-714-821-4641	mis.custserv@computer.org	Fax: +81-3-3408-3553
cs.books@computer.org		tokyo.ofc@computer.org

Publisher: Matt Loeb
Project Editor: Cheryl Baltes
Advertising/Promotions: Tom Fink
Printed in the United States of America

IEEE
COMPUTER
SOCIETY

IEEE

1197005 7

Contents

Introduction

• •

The nineteenth century was a period when fundamental advances were made in our understanding and exploitation of electromagnetic phenomena. For example, in 1820 the Danish scientist H. C. Oersted discovered that a suspended magnetic needle was deflected when a wire carrying an electric current was brought near it.

In 1832, American artist Samuel Morse abandoned his career and returned to studying science—a subject in which he first became interested while at college. In 1837 Morse exhibited in New York a working telegraph instrument that exploited Oersted's electromagnetic effect. Like most inventors, Morse found it difficult to attract government interest in his invention, but in 1844 a telegraph line was established between Washington and Baltimore. The age of electronic communication had begun.

Meanwhile, in the United Kingdom, Charles Babbage was struggling with the technology of brass gears to build computational engines. In 1823 he developed the Difference Engine and in 1833 proposed the Analytical Engine. He, too, experienced great difficulty in securing government support in developing his invention. Nevertheless, the age of computing was about to begin.

The following 150 years was a period of amazing creative invention and discovery. People such as Faraday, Boole, Maxwell, Edison, Hertz, Hollerith, Baird, Turing, Shannon, von Neumann, and Wilkes pioneered a technological revolution that has culminated in today's information society.

Today, digital computers are an integral part of our daily lives. Their ubiquity has become a significant feature of the last years of the twentieth century. So, too, has the telephone and the digital communication networks that allow us to interact with virtually anyone, anywhere in the world.

In recent years, computer graphics has evolved into the major disciplines of computer animation, image processing, visualization, and virtual reality. Such image-based applications require very high bandwidths, which fortunately have been anticipated by developments in fiber optics, satellites, and digital broadcasting systems.

As all of these technologies merge into one seamless digital medium, some very exciting tools are appearing that will transform the way we will work in the next century. In this book, we examine how the latest developments in virtual environments, computer animation, communication networks, and the Internet are being configured to create revolutionary tools and systems.

We begin with the paper "Data Flow Languages for Immersive Virtual Environments," by Anthony Steed. This paper addresses how behavior and interactions should be described when working with immersive virtual environments, particularly those that have a VRML description. The author describes recent research undertaken at University College London into a virtual environment dialogue architecture (VEDA). Underlying VEDA is a data flow model similar to that of the recent VRML 2.0 standard. This paper contrasts the two models and highlights the strengths of both.

The use of VRML is becoming well established for tasks such as data visualization and creating shared virtual environments. In the second paper, "VRML Interfaces to Information Systems," Christine Clark and Adrian Clark explore a different use of VRML, namely, to provide interfaces to information systems. Three particular examples are examined, and underlying all three interfaces is a programmed approach to virtual world creation, which is discussed with reference to the current VRML standard.

Although photo-realism has been a driving force behind many computer graphics research projects, it is very difficult to implement such techniques within a real-time environment. However, Stephen Boyd Davis and Helena Athoussaki question the emphasis on naturalism in virtual environments and argue that "such naturalism is often unthinking and based on a mistaken view of what VR can achieve." In their paper "VRML: A Designer's View," they indicate the advantages of nonnaturalistic approaches, using theatrical design and film as exemplars.

There is no doubt that one of the next major breakthroughs required to extend the human-computer interface is in the field of speech recognition. Considerable effort is being put into developing speech synthesizers, syntactic parsers, semantic and pragmatic analyzers, and dialogue managers. The ultimate goal is to control computers using natural language. In this next paper, Christophe Godéreaux, Pierre-Oliver El Guedj, Frédéric Revolta, and Pierre Nugues describe their research into a spoken dialogue interface in a virtual reality environment. Their paper "*Ulysse:* An Interactive Spoken Dialogue Interface to Navigate in Virtual Worlds" describes their prototype system *Ulysse* and how it has been integrated into a virtual environment system.

Information drilling and interactive data querying are sometimes also referred to as visual data mining and will become a major activity on the Web. Currently, the VRML standard allows users to view and navigate through 3-D information data worlds and hyperlink to new worlds. In his paper "Information Drill-Down Using Web Tools," Mikael Jern explains how HTML's Image Map and VRML's anchor code and multiple predefined viewpoints are used for information drilling. The author predicts that "over the next couple of years, we shall see 3-D visualization technology closely integrated with the data warehouse and multidimensional abstract and geospatial models."

Visualization and conceptualization are two generic processes associated with real world data for education and research. Visualization concerns the comprehension of data sets as a whole, while conceptualization concerns the application of theoretical models to such data sets. In the paper "Generic Uses of Real World Data in Virtual Environments," Roy Middleton and his colleagues M. Wright and G. Watson describe a tool kit for the development of virtual environment applications for education and research. A particular focus of the project is the use of real world data from the natural sciences, and this paper addresses some technical issues in their use. A major focus of the project is the completion of four pilot studies chosen from diverse academic disciplines: geology, genetics, meteorology, and veterinary science.

Multiuser computer systems that employ a common database must preserve the integrity of the database while it evolves within a shared user environment. Now that 3-D virtual environments are being shared by multiple networked users, special techniques are required to support such systems. Tao Lin and Kevin Smith describe in their paper "A Generic Functional Architecture for the Development of Multiuser 3-D Environments" how it is possible to decompose the functionality of a multiuser 3-D modelling system into functional tools and specify the interfaces between them.

Although there are some virtual environments that are static and may be inspected only by interactive navigation, there are many that are dynamic and must be capable of responding to different events. In their paper "Strategies for Mutability in Virtual Environments," Ben Anderson and Andrew McGrath propose five approaches to supporting changes in virtual environments: designer change, where the alterations are handcrafted by the original designer; user-driven changes, where users are able to contribute to the environments they inhabit; system-driven dynamics, where an environment automatically responds to changes in information associated with objects; ecological responsiveness, where a virtual environment adjusts to real world conditions, such as weather or time of day; and lastly, chaotic mutability, where an environment responds to random or periodic fluctuations.

The Internet is a packet-switched network. For example, when one sends an identical message, such as an e-mail, to multiple recipients, separate packets containing identical data are routed to each recipient. As audio and video streams generate data rates in excess of 100 Kbit/s, broadcasting to multiple recipients over a wide-area network is virtually impossible. However, if one can exploit the Internet's routing hardware and software, the overall data rate would be much lower. Adrian Clark's paper "Bringing the Mbone to Web Users" explains how the multicast backbone, or MBone, can be used to improve the performance of the Web using this technique.

Currently, most Web users surf the Net in the relative comfort of their office or home. However, as computers become even smaller and more mobile, users will demand seamless access to these services as they move about in cars, trains, boats, and planes. Unfortunately, cellular networks such as GSM and DCS-1800 operate at 9.6 Kbit/s and therefore pose a problem if they are to be used to distribute multimedia data. In the paper, "Handling of Dynamic 2-D/3-D Graphics in Narrow-Band Mobile Services," the authors C. Belz, H. Jung, L. Santos, R. Strack, and P. Latva-Rasku describe how it is possible to transmit vector graphics and animations over narrow-band transmission channels.

Users of early virtual reality systems would normally see a representation of their hand in the virtual world to help them interact with objects and issue navigational commands. With the advent of more powerful computers and the experience gained from working with virtual environments, it is now widely accepted that the user should be represented in greater detail. Indeed, when working with multiple-user systems, it seems only natural to see other users in some recognizable form. Modelling and animating such "avatars" is the theme of the next paper and arises from the pioneering work of Nadia and Daniel Thalmann in the world of computer animation. Their joint paper "Realistic Avatars and Autonomous Virtual Humans in VLNET Networked Virtual Environments," with Tolga Capin and Igor Pandzic, explores how avatars can be made more humanlike.

Many of the problems associated with computer animation are now becoming issues within virtual environments. Because computer animations are rendered in advance of viewing, they can incorporate complex algorithms for simulating physical phenomena. Virtual environments, on the other hand, must operate in real time and be able to compute such algorithms without introducing any lag. In the paper, "Interactive Cloth Simulation: Problems and Solutions," Pascal Volino and Nadia Magnenat Thalmann describe the problems of simulating virtual garments. They propose a new, efficient, cloth mechanical model that includes fast collision detection and constraint handling techniques.

Brian Wyvill and Andrew Guy's paper "The Blob Tree: Implicit Modelling and VRML" has an unusual title. The term *blob* was first introduced by Jim Blinn when he used implicit surfaces to model *Blobby Molecules*. Brian Wyvill has continued the research started by Blinn and now believes that implicit surfaces could play an important role in modelling virtual environments. In their paper, the authors describe an implicit modelling system that includes an interactive editor for building models defined by blending, warping, and Boolean operations. They also propose a definition of such models as an extension of VRML.

In a relatively short period of time, the Internet has become an important medium for commercial applications. However, if consumers are to be attracted away from their out-of-town supermarkets and coerced to purchase their products from virtual shops, virtual shopping must be made cheap, easy, and effective. To this end, the Computer Graphics Center at Darmstadt has been investigating the problems of modelling virtual supermarkets. In his paper "Automatic Generation of Virtual Worlds for Electronic Commerce Applications on the Internet," Klaus Bauer reports on the latest research results. He describes the advantages of 3-D environments for shopping applications on the Internet and describes the prototype of a software tool that automatically generates 3-D models of virtual supermarkets. An important feature is the system's ability to generate individualized supermarkets for each customer's specific needs and demands.

Developing the theme of the last paper, Nick Burton, Alistair Kilgour, and Hamish Taylor describe the problems of marketing an object on the Internet. In their paper "A Case Study in the Use of VRML 2.0 for Marketing a Product," they investigate the possible use of VRML 2.0 to enhance the Web-based marketing of electronic bagpipes. The case study raises many issues such as modelling, animation, interaction, and sound. Although they concluded that it is not cost-effective for inexpensive products and that VRML 2.0 is nonstandard across browsers and platforms, the research is nevertheless valid and valuable to anyone contemplating a similar project.

As companies and government agencies continue to introduce distributed systems into the workplace, there is a demand for more sophisticated software tools to support the different virtual organizations created. The aim of a virtual organization is to provide support for members of one or more actual organizations who are physically distributed or otherwise unable to meet. In their paper "A Virtual Environment for Collaborative Administration," David England and his collaborators, W. Prinz, K. Simarian, and O. Stahl, describe a virtual reality interface to a distributed cooperative system for government workers. Their aim is to improve the levels of mutual awareness among colleagues by reinforcing the concepts of shared documents and showing some measures of sharing.

Historically, theater sets required paper plans or low-tech models for implementation. However, these plans are hard to visualize, and the models are difficult to make and awkward to alter. Neither design method can accurately recreate the desired lighting conditions or give any feeling of movement or "sight lines" within a theatre space. To resolve these issues, Ian Palmer and Carlton Reeve investigated how theatre set designs could be communicated over the Internet to designers, directors, and producers. Their paper "Collaborative Theatre Set Design across Networks" reports on a project involving a theatre based on Bradford University's campus.

Today's museums face various challenges to the role they have traditionally performed. Some people view a visit to a museum as an educational experience, while others see it as a leisure experience. In his paper "Moving the Museum onto the Internet: The Use of Virtual Environments in Education about Ancient Egypt," William Mitchell describes how virtual environments provide the means of meeting these challenges. The paper describes two projects: the first is based upon the tomb of the Egyptian noble Menna, and the second is a virtual environment representing the pyramid builders' town of Kahun. The paper details how the models were created and evaluates how they were received by different groups of children.

Virtual reality systems have always interested the military sector, and the Naval Research Laboratory (NRL) in Washington is no exception. In recent years, the laboratory has monitored different VR systems and how they could be employed in various military scenarios. Although head-mounted displays provide a useful immersive environment, they can restrict the way virtual environments are presented to a group of people. To overcome such problems, the NRL developed a "VR Workbench" to display strategic information that could be viewed by a group of users. In their paper "The Virtual Reality Responsive

Workbench: Applications and Experiences," the authors, Lawrence Rosenblum, James Durbin, Robert Doyle, and David Tate provide an overview of VR display systems and then describe the technology behind their Workbench and its applications.

To conclude this volume, David Leevers reflects upon his own introduction to virtual reality systems and how they have influenced his understanding of the potential offered by this new technology. His paper "Inner Space: The Final Frontier," explores a number of issues common to all of the previous chapters. He concludes with the following statement:

> Not a moment too soon, the Web is becoming the nervous system of the planet, albeit a dry silicon one rather than a wet carbon one. Perhaps this is the time to update the Gaia hypothesis. The new CyberGaia is a global, web-footed amphibian who is equally at home in the world of bits as in the world of atoms, and whose own webbed feat is to provide the physical and the information infrastructure for a sustainable, balanced, and fair society.

We have come a long way since Oersted's discovery of the relationship between electricity and magnetism and Morse's invention of the telegraph. Yet, we have only started a new technological journey whose goal can still not be imagined. The following papers reveal a plethora of ideas that will influence computer systems of the twenty-first century.

## Chapter 1

Data Flow Languages for Immersive Virtual Environments

Anthony Steed

Department of Computer Science
University College London

Abstract

Current immersive virtual environment (IVE) systems use a wide variety of programming methodologies and scene description languages, and although many support importation of VRML 1.0 scenes, there is no agreement on how behavior and interactions should be described. Prototyping IVE behaviors can be a tedious process. To solve this, an immersive programming language, the virtual environment dialogue architecture (VEDA), was built to allow rapid and immediate changes to be made. Underlying VEDA is a data flow model similar to that of the recent VRML 2.0 standard. This paper contrasts the two models and highlights the strengths of both. Various approaches to creating an immersive VRML 2.0 programming environment are proposed from this comparison.

Introduction

A data flow model describes a system in terms of the data being passed between functions that transform its state. The tracing out of all such data flows through the system forms a directed graph, with nodes corresponding to functions and arcs indicating the possible routes for data to take. Such a model has a direct visual representation, and so-called *data flow visual programming languages* (Hils 1992) have proven useful and instructive in a number of application areas, such as scientific visualization.

The current virtual reality modelling language specification, version 2.0 (1996) uses a data flow model of operation to describe the virtual environment (VE). The nodes represent functions for describing behaviors, interactions, geometry, and appearance, and the arcs represent the manner in which node properties will be altered. A visual representation of the data flow of VRML 2.0 would certainly be possible, but as of yet, no tools exist that can directly use such a metaphor for immediate and interactive editing of scene graphs.

The utility of such a system has been shown through the design of implementation of a similar system for immersive virtual environments (IVEs). The virtual environment dialogue architecture (VEDA) system (Steed and Slater 1996; Steed 1996) was designed to allow immersed participants to rapidly prototype new behaviors and interactions without leaving the environment. Essentially, VEDA is a hybrid object hierarchy and data flow model with an immersive 3-D representation. VEDA's data flow model is similar to that of VRML 2.0.

This paper begins by describing current work on visual languages for virtual environments. Then it describes the motivation and design of the VEDA system. This is contrasted with the VRML 2.0 design later in the chapter. The final section proposes a number of approaches to creating a visual editor for VRML 2.0.

Current Technology

Data Flow Model

A data flow model is composed of nodes, that transform data, and arcs, which represent the passing of data between those nodes. Figure 1.1 shows an abstract representation of a data flow graph. Data flow starts from particular nodes, known as triggers or sensors, that are the external interface of the system. They might sense devices or react to a certain event in the VE, such as collision detection. Once a trigger is activated, data flows along all its connections to connected processing nodes. These nodes transform the data and then pass it on to further nodes that process or have an effect on the external world. Data flow stops at a node when the reception of data causes no further outputs. When constructing the entire graph of data flow, both *fan-in* and *fan-out* are usually allowed. Respectively, these are data from multiple sources arriving at the same receiver, and data from a source being sent to multiple receivers.

VRML 2.0 uses such a model to describe transformations of the properties of objects in the scene database. VRML 2.0 contains similar elements to those in VRML 1.0 (Bell, Parisi, and Pesce 1995) to describe the visible elements of a scene: geometry, appearance, lights, and transformations. The addition of a data flow model, through the use of sensors, interpolators, scripts, and routes, allows dynamic and interactive worlds to be built. Each VRML 2.0 node may have one or more inputs, or *eventIn* fields, where data can be received. The reception of data on certain fields triggers the calculation of an internal function, which may subsequently generate values on one or more *eventOut* fields, which are propagated to all connected nodes. The types of data that can be passed around include boolean values, colors, positions, orientations, and other nodes. In addition, nodes might have plain *fields,* which are not exposed to the data flow and are used for static state that

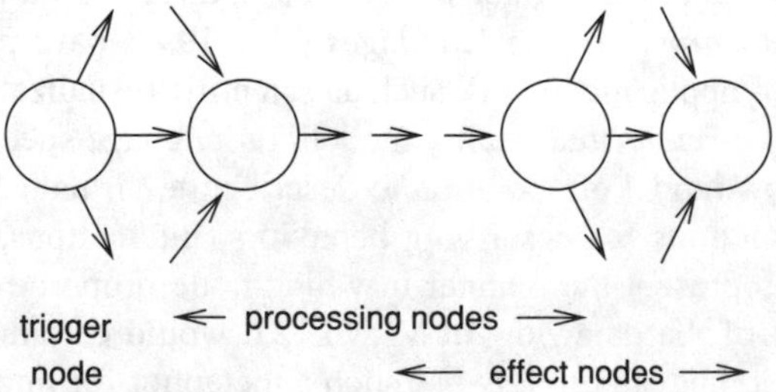

trigger   ← processing nodes →
node   ← effect nodes →

Figure 1.1 Components of a data flow model.

is initialized at start-up, and *exposedFields,* which serve as both eventIn and eventOuts. That is, data received has an effect on the node and is routed through to further nodes.

As an example, the fields of the *OrientationInterpolator* are given in Figure 1.2. The OrientationInterpolator's purpose is to generate values on the eventOut *value_changed,* whenever the eventIn *set_fraction* is changed. These values are chosen by interpolating between the values defined by the exposedField *keyValue* where the exposedField *key* gives the reference fraction values.

Figure 1.3 shows a sample VRML 2.0 file that uses the OrientationInterpolator, and Figure 1.4 the corresponding scene graph and data flow graph. The behavior is very simple; when the user clicks upon the red cube, it starts spinning. The TouchSensor senses the user clicking on the geometry and generates a time value on a data stream. This then flows to the TimeSensor node, which in turn generates a fraction value. This is then used by an OrientationInterpolator to start the generation of orientation values, which are routed to the position of the cube to provide an animation. The four elements of the data flow graph map onto the different types of data flow nodes in the abstract graph of Figure 1.1. In this case,

```
OrientationInterpolator {
eventIn    set_fraction
exposedField MFFloat key []
exposedField MFVec4f key value []
eventOut    MFVec3f value_changed
}
```

Figure 1.2   Fields of the OrientationInterpolator.

```
#VRML V2.0 utf8
Viewpoint {
  position 0 0 5
}
DEF BOX_POS Transform {
  children [
    Shape {
      geometry Box {}
    },
    DEF BOX_TOUCH TouchSensor {},
  ]
}
DEF BOX_RIMER TimeSensor {
  cycleInterval 5
  startTime - 1
}
DEF BOX_ENGINE OrientationInterpolator {
  key [ 0, .5, 1]
  keyValue [ 0 1 0 0, 0 1 0 3.14, 0 1 0 6.28]
}

ROUTE BOX_TOUCH.touchTime TO
BOX_TIMER.set_startTime
ROUTE BOX_TIMER.fraction TO BOX_ENGINE.set_fraction
ROUTE BOX_ENGINE.value_changed TO
BOX_POS.set_rotation
```

Figure 1.3   Sample VRML 2.0 file.

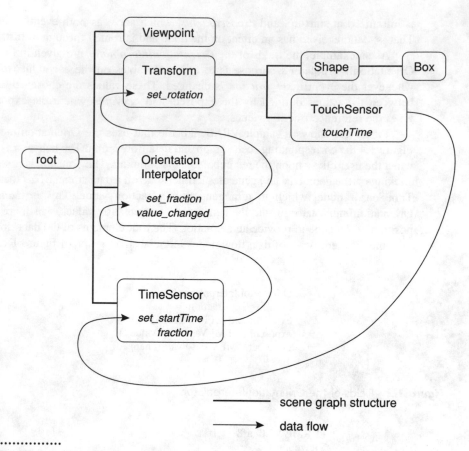

Figure 1.4 Scene graph of sample file.

the trigger node is the TouchSensor since it reacts to an event external to the scene graph (the user). The TimeSensor and OrientationInterpolator are the processing nodes, and the cube is the effect node, since the modification is visible to the user through the use of a browser.

VRML 2.0 is only one example of the data flow model, and many variations in the types of nodes and arcs allowed are possible (Hils 1992). Given a data flow model, there is a direct visualization of the data flow (see Figures 1.3 and 1.5), and a data flow visual language makes the obvious visualization into a programming environment. Many examples exist; IRIX Explorer (Foulser 1995) and AVS (Upson et al. 1989) are two well-known examples.

Visual Languages for Virtual Environments

In describing current visual programming systems for virtual reality, we have to distinguish between two classes:

- Visual languages to *describe* virtual environments
- Visual languages *within* virtual environments

Body Electric from VPL is an example of the first class of language (Kalawsky 1993, 212–219). Body Electric ran on a Macintosh adjacent to the main virtual environment generator. It used a node and arc-type diagram with the data flowing from the input devices through data "massage" units to drive events in the virtual environment. Since Body Electric ran on a separate system to the display generator, it was not accessible to the immersed participant.

Other more widely used 2-D data flow visual languages have been extended to support VEs. These include the AVS scientific visualization (Sherman 1993) and IBM's Data Explorer (Gillilan and Wood 1995). However, in both cases the VE presentation is an effect of a node within the data flow diagram rather than being the specification environment.

The second class of system involves a visual language, which, rather than being manipulated on a 2-D display, is manipulated within a 3-D environment. Such an approach was advocated by Glinert (1987) who proposed extending his BLOX method, which converts flat jigsawlike pieces to 3-D, using cubes that can be snapped together much like building blocks.

Several visual programming languages with 3-D representations have been built. They vary greatly in the style of interface and level of liveness at which they operate.

VPLA (Lytle 1995) is an interpreted language specified by a 3-D network editor that describes components of an animation in terms of hierarchical objects and actions on them, such as transformations, modelling operations, deformations, particle systems, and recursive procedures. However, VPLA is not "live" since it is used to produce RIB scripts that are subsequently rendered off-line.

Two virtual environment languages for programming that have been implemented are CUBE (Najork 1994) and Lingua Graphica (Stiles and Pontecorvo 1992). CUBE allows visualization of expressions in a functional language and some limited editing through a desktop-based interface. Lingua Graphica is a 3-D editor for a C++ based language. The Lingua Graphica workspace looks like a tool board with the tools being the various library functions, primitives, and predefined types. These can be juxtaposed with the spatial relationships between the objects corresponding to the syntax rules of the language. Once a procedure has been constructed, it can be written out directly in a text form that is compilable.

A visual language that belonged to both of these classes would both describe a virtual environment and exist within it. It would also ideally operate in a *live* mode. That is, the changes to the data flow could be made without leaving the environment and with immediate effect.

It is worth noting that it is not unusual for a mature IVE system to contain immersive tools with which to modify the environment. A tool such as dVISE 3.1 from Division Ltd. (1996) includes immersive and standard desktop menu systems that allow certain behaviors and properties of the objects to be modified. However, not every function of the underlying system is available to the immersed participant, and many programming tasks have to be conducted outside the environment.

Virtual Environment Dialogue Architecture

Model

The virtual environment dialogue architecture (VEDA) is a model for the programming of VEs and also an environment and set of techniques within which the behavior of the environment components can be manipulated. It uses a data flow model whose three types of functions map directly on to the three classes in Figure 1.1:

- **Devices** that return information about the participant and generate data streams
- **Filters** that process data samples on one or more incoming data streams and then generate new data streams
- **Tools** that take input streams and act upon the objects of the VE

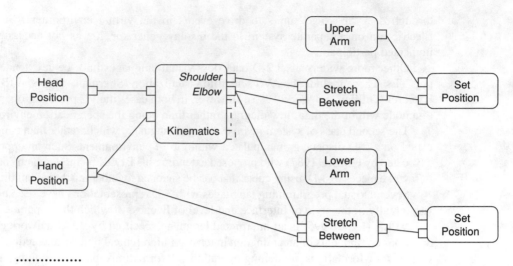

Figure 1.5   The abstract data flow model of the arm.

Figure 1.5 gives an example of how the arm of a participant is modelled using inverse kinematics. The data flow starts with two devices, one that returns the head position and one that returns the hand position. The kinematics function takes both of these as inputs and then generates further streams that correspond to other joint positions based upon a standard set of parameters for the body. The stretch-between function takes two positions and generates a data stream containing the transformation that would stretch a unit object between them. Finally, the setposition tool takes a position stream and applies this to an object, the id of which is passed across a second data stream.

Representation

The data flow model of the VE is presented immersively to the participant. This means that they can make changes to the live application without exiting the environment. Since the descriptions can be highly detailed, there are tools to hide and reveal parts of the data flow and to encapsulate several functions into a higher-level function.

An example of the immersive representation is shown in Figure 1.6. This representation implements a navigation metaphor called the *virtual treadmill* (Slater, Usoh, and Steed 1995). The functions represented are the head sensor on the left, the move function on the right, a gesture recognizer at the bottom, and the object referencing the participant at the top. The effect of this data flow is to move the participant in the direction they are looking when the head is making the gesture of walking in place. Data flow is represented by the pipes between input and output gates of functions. The shapes below the functions represent the types of the data and are colored, depending on whether they are input or output.

Basic Functions

The set of functions provided by VEDA was designed with the aim of providing interesting interactions and behaviors. These include the following:

- Gesture recognition—path and neural net based
- Object properties—color, scale, and position

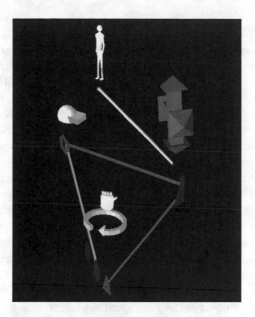

Figure 1.6 Immersive representation of the virtual treadmill navigation metaphor. (*See color plate on page 305.*)

- Object behaviors—collisions, animation, and constraints
- Logical and timing—toggle, double click, and so forth

These serve as the static parts of a scene graph, but VEDA was designed with the aim of allowing editing of the graph, so any part of the data flow representation can be picked and moved, functions can be copied and deleted, and the tube connectors can be dragged and snapped between function inputs and outputs.

Meta Objects

The immersive representation of the functions form the specification layer of the virtual environment, but the presentation layer certainly should not contain these objects if they clutter the space or detract from the purpose of the environment. The first way in which this is tackled is to provide level of detail tools within the environment. When applied to a function node, they can either hide or reveal the connecting tubes between two nodes.

VEDA also uses the concept of *meta objects* to provide complete hiding of function nodes and higher level abstractions of complicated behavior. A meta object can have any representation, so at a basic level it serves to allow arbitrary geometry to be included in the presentation of a world. A meta object also serves as a container, in that it can have a collection of other objects as children and a selection of their input and output streams as an interface. When combined with the level of detail tools, this allows whole sections of the data flow to be hidden or exposed, since the lowest level of detail for a meta object is for its children to be made invisible. Thus, an application environment will consist of a number of top-level meta objects that represent the form of the environment, and each one of these will be hiding within it the functionality of the data flow associated with that object.

Using these mechanisms, many higher-level objects have been built within VEDA that allow easy manipulation of new environments. These include virtual sliders, path tools, color tools, scale tools, and a virtual multimeter.

Motivation

The foremost reason for using an immersive representation of the scene behavior was so that editing could take place within the VE and not require the participant to repeatedly enter and leave the environment when making modifications. Apart from the obvious time penalty this involves, exiting the environment leads to a loss of *presence* of the participant in the environment, which might in turn lead to their forgetting exactly what the state of the environment was. Also, in many systems, editing the environment involves rerunning or, at worst, recompiling the environment, which leads to an application loss of state that could be time-consuming to reconstruct.

Another motivation is that the immersive 3-D nature of the environment should provide a good environment to display the network of data flow between objects. There is direct evidence to show that a 3-D node and arc diagram may offer substantial benefits to comprehension of the dialog structure over the 2-D diagram approach (Ware and Franck 1994).

From a design point of view, there might be another, more subtle benefit to immersively editing behavior. That is, since the participant has a sense of presence within the environment they are designing for, it may be the case that any techniques or design decisions that are effected would be more appropriate to that environment. Essentially, one would hope that since the participant is present within the VE, anything they would conceive of designing or altering would be consistent with the VE's overall style.

Comparison of VRML 2.0 and VEDA

Although VRML 2.0 and VEDA were both designed to describe virtual environments, their scopes are not identical, and thus the models and nodes that each supports differ on a number of points. VRML 2.0 is primarily a description of a file format for the transmission of interactive 3-D worlds across the Internet. VEDA was designed as an interactive VR tool kit for describing interfaces for both desktop and immersive systems. Thus, although there is a large overlap in nodes types, VEDA does not deal with URLs, scripting, or external interfaces, and VRML 2.0 does not support some forms of interface, a broad editing API, or a visual representation. The common ground is the similar descriptions of object properties such as color, position, and animation, and the basic data flow model of object behavior.

Models

Scene Graph

The data flow model is common to both systems, though the nodes of the data flow are different in each case. In VRML 2.0, the primary database structure comes from the scene graph, which is a property hierarchy, and the scripting and routing mechanisms are largely independent of this in that their behavior does not depend upon position. In VEDA, the primary database structure defines the relationship between meta objects and the components that make up their behavior. In particular, VEDA has no property hierarchy. Every VEDA function or meta object has a collection of properties: geometry, appearance, and position. Figure 1.7 shows that to change the position and scale of a scene element involves operating on a transform property in VRML 2.0, or on an object in VEDA.

Function Style

Figure 1.7 also illustrates the fact that VEDA effect nodes were designed to be function orientated rather than node orientated. In VEDA the data flow graph to change both the position and scale contains two functions, SetScale and SetPosition, which operate on a

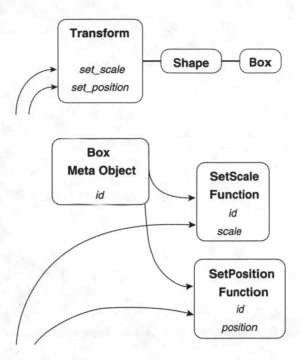

Figure 1.7 Contrasting data flow hierarchies for a simple task in VRML 2.0 and VEDA.

reference to the object, whereas in VRML 2.0, this is achieved with two data flow inputs to the Transform node. VEDA's approach was chosen to allow easier visual representation of the data flow, since the irrelevant part of the data flow could be hidden or revealed. It is possible in both systems to use prototypes or encapsulation to make data flow constructs that look like the corresponding ones from the other language.

Flow Model

VEDA was designed around a *data driven* model of data flow. This has the data flow starting at trigger nodes, which are updated at a continuous frequency. The alternative is the *demand driven* model that does a reverse, request-based polling of functions. In this case, data flow starts at nodes that have an effect in the displayed environment and propagates backwards until enough data has been accumulated in order to compute the correct state of the world. The advantage of the second approach is that it can be done in a lazy fashion, with only visible effects being computed. This means that some functions need not be evaluated, and if the time to propagate all the required data can be estimated, requests can be timed to be as close to the display refresh as possible. VRML 2.0 does not impose a model on the browser author, but in effect, VEDA had to, since external sensor updating and rendering were controlled by separate processes (see later section on Extensions). This meant it was impossible to do lazy propagation since there was no way of testing whether an object was being rendered, and it was also impossible to time the data flow to the rendering cycle.

Interaction Model

Since one of the motivations for building VEDA was to allow customization of the interaction metaphors, VEDA allows direct interaction with the devices that the participant is using. In VRML 2.0 the user can interact with an object only through Sensors, which

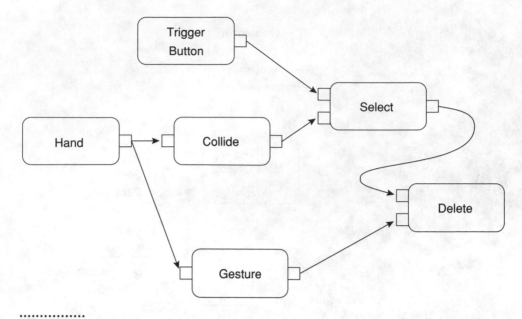

Figure 1.8   Abstract data flow for a delete mechanism.

impose an interaction style on the world since they also constrain the interaction in some way. In particular, there is no sense of the participant having a body in VRML, whereas VEDA considers the participant to be presented by a set of objects that are being updated by sensing devices, both position sensors and buttons. Thus, VEDA trigger nodes give the position and state of parts of the participant's body continuously rather than only when they interact.[1] The use of a body model makes the interaction style largely independent of the interaction devices, since as the example in Figure 1.5 showed, the body state can be inferred from few sensors. This allows fundamental changes to be made to the interaction structure within the data flow graph. For example, Figure 1.6 showed the definition of an interaction metaphor within VEDA based on where the participant was looking.

Editing Mechanisms

Another motivation for the design of VEDA was to allow live editing of the environment, so VEDA incorporates effect functions that copy and delete other nodes. This, in combination with collision detection, allows a simple tool-based approach to environment construction. The mechanisms by which the editing functions are used is in fact described within VEDA. Figure 1.8 illustrates the standard way in which the delete function is used. The objects to delete are a set selected by touching them with the hand and pulling on the trigger button. Once selected, they can be deleted by making the delete gesture with the hand. This means that within VEDA, it is possible to delete the delete function, but there are situations where you might want this, such as the last stage of reconfiguring an environment for use by a novice.

---

[1]Continuous extraction of the position of the viewpoint is actually possible in VRML 2.0 using a large ProximitySensor, but it is not possible to get the position of the hand.

Table 1.1   VEDA and VRML 2.0 Node Comparison

Node Class	VEDA Support	VRML 2.0 Support
Gestures	Training and recognition nodes	Possible with scripts
Logic	And, or, not, then,	Possible with scripts
Collision Detection	Supported	Viewer collision, but no object collision
Constraints	Supported	Possible with scripts and prototyping
Animation	Path following and description	Interpolators give a flexible approach
Sensors	Device and collision sensors, no visibility sensor	Constrained interaction sensors
Appearance and geometry	Properties excluding texture and geometry	Full support
Sound	Play recorded sounds	Spatialized playing of recorded sounds
Lighting	Is an object property	Supported
Node editing	Delete, copy, selection	Indirect through scripting

Nodes

The previous section described the differences between the models and introduced some of the nodes in areas of nonoverlapping scope. Table 1.1 summarizes the node differences where there is common support.

Apart from the differences mentioned already, such as the role of sensors in both systems and the lack of collision detection in VRML 2.0, there is better support in VRML 2.0 for changing properties. This is partly due to the property graph structure of VRML 2.0 where each property, such as texture and geometry, is a candidate for inclusion in the data flow. Part of the reason VEDA does not support these data flow types is that the platform it was written on supported neither texture mapping nor dynamic geometry editing. However, given the function-oriented approach of VEDA, such functions are simple to add as long as the underlying database API supports them.

Extensions

One of the most interesting proposed extensions to VRML 2.0 is the External Authoring Interface (Marrin 1996). This will allow external control of a VRML 2.0 scene through an API within the browser. This will allow other applications to alter the scene graph independently of the data flow model. In effect, such a layer is inappropriate for VEDA since it acts as an external process, operating on an object store through an API. VEDA was written as a dVS application and is only one of a number of actors sharing a database (Division Ltd. 1994). Figure 1.9 illustrates the separation of the two, with the clear interfaces between the database and event mechanisms.

The VEDA equivalent of a process using VRML 2.0's external API is simply another dVS client. As a consequence of this, VEDA automatically supports multiple users since in the dVS model, a second user is simply a second set of trackers and renderers. However, since the data flow is not represented in the dVS database, there is no capability for two clients to interact except through surface interactions; that is, interactions that can be detected from the current state of the scene rather than any application level semantics.

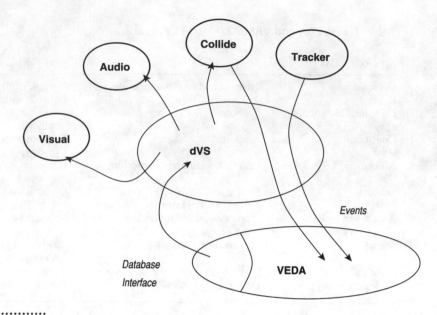

Figure 1.9   Relationship between VEDA and dVS.

Towards a Visual Representation of VRML 2.0

The experience with VEDA has shown that it is useful to have the capability to make changes to the world while the participant is immersed. This allows aspects of the behavior of the environment to be demonstrated in situ and take full advantage of the full body interaction style.

The two data flow systems have a number of parallels in the node classes they provide, but they differ in some important areas: sensor information and collision information, which were discussed previously. The second could be prototyped in VRML 2.0, but the first would either require access through some sort of browser API to devise information or through new nodes that modelled the participant's body. The body model approach has advantages for portability across browsers, but the exposed device approach allows the browser users greater scope for customizing their representation and behavior in the environment.

As it stands, VEDA would require a handful of new nodes in order to be able to support the import of VRML 2.0 worlds. Export is more problematic since the mapping of gestures and interaction techniques into VRML 2.0 could not be made without new browser-dependent nodes. However, since one of the aims of a VEDA style system would be to provide editing for the end-user rather than just scene authors, it would be necessary to provide the required functionality across different browsers, so the application itself must be written in VRML 2.0.

As noted earlier, VEDA is simply a dVS client and interacts with it through a database API. Since the separation is quite distinct, it would be simple to write a similar application that used the proposed external authoring interface VRML 2.0. This would avoid the problem of accessing tracking devices since the external applet could interface to them directly. Either the browser or the scene itself would have to have a hook that allowed the presentation level to be augmented with the VEDA tools, and these could be downloaded as required.

Conclusions

A number of systems have started looking at applying a visual approach to the description of virtual environments. The VEDA system is the first to combine a data flow model with a live immersive programming environment. It allows the participant to alter the behavior of a running application by examining and editing the underlying data flow model.

While the recent VRML 2.0 standard also supports a data flow model of execution, it does not support or define a visual representation. The interaction model lacks the flexibility to be used to describe novel manipulation or navigation techniques, but through the use of VRML's external API, this can be rectified to some degree. The external API also gives us the ability to introduce editing tools into the environment, and thus a VEDA-like approach to environment design could be supported as a VRML 2.0 scene.

References

G. Bell, A. Parisi, and M. Pesce, "The Virtual Reality Modeling Language, Version 1.0 Specification," Web document, available at http://vrml.wired.com/vrml.tech/vrml10-3.html (May 26 1995).

dVISE for Unix Workstation, User Guide, Division Ltd., Almondsbury, Bristol, U.K., 1996.

dVS Technical Overview Version 2.0.4, Division Ltd., Almondsbury, Bristol, U.K., 1994.

D. Foulser, "IRIS Explorer: A Framework for Investigation," *Computer Graphics,* May 1995, pp. 13–16.

R.E. Gillilan and F. Wood, "Visualization, Virtual Reality, and Animation within the Data Flow Model of Computing," *Computer Graphics,* May 1995, pp. 55–58.

E. Glinert, "Out of Flatland: Towards 3-D Visual Programming," *Proc. 2d Fall Joint Computer Conference,* IEEE Computer Soc. Press, 1987, pp. 292–299.

D. Hils, "Visual Languages and Computing Survey: Data Flow Visual Programming Languages," *J. Visual Languages and Computing,* Vol. 3 No. 1, 1992, pp. 69–101.

R.S. Kalawsky, *The Science of Virtual Reality and Virtual Environments,* Addison-Wesley, 1993.

W. Lytle, *Vpla: Visual Programming Language for Animation,* technical sketch, SIGGRAPH95, 1995.

C. Marrin, "Proposal for a VRML 2.0 Informative Annex, External Authoring Interface," http://vrml.sgi.com/moving-worlds/spec/ExternalInterface.html (Sept. 12, 1996).

M. Najork, *Programming in Three Dimensions,* PhD thesis, Univ. of Illinois, Urbana-Champaign, 1994.

W. Sherman, "Integrating Virtual Environments into the Dataflow Paradigm," *Fourth Eurographics Workshop on ViSC,* Abingdon, UK, April 1993.

M. Slater, M. Usoh, and A. Steed, "Taking Steps: The Influence of a Walking Metaphor on Presence in Virtual Reality," *ACM Transactions on Computer Human Interaction,* Vol. 2, No. 3, 1995, pp. 201–219.

A. Steed, *Defining Interaction within Virtual Environments,* PhD thesis, Queen Mary and Westfield College, Univ. of London, Dept. Computer Science, 1996.

A. Steed and M. Slater, "A Dataflow Representation for Defining Behaviours within Virtual Environments," *Proc. VRAIS'96,* IEEE Computer Soc. Press, 1996, pp. 163–167.

R. Stiles and M. Pontecorvo, "Lingua Graphica: A Visual Language for Virtual Environment," *Proc. 1992 IEEE Workshop Visual Languages,* IEEE Computer Soc. Press, 1992, pp. 225–227.

C. Upson et al., "The Application Visualization System: A Computational Environment for Scientific Visualization," *IEEE Computer Graphics and Applications,* Sept. 1989, pp. 30–42.

"The Virtual Reality Modeling Language Specification, Version 2.0," http://vag.vrml.org/VRML2.0/FINAL/ (Aug. 4, 1996).

C. Ware and G. Franck, "Viewing a Graph in a Virtual Reality Display Is Three Times as Good as a 2D Diagram," *Proc. 1994 IEEE Symposium Visual Languages,* 1994, pp. 182–183.

Chapter 2

VRML Interfaces to Information Systems

Christine Clark and Adrian F. Clark

VASE Laboratory
Electronic Systems Engineering
University of Essex

The use of VRML, the Virtual Reality Modelling Language, is becoming well-established for tasks such as data visualization and in creating shared virtual environments. This paper explores a different use of VRML, namely, to provide interfaces to information systems. Three particular examples are examined: a model of the University of Essex campus, which interfaces to departments' Web pages; a model of a theatre, which interfaces to a booking system; and a developing VRML interface to the classic *Adventure* game. Underlying all three interfaces is a programmed approach to virtual world creation, which is discussed with reference to the current VRML standard.

Introduction

VRML, the *Virtual Reality Modelling Language,* has evolved from nothing to almost an International Standard in a little over two years, and it is almost certain to become the norm for describing 3-D environments, both on the Internet and elsewhere. With VRML 2, the language's ability to allow objects within an environment to exhibit behavior is starting to be exploited in shared virtual environments; and, of course, its use for data visualization is already well established. This paper explores a somewhat different use of VRML, namely, to provide easy-to-navigate interfaces to information systems. This is done by discussing three exemplars:

- A model of the University of Essex that interfaces to the university's information system
- A model of a local building that is to be converted into a theatre, the model interfacing to a seat-booking system
- The use of VRML 2 to resurrect the classic 1970s *Adventure* game

Underlying all three interfaces is a developing approach to generating the VRML descriptions by program.

Figure 2.1   Aerial view of the University of Essex campus. (*See color plate on page 305.*)

The following three sections describe the interfaces and how they were constructed. The main features of our program-based approach are then discussed. Finally, some conclusions are drawn.

Interface to a Campus Information System

Most universities have a substantial investment in their information systems, which are now invariably Web based. However, the best way of navigating through these seas of information is still not clear, as even a cursory examination of universities' home pages will demonstrate—there is a vast range of styles. With the advent of VRML 1 in 1994, the authors speculated whether a 3-D interface might be an interesting navigation scheme, as well as an excellent vehicle for learning VRML!

Fortunately, Essex is geographically quite compact, the campus lying in an area of parkland (Figure 2.1); this simplifies the problem when compared to, say, a city-center site. Since the model was intended from the outset to be a Web navigation aid, its most important characteristic is speed, both of downloading and of rendering. This means that many features that contribute to realism (windows, trees, textures, etc.) have to be dispensed with. This is actually not too much of a problem, since Essex is a "red-brick"[1] university, rather boxlike on the whole. On the other hand, the buildings lie in a valley, and

[1]Actually, poured concrete.

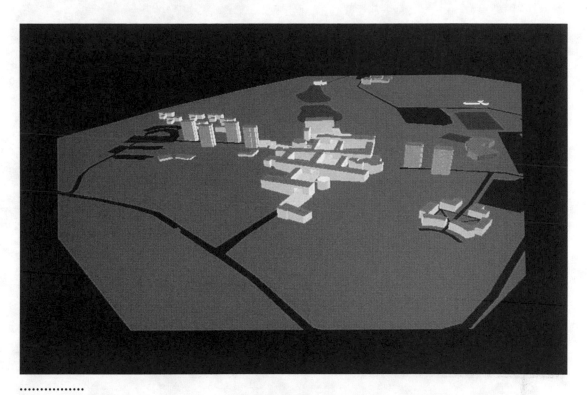

Figure 2.2   View of the University of Essex VRML campus model. (*See color plate on page 305.*)

the model would have to reproduce that in order to look at all sensible. The model was logically, though not necessarily chronologically, constructed as described in the following paragraphs. The model is available on the Web as

```
http://peipa.essex.ac.uk/vase/demo/campus/
```

The entire process took about 18 person-days, including learning VRML, programming, and chasing up and measuring drawings from our Estates section.

The first step was to construct a digital elevation model (DEM) of the campus. Study of a 1:1,250 map of Wivenhoe Park showed that a relatively coarse DEM would suffice: 20 x 16 samples. Ground height above sea level was estimated at each sample by interpolating between contour lines. A program was written in the Tcl programming language (Ousterhout 1994; Welch 1995) (discussed later) to convert these data to a VRML model. The obvious solution was to generate a set of quadrilaterals; but these would not necessarily be planar, so each set of four points was used to produce two triangles.

The main campus buildings were somewhat harder to model. They are constructed around a series of squares connected by flights of steps. The approach that was taken was to read points representing the corners of buildings from the architect's plans and to enter them manually. Each point entered includes the building's height. These data were also processed by the Tcl script to yield the sides and roof of the various buildings. The Tcl script ensures that the buildings stand on the ground by interpolating (bilinearly) between DEM samples. Some significant simplifications were made at this stage: buildings are straight-sided on the model, the cloisterlike walkways were omitted around the squares, and stairs were replaced by ramps; but the essential geometry was retained, as Figure 2.2—which approximates the aerial photograph of Figure 2.1—demonstrates.

With the buildings in place, the next step was to incorporate roads, paths, sports fields, and the lakes. Our experiments at that time showed that texture mapping was slow and not supported by all browsers (this situation has improved subsequently, of course), and so we made all these features from polygons that float above the ground—something that no one has ever commented on, despite the model having been visited several thousand times.

The final stage was to associate particular parts of buildings with Web pages. After a couple of experiments, the approach that was found to work best was to associate the home page of departments with the relevant building. Fortunately, departments at Essex normally inhabit a vertical slice of a building: were one department to lie on a floor above another, the building would have had to be split into floors, making navigation somewhat more difficult.

The majority of Web pages linked to the buildings are, of course, HTML pages; however, the link to the authors' laboratory ends in another VRML model, this time of the laboratory itself (Figure 2.3). This model was again produced manually, but with the benefit of experience, a somewhat more general set of Tcl scripts were used to convert measurements into VRML: it took about three days. Although the model is a genuine reproduction of the actual room and its layout, many of the objects within it act as icons:

- The computers are links to the Web pages of individual projects taking place within the laboratory
- The lab's library cupboard is a link to its Web-based catalogue and issues database
- The whiteboard links to the lab's information services: diary, mail message archive, and so forth

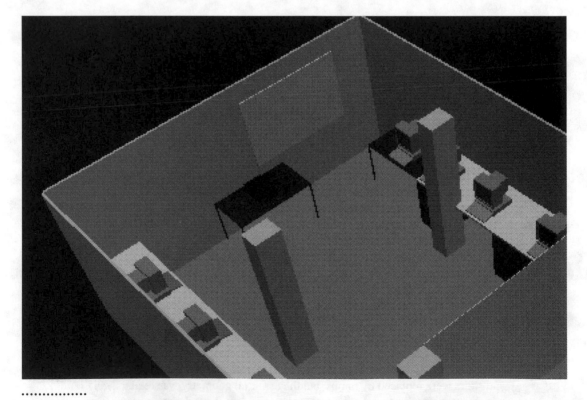

Figure 2.3   View of the VASE laboratory model. (*See color plate on page 306.*)

- The telephone links to the university's X.500 database, allowing us to find contact information for any person at the university (or, by following a couple of HTML links, in any X.500 database around the world)

We have monitored the feedback that we have received in the 18 months since the campus model was completed and have found it to be generally positive. (Many of the negative comments actually relate to the difficulty of navigation within VRML browsers.) People have generally found the model to be genuinely useful in conveying an idea of the geography of the campus—indeed, we have had comments on several occasions to the effect that it has helped visitors find departments—and that it provides a refreshing alternative to scanning through HTML pages.

When we convert the model to VRML 2, which we shall probably do over the summer of 1998 when VRML 2 browsers have supplanted VRML 1 browsers, we have a number of enhancements lined up, including a guided tour of the campus with links to photographs of buildings and the parkland.

A Theatre Model and Booking System

A somewhat different type of interface to an information system is demonstrated by our second example. A local theatre group hopes to convert a disused building into a theatre by means of a government grant. We were approached to produce a model as a visualization aid for the local community; however, since we were interested in doing something more than just modelling the building, we are using the model as an interface to a seat-booking system. The theatre model is on the Web at

```
http://peipa.essex.ac.uk/vase/demos/theatre/
```

As with the campus model discussed previously, the theatre model was built using measurements taken from the architect's drawings (Figure 2.4). The existing fabric of the building is shaded red and proposed constructions are colored white to help visualize the modifications. Other features, such as the seats, were also modelled by hand and are replicated within the model using VRML's DEF and USE constructs for efficiency. In preparing the theatre model, an effort was made to write some general Tcl code for creating VRML primitives, an approach that has paid dividends subsequently. Based on these primitives, procedures for generating higher-level objects (walls, seats, staircases, etc.) can be written quite quickly. The use of a Tcl script for generating the VRML also allows us to produce models with or without roof, textures, and so forth.

The model has found a number of uses:

- Interactive inspection of the VRML model brought to light some minor inconsistencies and errors in the architect's drawings.
- Direct manipulation of the VRML model by the potential users of the theatre gave them a fair idea of how the auditorium will appear (and actually prompted them to redesign the seating arrangement) and how easy it will be to manhandle scenery between the stage and the store.
- At the request of the local planning department, we recorded some "walk-throughs" of the model onto a video. This was then used in a touring planning exhibition, giving people a better idea of the way the theatre will appear than any simple drawings could.
- Finally, the fact that the model is program generated makes using it as an interface quite straightforward. In particular, we have used it as an easy-to-use interface to a prototype seat-booking system.

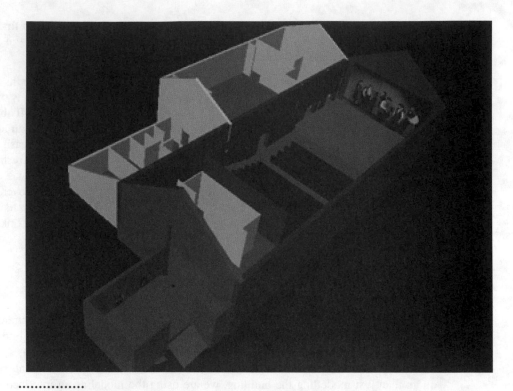

Figure 2.4   Model of the proposed Wivenhoe Theatre: Roof and first floor removed. (*See color plate on page 306.*)

When a Web user wishes to book a seat, he or she selects the desired performance. This causes the model to be loaded into the user's browser; however, rather than downloading a fixed model, the seats are colorcoded: blue for those that have been booked and green for those that are available. This gives the customer an immediate visual overview of the auditorium and the available spaces within it. By clicking the mouse pointer of a particular seat the user's viewpoint changes so that he or she has a view of the stage from that seat. (Figure 2.5). (Since this is a VRML 1 world and VRML 1 browsers are somewhat lacking in sophistication, this is achieved by reloading the world with a different viewpoint; with the new features offered by VRML 2, it will be possible to move the viewpoint without necessitating reloading.) The stage has some "scenery" texture-mapped onto it; above that are two "buttons" that link back to the auditorium view or onward to a fill-in form to complete the booking transaction.

Although the VRML-based seat-booking system was designed for use within the theatre, we have found that it works perfectly well across the Web, with two provisos: the download time is significant because of the texture-mapped scenery; and the Tcl code that implements the booking system on the Web server does not currently enforce mutually exclusive access to the database.

When we move the model to VRML 2, we are considering using an approach based around a downloaded Java applet: this can generate the VRML model within the Web browser, control viewpoints, and handle forms entry. Information regarding the state of each seat (booked or available) can then be communicated in a very compact form. How well this approach will work depends on the level of Java support in VRML 2 browsers; our initial experiments suggest that it will not work with all browsers.

Figure 2.5   Model of the proposed Wivenhoe Theatre: View of the stage from a seat.
(*See color plate on page 306.*)

The Programmed Approach to Virtual World Construction

When people consider creating 3-D objects and environments, they normally think of interactive, mouse-driven tools. Although there are many such tools, few of them are well-suited to creating VRML worlds, where objects may exhibit behaviors—though SGI's *Cosmo Worlds* and Sony's recently announced *Community Place Conductor* both show promise. These tools are geared towards rapid prototyping of objects that look more or less right. However, the models described in this paper either have to mimic real artifacts or are generated on-the-fly from information stored in non-VRML forms; interactive, manually guided tools are quite inappropriate for these tasks. Conversely, the authors' programmed approach fits well with these requirements. Furthermore, since the Tk permits rapid construction of GUIs in Tcl, it is feasible to write a GUI for virtual world construction that makes use of the Tcl procedures developed for programmed use.

It can be argued that the PROTO mechanism of VRML 2, coupled with a script, can be used to define higher-level objects. While this is true in principle, the current conformance requirements of the VRML standard (ISO 1996) do not place constraints on the language use for scripting—or even that scripting must be available! Thus, we find that the two major VRML 2 browsers—Sony's *Community Place* and SGI's *Cosmo Player*—employ different scripting languages (Java and Javascript, respectively). This is far from ideal and means that PROTO nodes are useful only for specifying behavior-free, precalculated geometry in practice.

For creating static geometry, the authors' Tcl-based approach is quite flexible. For example, we are currently working on a VRML recreation of a temple in Roman Colchester with the Colchester Archæological Trust. Roman temples invariably have many columns, and so the obvious VRML 2 solution is to prototype the column and then use it, using either PROTO or DEF/USE. Bearing in mind the above comments regarding scripting and PROTOs, we have instead written a generic column-generator in Tcl. This script is invoked as follows:

```
column -width w -height h -flutes f -pediment p -capital c
-accuracy a
```

where p and c are one of none, doric, ionic, corinthian; w and h are the width and height; f is the number of flutes; and a is the required accuracy. The script generates a cylinder if the number of flutes is zero; otherwise, it generates a single flute and rotates that $f - 1$ times to produce a complete column. The resulting column may then be incorporated into a scene via DEF/USE. A couple of example columns are illustrated in Figure 2.6 and a model of a Roman temple built from them in Figure 2.7.

The temple is actually a visualization of the Roman Temple to Claudius, built when Colchester was the capital of Britain, and is part of ongoing work being carried out in association with the Colchester Archæological Trust. (The actual temple would have had round columns, not fluted ones as shown in the figure.) Temples such as these were generally constructed in accordance with some rules documented by Vetruvius, with most measurements being described in terms of the column diameter. Hence, the model of the temple may be constructed knowing little more than the column diameter and the number of columns across the front: most other measurements are calculated from them by another Tcl script.

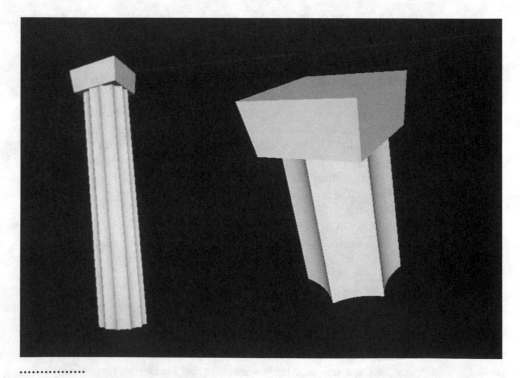

Figure 2.6 Examples of columns from column-generating Tcl script.

This approach can also bring other benefits. We have demonstrated on some of our models that by replacing the procedures that generate primitive shapes, a single-scene description may be processed to yield either VRML or input to the *Persistence of Vision* ray-tracer (POV 1996), which is useful for making high-quality animations of scenes. We are now working towards revising our software to support other rendering systems, including Renderman (Upstill 1989), Criterion's Renderware (Criterion Softward Ltd. 1995), and OpenGL/Mesa via the TkSM extension to Tcl.

At the same time, we are investigating how the same Tcl software may be used for specifying behaviors in VRML 2. The intention is not to introduce yet another incompatible scripting language, but to generate Java or Javascript that is functionally equivalent. Until the VRML community imposes a *requirement* on browsers to support a specific scripting language, macrogenerating the same script in multiple languages may be the only way to produce portable VRML.

A VRML Interface to the Adventure Game

As a final example, we consider an interface to a somewhat different type of information system—a game. Those readers who are old enough to have used a card punch will almost certainly have seen, and have probably spent many hours playing, the classic 1970s *Adventure* game written by Willie Crowther and Don Woods. For the uninitiated, the game

Figure 2.7   Roman temple model built using column-generating script. (*See color plate on page 307.*)

consists of a series of rooms in Colossal Cave (which actually started out as a reasonably accurate model of a real cave system of the same name) where the player was required to solve a number of problems as he or she found his or her way through the cave system. The user interacts with the game via typed commands; an excerpt from a typical dialogue is shown in Figure 2.8.

```
You are standing at the end of a road before a small brick building.
Around you is a forest. A small stream flows out of the building and
down a gully.
> enter
You are inside a building, a well house for a large spring.
There are some keys on the ground here.
There is a shiny brass lamp nearby.
There is food here.
There is a bottle of water here.
> take keys
OK
> take lamp
OK
> take food
OK
> take bottle
OK
> go out
You're at end of road again.
...
You are in a 20-foot depression floored with bare dirt. Set into the
dirt is a strong steel grate mounted in concrete. A dry streambed
leads into the depression. The grate is locked.
> open grate with keys
The grate is now unlocked.
> down
You are in a small chamber beneath a 3x3 steel grate to the surface.
A low crawl over cobbles leads inward to the west.
The grate is open.
> turn on light
Your lamp is now on.
> w
You are crawling over cobbles in a low passage. There is a dim light
at the east end of the passage.
There is a small wicker cage discarded nearby.
> take cage
ok
```

Figure 2.8 Typical dialogue from the *Adventure* game.

Adventure was chosen deliberately: it is obvious that shoot-'em-up games such as *Doom* or *Descent* can be mapped reasonably easily onto the capabilities of VRML 2, but games that require a greater range of interactions than simply motion within a 3-D space and shooting a weapon are likely to be somewhat harder to migrate.

The source code of *Adventure* is fairly widely available, both in its original Fortran and in C. Accompanying the source is a fairly substantial database that includes long and short descriptions of rooms; their interconnections, some of which have a degree of randomness; the various objects (keys, cage, etc.); the verbs; and so on. This "database" approach suggests that conversion to use VRML might be feasible, and though the original program had many special cases handled by custom code, the fact that *Adventure* was subsequently rewritten to work with "*Adventure* interpreters" such as Infocom, means that a generic solution is possible.

How, then, can a VRML-based *Adventure* be constructed? The approach that we initially took, for the conversion is still in progress at the time of writing, is to make the game run under the control of Sony's *Community Place Bureau* server. This supports shared virtual environments, which will allow us to bring a new dimension—collaboration—to a previously single-user game.

The room descriptions contained in the database have been converted to invocations of a Tcl procedure:

```
room -name n -long t -short t -size {hwd}
-floor a -ceiling a -n.wall a ...
-n.exit e -ne.exit e ...
```

where *n* represents the room name, the *t*'s are long or short textual descriptions of the room, {*hwd*} are the dimensions of the room, *a* represents a procedure that determines the room's appearance (via a texture map), and *e* represents the room that an exit moves the player to. The room procedure is capable of generating both textual and VRML-based descriptions of the appearance of the room, so that, in principle, players may use either mode of interaction. As well as generating the VRML code for the room itself and for the tunnels that form the exits from it, the room procedure generates TouchSensors at each exit to implement the link.

Our initial implementation downloaded a description of each room as the player entered it, but we found that too slow. The approach that we are now investigating is to have all the room invocation generate Java data objects that contain the room descriptions, and then to compile them into a single "*Adventure* applet" that is downloaded only once. The VRML descriptions are then generated as required by the applet, giving an improved response. However, the problem with this approach is that it is not yet possible to delete objects from the scene with Sony's *Community Place* browser. Our expedient is to move objects below the floor, but they still consume memory—and presumably impact a little on rendering time.

As we have seen, generating the VRML descriptions is fairly straightforward; however, the most important features of the game are the way in which the player interacts with the objects. The objects in the rooms are also represented as VRML and have TouchSensors by which the player picks them up. The text-based *Adventure* allows the user to inspect his or her possessions at any time; the obvious way to do this is to have some buttons that are permanently placed to one edge of the field of view using a Billboard. However, although a Billboard can be rotated to face the player all the time, it is subject to translations and hence can disappear from the player's view as he or she moves around a room. Notwithstanding these difficulties, selecting the **Inventory** button moves the player to a "room" in which the objects lie; clicking the mouse on an object will then cause it to be dropped. The player may use the browser's **Back** button to return to the game.

These interactions are achieved by means of Java code downloaded into the VRML browser. However, since the various VRML 2 browsers provide different Java interfaces, we have not been able to produce a single interface to all browsers.

Other interactions are more problematic. *Adventure* involves performing certain actions with particular objects, such as waving the rod. While it is possible to conceive of performing such an action by picking the rod (i.e., activating its `TouchSensor` there is no way that the Java code can monitor how the rod is subsequently moved. This is not really a shortcoming in VRML *per se,* since VRML is much more concerned with the appearance of virtual worlds; however, it is indicative of shortcomings in VRML browsers that will have to be addressed before a wider range of interaction modalities can be achieved. Perhaps the best approach for browser writers would be to provide and document programmer interfaces for input devices, analogous to the external application interface that VRML 2 browsers are now expected to provide. This would at least permit the controlling application to monitor actions such as mouse movements.

Conclusions

This paper has demonstrated that VRML can be used to provide innovative interfaces to existing information systems. In developing several working VRML interfaces, the authors have developed an approach to generating the VRML descriptions based around scripts written in the Tcl programming language. This proves to be invaluable when the VRML interface contains information extracted from non-VRML data, such as a database. The same program-based approach holds promise as a possible solution to the current dichotomy over scripting languages in VRML 2.

In exploring the conversion of the *Adventure* game to VRML 2, our preliminary work indicates that generating VRML of the rooms is fairly straightforward but that many of the tasks the player is expected to perform (e.g., dropping the cage, waving the wand) do not translate easily to the limited interaction capabilities of VRML 2. We have managed to work around some of the problems but many still remain. It is clear that VRML browsers need further development before games such as *Adventure,* which requires a greater range of interaction modalities than existing VR-based games, can be accommodated. A programmer interface for input devices would certainly make greater degrees of interaction possible.

Acknowledgments

The authors would like to thank Kashaf Khan and Mark Bird, who are carrying out most of the experimental work in trying to convert the *Adventure* game to VRML.

References

J.K. Ousterhout, *An Introduction to Tcl and Tk,* Addison-Wesley, 1994.
Persistence of Vision Ray-Tracer: User's Documentation, (POV), tech. report, available from http://www.povray.org/ (1996).
Renderware Reference Manual, Criterion Software Ltd., 1995.

S. Upstill, *The Renderman Companion,* Addison-Wesley, 1989.

Virtual Reality Modelling Language: Committee Draft, tech. report, International Standards Organization (ISO), 1996.

B.B. Welch, *Practical Programming in Tcl and Tk,* Prentice-Hall, 1995.

• About the Authors •

Dr. Christine Clark is a Senior Research Officer in the Department of Electronic Systems Engineering at the University of Essex. She obtained a BSc in Physics at the University of Newcastle upon Tyne and then worked for British Aerospace. She subsequently returned to academia and obtained MSc and PhD degrees in Remote Sensing from UCL. She has worked on the identification of remotely-sensed spectra by neural networks and genetic algorithms, neural approaches to automated cartography, image presentation and processing for the production of teaching materials on remote sensing for schools, and automatic identification of ships in infrared imagery. Since arriving at the University of Essex, she has become involved in virtual reality, and she helped set up the *Virtual Applications, Systems and Environments* Laboratory. She is the main creator of the lab's virtual worlds.

Dr. Adrian F. Clark obtained a BSc degree in Physics from the University of Newcastle upon Tyne in 1979 and a PhD in image processing from the University of London in 1983. He has been with the Department of Electronic Systems Engineering at the University of Essex since 1988, where he is now a Senior Lecturer. He has worked on image restoration, analysis, compression and synthesis, and on algorithms and software techniques for image processing, using both serial and parallel hardware. Presently, he has become involved in networked virtual reality through Essex's *Virtual Applications, Systems and Environments* Laboratory, which he heads. Dr. Clark acted as Convenor of a British panel of experts involved in the development of an International Standard for Image Processing. He is a former Secretary of the British Machine Vision Association and has chaired Technical Committee 5 (Benchmarking and Software) of the International Association for Pattern Recognition.

Chapter 3

VRML: A Designer's View

Stephen Boyd Davis

Centre for Electronic Arts, Middlesex University

Helena Athoussaki

BT Laboratories

The authors question the emphasis on naturalism in virtual environments. They argue that such naturalism is often unthinking and based on a mistaken view of what VR can achieve. Using theatrical design and film as exemplars, they indicate the advantages of nonnaturalistic approaches, both in the models which are built and in how they are projected and rendered for the user. Attention is drawn to the value in providing means of authorial control over the parameters of viewing, and this control is discussed in light of the issue of narrative versus interaction. Recommendations are made concerning both the design of virtual environments and the facilities provided by the VRML specification.

Replication or Representation?

A key point underpins this paper. It is that the construction of virtual environments should be seen as an intentional activity, based on thoughtful, well-informed and inventive decision making. It is a design process in the best and broadest sense of the term. We argue that the construction of virtual environments is best seen as the construction of meaningful forms and experiences rather than as replication of the real world, and we question often unspoken assumptions about naturalism.

Some believe VR should be limited to those environments which mimic external realities, but there are problems with this approach. At first sight there might seem to be a clear distinction between virtual reality—the reconstruction of slices of life in the machine—and the building of artificial environments, where less reference is made to the characteristics of the observed world, as for example, a navigable space derived from scientific data. But, in fact, the difference is one of degree. Even imitative representations (what we might call hard VR) involve (and always will involve) very significant filtering and "falsification."

Part of the problem is that VR is seen by some as by definition concerned with the replication of reality. Work by many experts has brought in recent years the ability to specify in computer graphics and simulation systems such diverse attributes as dependencies between objects, object qualities such as brittleness, flexibility, and anisotropy, and environment qualities such as gravity and wind speed. However, it would be a misunderstanding to suppose that these facilities, however marvellous, are converging on the goal of replicating reality. A moment's thought will show that the only model that would approach the goal of replication would be boundless in size and of infinitely small granularity, modelling the behavior of subatomic particles! We must acknowledge that VR is a system of representation, involving selection and abstraction (Davis, Lansdown, and Huxor 1996). It is easy to imagine a time when, under highly controlled conditions, it will be possible to convince a user that a virtual environment is in fact real, but this illusion will always break when those conditions are suspended, or the user attempts to investigate more deeply those phenomena with which they seem to be presented. We can model aspects of reality, but that is no different in principle from any other act of representation. As creators of virtual environments, we have both the obligations and the freedoms that any designer has in constructing a representation. But rather than regretting that we cannot replicate reality, we can enjoy the freedom to make choices, as all designers do.

Designers and VRML: What Designers Are Not Complaining About

We do not wish to complain that current VRML models are too simple or that interaction in most browsers is not fluid and immediate. We are happy to work within the limitations of what the technology can provide, reducing the number of polygons, using flat shading, minimizing texture maps and so forth. This is something designers in all media do all the time. Good designers have no problem designing within limits—it is what they are trained to do. No one complains that a pencil is not a box of paints. As designers we do not even mind that what we design will appear differently on different machines: even this is not unprecedented in other media—a television designer cannot accurately predict the color gamut on a given display, or even how much of the image will be clipped at the edges. Nor are we making that regular complaint that VR is technology driven, or a solution looking for a problem. Some of the most exciting, practical, and conceptual inventions have been driven by technology, or even "idle" speculation, not by need. We do not care where innovative ideas come from—where they are heading is the issue.

What Designers Are Not Trying to Do

Designers do not try to copy the world. Very rarely, painters have believed that their task was to depict everything within view. Ruskin instructed painters to "Paint all you see, selecting nothing, rejecting nothing" and his view of the painter's eye as a passive data-collection device echoes his contemporaries' views about the nature of material evidence. Representation became a matter of choosing where to put the frame around the world. Photography was an answer to a Victorian prayer. But the remarkable thing about this view is that it is so unusual. The history and practice of representation provides few examples of unselective monitoring of the pattern of light on the retina (Gombrich 1980). There are of course many aspects of reality that are not captured in most virtual environments and which, in some cases, cannot be specified within the given system. Every system (including VRML 2) supports the specification of the obvious static attributes, such as object extent and position, plus dynamic attributes essential for animation, such as translation and rotation. Most systems have now added to the simple specification of color that of transparency, translucency, reflectivity, and surface texture. It is certain that progress will be made by the most humble systems towards the complexities of the most advanced. For example, VR systems have hitherto tended to model passive visual qualities, rather than those of activity, behavior and process. However, the open architecture of VRML permits the incorporation of properties that are not specifiable within the base system, and these might include more "organic" attributes such as tendencies to fall, to continue along a trajectory, to change color, decompose, grow, subside, or disperse. More complex behaviors will be incorporated, such as the simulation of animal or even human behaviors and propensities necessary for the creation of the virtual actor (Thalmann 1987, 1993).

Theatricalities: Realism and Nonrealism

If it is accepted that VR can, and in many cases should, transcend the borders of physical realism and nonrealism, an obvious place to seek inspiration is the theatre. As with painting, there has been a significant transition throughout this century from a view of the theatrical set as a realist pictorial view through a window, to that of the stage as an arena where the set designer can do anything at all, take whatever freedoms she or he chooses, in the service of the drama. Ironically, this view of theatrical design has now itself been cited by Laurel (1991) as a model for interface design—a plea to interface designers to view their task as the construction of environments suitably equipped with "props" in which the user's actions will take place.

At the beginning of the twentieth century, the art of the theatrical designer was little theorized. There was a general assumption that, in addition to providing the entrances and exits, the furniture and the props specified by the script, it was the job of the theatre designer to represent as accurately as possible the architecture, furniture, and so forth of the period described in the play. That the set should do something other than represent the surfaces of a real scene was not considered. However, there was, for example, in the productions of Irving, an insistence on unity in the overall conception, which already might begin to guide our thoughts about VR (Bablet 1962). In particular, contemporaries praised Irving's use of electric lighting to pull together the disparate elements of the scene (and no doubt also to hide in shadow some of the mechanical aspects of the set). As VR systems become more capable of handling complex rendering, including the effect of multiple light sources, we shall almost certainly want to use lighting more selectively, and perhaps partly under the user's control, to make the environment "speak" more effectively.

Edward Gordon Craig (1872–1966) was a follower of Irving who from 1897 took theatrical design strongly away from the naturalism of his predecessors. Why did he do this? What was it about a wholly naturalistic set that he felt did not serve his purpose, and how might his thoughts guide us in our choices about virtual realities? Craig's reasons were based on a desire for what we might now call media integration, concentration, and action.

Media Integration

By avoiding naturalism, the mimicry of reality, Craig found that he could better integrate all the elements of the production—in his case: actors, color, music, movement (Bablet 1962, 13). One of the problems of virtual environments is that we shall probably want to embed other media in them: for example, text. If we took the naturalistic route, we would be obliged to put text into virtual books on virtual shelves, which seems to forfeit many of the advantages of nonphysical media. To watch a movie, we would have to operate a virtual VCR. But by avoiding mimicry, we may be better able to create "realities" that provide a seamless integration of the different media types, while at the same time keeping the strengths of each constituent medium.

Concentration

By being selective and nonnaturalistic, Craig also found that he could focus attention where he wished. One of Craig's problems with naturalism (at least, with unthinking naturalism), was its inadequacy to get beyond realism. There were several aspects to this. For example, in creating a scenario for Shakespeare, a naturalistic set could undermine the intentional nonrealism of the words spoken (ibid., 45). Perhaps this seems an arcane problem, but in fact it is just a special case of the tension between symbolism and naturalism, and symbolism is by no means confined to the arts. When we consider something as simple as a graphical user interface and consider how such a symbolic environment might be concretized as a virtual environment, it is clear that not all the advantages lie with a naturalistic solution. The nonnatural symbolic nature of the GUI may actually be helpful, for example, in making clear that this is only one of many possible metaphors for the operation of the computer or in its ability to take liberties with scale so that all objects however near or distant have the same size.

Another Craigian idea valuable in thinking about virtual environments is the greater ability of nonnaturalistic environments to suggest the unseen (ibid., 134). In current practice, VR systems deal predominantly with surface appearances, and the systems in existence tend to emphasize the visual over the behavioral aspects of what they model—depiction rather than simulation. Craig's objection to naturalism in the theatre was partly based on his fear that it created a surface layer, which distracts attention from the real messages of the play. Ironically, it is the limitations of current VR technology that have tended to enforce heavily simplified and stylized worlds upon us, so that we cannot currently begin to create virtual worlds containing the rich minutiae of the real world. A Craigian reduction to elementals has been forced on us by the technology! But when the naturalistic detail of the real world becomes more achievable in VR, we shall need to decide whether this is, for some purposes, both more and less than what we want: more because details may distract us from what is being said; less because we may need powerful symbols, and symbols normally work precisely by eliminating particularities of surface appearance. A historical example might be the work of the pre-Raphaelite painters, whose work seems now to lack any clarity of meaning but which preserves in minute detail the (irrelevant) surface characteristics of what the painter observed. Reality is full of clutter, and we do not necessarily want our virtual environments to be the same.

Action

We remarked that the theatre has been adopted by Laurel as a metaphor for the human interface. One of the points on which she and Craig concur is in emphasising that the stage is not in itself the point, that it is a location for action. Traditionally, in the theatre and in VR, there is an assumption that people (or avatars) move, while environments remain static. For Craig, the emphasis on the set as a site for action led to his facilitating in his designs the process of change, often by the reconfiguration of modules. He likened the set to a face, which

TimeSurfer is a project undertaken for BT Labs at the Centre for Electronic Arts. The user chooses a configuration of avatar and depending on that choice, enters one of a variety of spaces. Here we see the Hellenic space.
Design: Helena Athoussaki and Taline Olmessekian.

always has the same parts laid out in a recognizable pattern, but which alters its expression with any change in one of its features (Bablet 1962, 126). Modern theatre design, following Craig's tradition, has made extensive use of the "stage as actor," using revolves, flying stages, and dynamic lighting. In particular, scenes transform themselves fluidly into other scenes: "it was important to conjure up a . . . world that could be at times interior, exterior, hospital, hotel, prison . . ." (Anthony McDonald describing his 1984 set for Orlando, quoted in Goodwin 1989). While real buildings and landscapes are not generally reconfigurable, virtual environments will surely want to reject adherence to this limitation of the real world. Incidentally, Craig became increasingly fascinated by the possibilities of a purely mechanical performance without the irritating distraction of real actors! He began to feel that sophisticated marionettes would be preferable. He would surely have loved VR.

From the Model to the View

It might be thought that we are advocating a wholesale antinaturalist approach to VR, such as Craig undoubtedly adopted in the theatre, but this is not the case. The point being made is that in deciding on the role of naturalism in virtual environments, some of the reasons that Craig had for objecting to naturalism may be worth considering for our own purposes. Above all, Craig was a functionalist in that he saw the purpose of the design as to serve the idea (not the other way round). In turning now to the very process of viewing itself, our argument will be that the cinema—while apparently a highly naturalistic form—also adopts an exceptionally stylized, indeed artificial, approach, and that once again VR has much to learn.

The VRML specification is largely given over to enabling authorial decisions about the model, and not about the view. Of course, model data do not exist visually, but numerically and textually. It is rendering that makes the model visible, and this must by definition incorporate a characterization of the process of seeing. Though at a technical level all designers of VR and other visual modelling systems recognize the important difference between features of the model and features of the view, there is a tendency for the implications of this distinction to be overlooked: the viewing process is largely unquestioned.

A View of Viewing

If we suppose a certain view of a model specified according to particular perspectival parameters, then this still leaves unspecified the actual marks that render that model visually. While one option is to render the projected scene using marks that will approximate a photographic view, an alternative is to render the scene using "expressive" algorithms that, for example, permit the accentuation of certain features in a more arbitrary way. These issues have been discussed by Lansdown and Schofield (1995), but we will not pursue them in depth here since we wish to concentrate on other aspects.

Since the Renaissance, when the rules of perspective construction in painting were developed and codified, most notably by Alberti, Viator, and DŸrer (Ivins 1938), the viewing of the three-dimensional world came to be seen as unproblematic, so that by the nineteenth century, it was assumed that the human perception of reality was fully understood. Painting was a matter of replicating what the eye sees when looking on the real scene. Since then, advances in the understanding of perception as an integrated eye-brain system (summarized by Lansdown in Davis, Lansdown, and Huxor 1996) rather than as a mechanical optical device have made us more wary of saying that we know what we see. For example, it seems likely that the design of early cinematography lenses was based on a view that the human eye sees everything in focus at once—a view that no one now holds. A different opinion about that perceptual process would have led to a different concept of photo-realism.

When three-dimensional cartesian data are stored in a computer system, this implies almost nothing about what the user will see. Assuming that vision (rather than sound or tactile stimulus) is the means of perceiving the three-dimensional model, then it is unavoidable that the three-dimensional data will be mapped onto two-dimensional surfaces, whether a conventional desktop monitor, projection screen, miniature LCDs in a headset, or even the retina itself. A 3-D to 2-D transformation must be applied, and what this transformation may be is dictated by the specification of the viewing system, not by the model. In a graphics rendering system, any consistent model of perspective can be applied (and with some difficulty, we could even make it inconsistent if we chose to). VR systems are normally assumed to map the three-dimensional data using three-point perspective (the system we normally think of as "perspective"). We cannot deal here with the question of whether three-point perspective is, in a special sense, correct (see Hagen 1986). The issue here is a more humble one: to consider the role of viewing in its own right and to ask for its specification as part of the VRML system.

Some Characteristics of Viewing

If we consider the act of viewing, we can begin to enumerate its characteristics. In doing so, we ought really to distinguish clearly between phenomena associated with the eye, the camera, and algorithmic projections, but in the interests of brevity, some of these distinctions are given less attention than they perhaps deserve. We shall need to deal with the effects of the parameters viewpoint, depth of field, and angle of view and also with the way these are altered and enhanced by being experienced over time.

Viewpoint

Viewpoint comprises two questions in one: from where is the scene viewed, and toward what is that view directed? It is fundamental to computer animation (for example for film and TV) that the author specifies the eyepoint, and vestiges of this have been transferred to VRML, with its ability to specify an initial view and a series of sequenced viewing stations thereafter. We shall see later that the inclusion of this facility opens a doorway on many more facilities that might be provided.

Depth of Field

We have mentioned that at the time of the earliest photography, it was believed that the entire visual field was in focus. Now this view is modified in two ways. First, the eyeball itself is known to change shape under muscular control in order to alter focusing distance: objects at different distances cannot all be in focus at the same time. Second, we know that only that portion of the scene that is opposite the fovea (a very small part of the retina) is clearly resolved and that it is through rapid movements of the eye (saccades) that this part of the eye is directed at different parts of the scene. So both in depth and across the scene, it is impossible for all parts of the scene to be equally resolved.

It seems certain that Rembrandt's self-portraits owe a part of their feeling of presence to his depiction of the differential focus of the planes in viewing the face, so that the eyes, the hypnotic subject for any painter observing his own face in a mirror, are in sharp focus and the tip of the nose and the distant parts of the head are relatively defocused. Baxandall (1985, 80) claims that Chardin uses selective sharpening and softening of edges in his paintings in order to imitate the effect of the eye taking certain trajectories across the scene. The painter hopes to lead the viewer's eye through the painted image by increasing the sharpness of certain edges and points on a particular trajectory.

The issue of depth of field really has two aspects: focal plane and focal range. We can say what is in focus and also to what extent objects at different distances are unfocused. The uses of these parameters in cinematography are further discussed later in this chapter.

Angle of View

We know that the eye is constantly making rapid saccadic movement, punctuated by short pauses (0.2–0.5 seconds) for attention. Thus, the center of attention moves around involuntarily, in addition to deliberate movements of the eye and head. The range of these movements when looking at a given scene is dependent on what is being looked at: it is emphatically not a rasterlike scanning process, tending instead to concentrate on those parts of the image that are the most informative. But even when the eye is directed at some part of the scene, it is capable of providing a much wider view in an unfocused form. This is a respect in which the eye is notably unlike a camera (or a computer monitor): the eye has no fixed boundary to its view, whereas mechanical devices have. The latter are, by their nature, framing devices (though we may be able to overcome these limitations using displays larger than the maximum possible angle of view). In addition, any mechanical lens will present the scene very differently depending on its focal length. This is familiar from the strangely foreshortened image that a cricket wicket presents when viewed using a telephoto lens, and from the almost unacceptably distorted image produced by a fish-eye lens. In terms of projection systems, these projections belong to the family of convergent perspective views, but we should also include nonconvergent projections such as the isometric and axonometric. In the isometric, all lines are drawn to scale, rather than decreasing in size with distance, horizontals are inclined at 30i, and all planes are equally distorted. Approximations to such views have been used in Roman, Byzantine, Persian, and Chinese paintings and extensively in Japanese woodcuts (Dubery and Willats 1972). In axonometric projection, horizontal surfaces are drawn in rotated plan view, and the necessary verticals and horizontals are then appended to them (ibid.). This projection has attracted architects throughout the twentieth century because of its ability to combine an undistorted plan with an evocation of the character of the interrelated spaces.

Another way in which we should distinguish between the eye and camera (on the one hand) and painted and synthetic views (on the other) is in relation to the vanishing point. This is actually another of the issues of framing. The eye and the camera are both in the same bind in that wherever they "look," it is towards that point that all lines are perceived as converging. One of the views of converging railway lines in Figure 3.1 cannot be achieved with a normal camera.

Distortion

Distortion is a loaded word. It implies falsehood. We think we know the difference between a visually distorted and an undistorted image. However, "distortion" can be a more subtle affair, which assists rather than impedes the user's experience. All the projection schemes just discussed are geometrically consistent renderings of three-dimensional scenes, but other transformations can be imposed for a variety of purposes. Sarkar and Brown (1994) give an account of useful synthetic "fish-eye" views of information on computer displays, which are not optically consistent but which are algorithmic. Within the

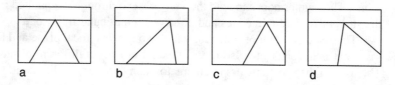

a   b   c   d

Figure 3.1   The archetypal view of railway lines converging at the horizon is shown in (*a*). In (*b*) we are still standing in the middle of the track, but it recedes towards a point that is not directly ahead of us. (*c*) shows the track receding from a point to our right towards a place that faces us on the horizon. But what does (*d*) depict? It is not a view that could appear in any uncropped photograph or in a film, yet such views have been used in painting ever since the Renaissance, partly for their ability to wrong-foot the observer, making the observer reappraise the image. Technically of course it is easy to produce. Cropping a photograph asymmetrically will produce such an effect, and digitally we can project an image that is then clipped to a noncentral viewport.

restrictive space offered by a conventional display, their aim was to reconcile the demands of providing maximum contextual information with the largest, clearest possible view of the current topic of interest. The solution adopted was to provide a central zone displayed in the standard way and to progressively compress the image nearer to the perimeter of the screen. Of course, one of the advantages of digital projections is that we can instantly abandon one that does not serve our purpose, or adopt a view that, for example, compresses only at two edges rather than all four.

Another form of distortion is one that is explicitly semantic. A famous two-dimensional example is the London Underground diagram. Garland (1994) documents the wealth of human decision making that since 1931 has gone into designing the deviations of placement of stations and the distortions to the geometry of the connecting lines, with the principal aim of allowing the topology of the lines to take priority over their topography.

The history of painting again provides insights: we cite one example given by Dubery and Willats (1972). If we survey a wide angle on a scene, using "undistorted," three-point perspective, the resulting image tends to have what we subjectively regard as distortions at the edges—things seem overlarge, or stretched, at the perimeter of the view (Dubery and Willats 1972, 87–89). However, if we distort the objects as though they were projected onto a curved picture-plane, we create what appears to be a more "natural" view (ibid., 89–90). In addition, we gain the advantage, as van Gogh does in his painting *Vincent's Room,* that viewers "sense" their own location in relation to the scene—a key requirement of virtuality. *Vincent's Room* feels like a small, intimate space because of the "distortions" imposed on it.

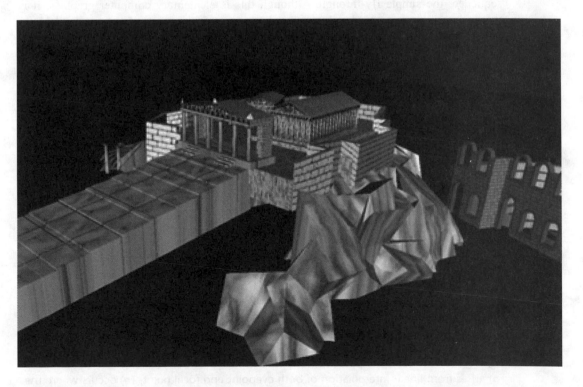

TimeSurfer, for BT Labs at the Centre for Electronic Arts. An overview of the Hellenic space. Modelling in 3-D Studio Rev. 4 on PC, StrataStudio Pro 1.75 on Macintosh, Cosmo Worlds on SGI.
Design: Helena Athoussaki and Taline Olmessekian.

Viewing in Film

It is worth looking in some detail at how film makes use of authorial control over viewing. A naive model of filmmaking supposes it to be the bringing together of protagonists and script, and filming as "transparently" as possible the events that unfold. But this is to ignore the crucial contribution made by cinematography: in some ways, filmmaking is control of viewing. The filmmaker decides what is revealed and what is concealed, and how those things which are revealed are framed, focused, enlarged, reduced, and so forth.

Originally the aim of filmmakers was merely to record moving objects. The earliest experiments of the Lumière brothers were essentially moving equivalents of still photographs, which had no narrative power other than that inherent in the subject matter (Reisz 1968). But since then, authorial control of viewing has become a key element in the power of cinema to shape the viewer's perceptions. Subsequent innovations have enhanced the ability to communicate—in an informational sense—and to incite in the viewer attitudes to what is shown. The former is most obviously relevant to documentary, the latter to the fiction film, though there is considerable overlap between these two genres.

Control of Viewing Trajectory in Film

Of course, a sequence of viewpoints (for example, in a VRML world) is more than the sum of its parts, since it implies the intermediate views encountered in moving from one viewpoint to another. We should consider the question of the "motorized monocular" viewpoint sequence, the simple fly-through. Although this is elementary computer graphics, it is worth describing in some detail, to remind ourselves of the range of options available in systems that allow aspects of a viewpoint path to be specified. Since any two views each comprise an eyepoint and a focal point, we have a number of possibilities as we move from one viewpoint to another. Once half the time from a view 1 to a view 2 has elapsed, where are we and what are we looking at? Are both eyepoint and focal point simply interpolated linearly? Probably not, since this will lead to peculiar "twitches" of view when we leave one viewpoint and set off towards another. But if the paths are curves, what shape are they, and do the two points move along their curves with equal velocity (Figure 3.2)?

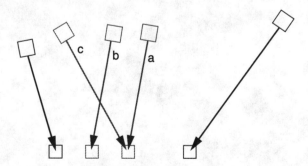

Figure 3.2 Three of the possible viewpoints at half-time between two known views: (*a*) arises from linear interpolation of both eyepoint and focal point; (*b*) occurs where the trajectories of both eyepoint and focal point are straight lines, but the rate of change is nonlinear; (*c*) assumes that the path of the eyepoint is curved.

The crudest form of trajectory control is simple omission of that which the filmmaker considers irrelevant: in other words, editing. From early days this has been used in a number of different ways. For example: to omit a period of time between interesting moments (as the acts in a stage play conventionally had done); to move attention from part of a scene of action to another without having to shift the camera view (but where time represents approximately the actual time which is elapsing in the scene); to alternate between two locations showing events which are occurring contemporaneously (thus taking liberties with the flow of time). These techniques were all introduced in the five years after 1895, when the Lumières showed film publicly for the first time (Figure 3.3).

The earliest films (for example, by Méliès in 1899) made no use of moving cameras (and the zoom lens was then unknown) so that the construction of a viewing trajectory was impossible: the camera was pointed at the action, and editing was the only means of taking the viewer from one shot to another. For example, in *Life Rescue at Long Branch,* made at Edison in 1901, a transition is made from a long shot of a beach resuscitation to a closer shot of the same, by cutting the two shots together (Salt 1990). By the 1920s, however, thanks to the work of Griffith and other pioneers, an innovative director like Abel Gance, in *Napoleon,* was using all of the following: cameras travelling on horizontal and vertical rails; a hand-held camera in a harness on the cameraman's shoulders; a camera on a portable "guillotine" so that it could rise and fall at the same time as being moved along; cameras on cars; even a camera mounted on a giant pendulum that could swing towards and away from its subject in deliberately giddying sweeps (Brownlow 1983). Nowadays, the majority of these techniques—tracking, dolly, crane, and helicopter shots—are part of cinema's everyday visual grammar. All are more expensive to shoot than static camera work, and all are used because they lend both clarity and power to the film.

One of the most powerful techniques that a filmmaker can use is to deny the viewer the sight that the viewer most craves (it is clear that we must soon discuss how this relates to an interactive medium under complete user-control). One particularly striking example will suffice: in *Rosemary's Baby,* directed by Polanski in 1968, Ruth goes to use the phone in an adjoining room. Polanski uses the doorway as a frame for this action but does it so that Ruth's face is concealed from view—the audience yearns to see her expressions, but cannot, and is thereby more deeply engaged in the action than they would have been if everything had been revealed. Earlier, in 1958, the opening sequence of Welles' *Touch of Evil* had played several similar tricks, denying viewers the chance to keep track of the very thing they most wanted to see by filming the main characters taking a journey around several blocks and down several streets, so ensuring that our view of them would be periodically interrupted by buildings, traffic, passing handcarts, and other obstructions. Again, the effect is of increased, not decreased, engagement.

Viewpoint	Viewing trajectory
Depth of field	Focus-pull
Viewing angle	Zoom

Figure 3.3 Time-based attributes of viewing. The addition of the time component to the attributes of viewing provides three important features of cinematic expression.

Focus in Film

The photographer can choose both focal plane and focal range. With the element of time, the cinematographer can manipulate these dynamically. Commonly, the intention is to transfer clear definition from one actor or significant object to another, as an analogue to the process of shifting one's attention (either deliberately or through the act of noticing). Simulating differential focus is computationally expensive and will not be incorporated early in accessible VR systems, but is complementary to the other aspects discussed here, which have no significant computational overhead.

One of the obvious uses of control of view in film is in "pulling focus," where there is a transition from one plane of the subject matter to another. This is a powerful authorial technique for forcing the viewer to attend first to one thing, then another. Towards the resolution of *Who's Afraid of Virginia Woolf?* (Nicholls 1966), the camera zooms or tracks towards Burton and Taylor, in an interior, closing in on their clasped hands, but the focus is then shifted to a glimpse of daylight in the world outside. Not only is there the obvious fact that something out of focus is more difficult to discern, but psychologically it is impossible to resist having one's attention captured in this way. Cause and effect are in reverse: normally shifting one's attention in depth leads to a change in focal plane; here a change in focal plane leads to a shift in attention. The filmmaker's will is irresistible: the viewer cannot choose to ignore this authorial instruction.

Zoom Control in Film

At one level, zooming is just a matter of changing gradually the degree of enlargement that a lens provides, which seems tantamount to moving the camera nearer or farther from the subject. However, the effects that it produces are more interesting than this implies, since the relationship between objects at different depths in the scene are not in fact the same if we compare viewing through a long lens and viewing close up as shown in Figure 3.4.

Gance's *Napoleon* (1927) used a wider variety of lenses than had been used before in a single film, from 275mm to 20mm (Brownlow 1983, 54). Subsequently, the use of a great variety of lenses, and of adjustable lenses that can be zoomed from one focal length to another, has become commonplace. Zooming in on a subject serves the practical need of revealing greater detail within a small part of a scene without a disruptive cut (which

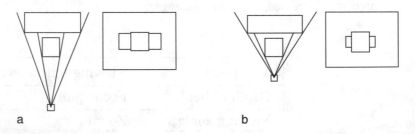

a   b

Figure 3.4 Long lenses and close-ups. An enlarged view from a distance (*a*) is clearly not the same as a close view (*b*). Both the relative distances of the objects from the lens and the degree to which objects occlude one another are different.

would be inevitable if two or more different lenses were used), but it is also a convenient analogue of the psychological process of concentrating one's attention on part of a scene (similar techniques are used in sound for film, where a sound to which we are meant to attend is increased in volume relative to the background, imitating our natural ability to discern those sounds on which we are concentrating). When combined with other techniques such as tracking (physically moving the camera), zooming can produce some remarkable effects. For example, in *Goodfellas* (Scorsese 1990), two characters sit facing one another at a cafe table. We view them in profile with a window behind, through which we see the street, its buildings, and traffic. Using a combination of tracking back while zooming in, a strange effect occurs: though the two characters remain the same size, the street scene behind them gradually moves nearer as they speak. The effect is to make the street ominous, insistent, and at the same time, more schematic, less real.

A point to emphasize is that the film conventions that we now find so convincing, unremarkable, and indeed largely transparent, themselves took time to evolve and had to be learned by audiences (Musser 1991, 392–407). We can only wonder what new forms of expression may be latent within VR. After all, the early cinema was confined to simple documentary sequences and technical showing-off of various kinds: no one could have predicted the major role of cinema and its offspring, television, in our culture.

Control of View in Virtual Reality

What relevance do the tricks and techniques of an authorial medium like film (or painting for that matter) have to the user-controlled interactive world of virtual reality? Surely we are missing the point of an inherently interactive medium. In the construction of virtual objects and environments, what business has the author in trying to dictate how the viewer sees?

It is true that one approach to VR is to regard it as equivalent to architecture or sculpture. When sculptors or architects make three-dimensional physical structures, they must reconcile themselves to working within the laws of vision as they find them: none of the film techniques we have described are available to them (though many architects have made attempts to harness optical illusion to their benefit, as close to controlling vision as they are able to get). Perhaps for the author of a virtual environment to interfere in the parameters of vision is equivalent to a sculptor altering the geometry of his spectator's eye, and is inexcusable.

One of the greatest constraints on the development of a new medium must be the set of preconceptions that is brought to it by early practitioners. Seeing the authorship of virtual environments as akin to sculpture/architecture is one approach but is not the only one available. Again, theatre may provide useful insights. Theatre design has a productive and healthy attitude to illusion. The twin millstones of twentieth-century design and architecture, functionalism and "truth to materials," are largely absent from theatre design. Theatre designers know they are dealing primarily in ideas, not functions, and this surely is the purpose of the majority of virtual environments, too. Theatrical design is free to resort to any illusion that serves its purpose. Even though constrained by the laws of vision, theatre design does all it can to subvert those laws: if it could change them, it would.

But let us return to the argument that in an interactive medium, the author has no business controlling the act of viewing. We could say that it is the user's prerogative, not the author's. Have we not provided browsers where the user can modify many aspects of the view at will? But we should acknowledge that VRML has already trespassed into "interfering" with viewing, albeit in a rather half-hearted way: we have already started down the path of giving authors control over how their work is seen. What are viewing nodes and the

Community Network is an experimental shared space created at BT Labs, using Cosmo Worlds on Silicon Graphics workstations for use with the Sony Community Place browser on Windows95. Shared spaces are implemented using Sony's Community Place server. The shared space is intended to avoid representing any recognizable locations, but to embody a Suffolk flavor in its design.

paths between them if not an authorial device to control the process of viewing? (Of course the user is also free to ignore the viewing paths provided by the author, but that is not the point.) Authors can control the users' viewpoint to make it look as if they are being lifted, or flying, under program control (for example, allowing the user to ride in an open lift, their "eyes" moving through the floors). And naturally, the idea of a viewpoint node offers a field of view parameter, which allows the view angle to be set like a zoom lens. So it is clear that we are already well on the road to interfering actively in the construction of the viewing process, not just the model. It would simply be untrue to say that VR (as represented by VRML) concerns itself only with what is in the world and not with how that world is seen.

We should also ask: Where does the virtual world stop and our viewing of it begin? If we consider the controls we have over levels-of-detail (LoDs), the distinction starts to break down. We use LoDs as an analogue of something that arises from viewing: those parts of the world that we are unable to clearly discern, being too far away, are represented by simple objects, but as we approach them and would be able to discern more detail, we swap-in a model that provides that detail. The analogue of an artifact of the viewing process is achieved by altering the world itself.

Is Control of Viewpoint Relevant to Interactive Environments?

We can discern two opposed schemas: that of the cinema, in which authors have absolute control, and that of virtual environments, in which the author's task is over once the model is built. But we hope we have indicated that VRML is not at either extreme. In fact, it tries,

Community Network is an experimental shared space created at BT Labs, using Cosmo Worlds on Silicon Graphics workstations for use with the Sony Community Place browser on Windows95. Shared spaces are implemented using Sony's Community Place server. The shared space is intended to avoid representing any recognizable locations, but to embody a Suffolk flavor in its design. (*See color plate on page 307.*)

rightly, to have the best of both worlds: to offer a world that can be browsed at liberty in any way the user chooses; and also to offer a series of authorial viewpoints on that world. As such, it offers a new site for an old issue—one that has been discussed widely in interactive multimedia: the tension between narrative and interaction.

The problem is—and it's terribly obvious really—that most successful communication involves a great deal of craftsmanship and authorship and point of view and storytelling and narrative. Every successful form, be it a novel or a feature film or a play or a comic, needs a skilled storyteller to weave together a spell in the mind of the audience, suspend their disbelief, and take them on a carefully planned emotional roller coaster through the story. Every successful form of communication involves protagonists; a set of conflicts and experiences; and at the end, some sort of resolution so the thing has a satisfying shape. Interaction largely destroys all that. By giving the audience control over the raw material, you give them precisely what they don't want. They don't want a load of bricks, they want a finished construction, a built house (Max Whitby, Multi Media Corporation, quoted by Cameron 1993).

This is an apposite symbol, because building houses is just what we seem to do in the construction of virtual environments. But perhaps our houses are just piles of bricks: perhaps it is the element of narrative that makes a (metaphorical) pile of bricks into a true house. We hope it is clear that we are not arguing that the user should be forbidden the freedom to view, move, interact as they please, but that oftentimes there will be a need for the authors to provide more.

Whitby seems to be referring primarily to fiction, where the element of narrative is traditionally given priority. In the context of hypertext and multimedia, many have wrestled with the relationship between the strengths of narrative and the freedoms of interaction (for example, Landow 1992; Rees 1994). But the issue is equally relevant to nonfiction works. A recent project has begun investigating the advantages of a narrative component in educational multimedia: "A narrative structure, whether provided by the design of a CD-ROM or the teacher, can hold everything together at a conceptual level" (Plowman 1996). A definite shift is detectable here. Whereas only a few years ago it was the complete freedom of interactive multimedia that gave it its appeal—unlimited choices, surfing multiply connected information spaces, the open work, readers as writers, and so on—now there is a renewed appreciation of the author's guiding hand.

Conclusions

We acknowledge that controlling the user's style of viewing can be said to be against the whole spirit of VRML, an unwelcome intrusion of authorial control into what is properly the user's domain. However, we believe that three factors will tend to lead to greater interest in authorial control:

- *Precedent*   In media where technology has facilitated authorial control of viewpoint, it has been enthusiastically used. The example we have cited at some length is film.
- *Motive*   In multimedia, there has been a shift from a wholesale embracing of the infinitely open work, to a new appreciation of narrative in both fiction and nonfiction. We have cited some of the benefits of authorial control over the parameters of viewing.
- *Means*   VRML already provides viewing nodes, with trajectories between them, and associated angles of view. Using program control, the user's "eyes" can be transported from place to place. The barrier between authorial building and authorial control over view has therefore already been crossed.

We believe that designers of virtual environments should recognize the merits of nonnaturalistic models and consider whether such models would allow them to better convey their ideas. They should also take advantage of the (currently limited) facilities for controlling viewpoint. Designers of virtual environments should take pains to learn the basic psychology and physiology of viewing and the practices of the filmmaker, or (far more probably) make use of the expertise of others. This would enable them to offer users the option of traversing and viewing models through "authored eyes," rather than making the model itself bear the full burden of engaging, informing, and perhaps entertaining the user. To this end, the design of VR systems and specification languages should provide greater opportunity to specify all aspects of the process of viewing.

References

D. Bablet, *Edward Gordon Craig,* published in French, 1962; 1966 translation published by Heinemann.

M. Baxandall, *Patterns of Intention—On the Historical Explanation of Pictures,* Yale Univ., Conn., 1985.

K. Brownlow, *Napoleon—Abel Gance's Classic Film,* Jonathan Cape, London, 1983.

A. Cameron, "Dissimulations—Illusions of Interactivity," University of Westminster, Hypermedia Research Centre http://www.wmin.ac.uk/media/HRC/ (1993).

S.B. Davis, J. Lansdown, and A. Huxor, "The Design of Virtual Environments Report of the Support Initiative for Multimedia Applications," ISSN: 1356–7370, Loughborough University, July 1996.

F. Dubery and J. Willats, *Perspective and Other Drawing Systems,* Herbert Press, London, 1972.

K. Garland, *Mr. Beck's Underground Map,* Capital Transport Publishing, Middlesex, UK, 1994.

E.H. Gombrich, "Standards of Truth: The Arrested Image and Moving Eye," *The Language of Images,* W.J.T. Mitchell, ed., Univ. of Chicago Press, Chicago, 1980, pp. 181–217.

J. Goodwin, *British Theatre Design: The Modern Age,* Weidenfeld and Nicholson, London, 1989.

M. Hagen, *Varieties of Realism: Geometries of Representational Art,* Cambridge Univ. Press, 1986.

W.M. Ivins, *On the Rationalization of Sight,* Metropolitan Museum of Art, 1938; reprinted, Da Capo Press, New York, 1975.

G.P. Landow, *Hypertext—The Convergence of Contemporary Critical Theory and Technology,* John Hopkins Univ. Press, Baltimore, 1992.

J. Lansdown and S. Schofield, "Expressive Rendering: A Review of Non-photorealistic Techniques," *IEEE Computer Graphics and Applications,* 1995.

B. Laurel, *Computers as Theatre,* Addison-Wesley, 1991.

C. Musser, *Before the Nickelodeon—Edwin S. Porter and the Edison Manufacturing Company,* Univ. of California Press, 1991.

L. Plowman, "Narrative, Interactivity and the Secret World of Multimedia," *The English and Media Magazine,* No. 35, Autumn 1996, pp. 44–48.

G. Rees, *Tree Fiction,* 1994.

K. Reisz, *The Technique of Film Editing,* 2 ed., Focal Press, London and Boston; reprinted, Butterworth, 1982.

B. Salt, "Film Form 1900–1906," *Early Cinema—Space, Frame, Narrative,* T. Elsaesser and A. Barker, eds., BFI Publishing, London, 1990.

M. Sarkar, and M.H. Brown, "Graphical Fisheye Views," *Communications of the ACM,* Vol. 37, No. 12, December 1994.

D. Thalmann and N.M. Thalmann, "The Direction of Synthetic Actors in the Film Rendez-vous Montreal," *IEEE,* Vol 7, No. 12, 1987, pp. 9–19.

D. Thalmann and N.M. Thalmann, "The World of Virtual Actors," *Virtual Worlds and Multimedia,* D. Thalmann and N.M. Thalmann, eds., Wiley, Chichester, 1993, pp. 113–126.

Films:

A. Gance, (director and sc.) *Napoleon,* WESTI/Society Generale des Films, France, 1927.

R. Montgomery (director), *The Lady in the Lake,* MGM, 1946.

M. Nicholls (director), *Who's Afraid of Virginia Woolf?,* Warner, 1966.

R. Polanski (director and sc.), *Rosemary's Baby,* Paramount, 1968.

M. Scorsese, *Goodfellas.*

O. Welles (director and sc.), *Touch of Evil,* Universal Int'l, 1958.

• About the Authors •

Stephen Boyd Davis is Principal Lecturer in Design for Interactive Media at Middlesex University's Centre for Electronic Arts. His specialist interests include novel forms of interaction, users' interpretation of visual designs, and design education. The University's Centre for Electronic Arts undertakes projects with BT Labs.

Helena Athoussaki is an Interaction Designer working in the Human Factors unit of BT Labs research department. She joined BT in October 1996 after receiving her MA in Design For Interactive Media from Middlesex University Centre for Electronic Arts. She is currently working on 3-D community intranet applications and immersive VR projects.

## Chapter 4

Ulysse: **An Interactive Spoken Dialogue Interface to Navigate in Virtual Worlds**

Lexical, syntactic, and semantic issues

Christophe Godéreaux, Pierre-Olivier El Guedj,
Frédéric Revolta, and Pierre Nugues
Institut des Sciences de la Matière et du Rayonnement

Abstract

We describe a spoken dialogue interface in a virtual reality environment. This prototype—*Ulysse*—accepts utterances from a user enabling him or her to navigate into relatively complex virtual worlds. The paper first describes *Ulysse*'s architecture, which includes a speech recognition device together with a speech synthesizer. *Ulysse* consists of four principal modules: a chart parser for spoken words, a semantic analyzer, a pragmatic analyzer, and a dialogue manager. Then it describes each of these principal modules by indicating their function in the processing of multiwords found in the corpus. The study of multiwords is primordial for a human-machine dialogue system. *Ulysse* has been integrated in a virtual reality environment and demonstrated.

Introduction

Speech interfaces are beginning to rise in simulation or virtual environments to ease interaction (Karlgren et al. 1995; Bolt 1980; Ball et al. 1995; Everett et al. 1995). Speech is considered more natural and in certain cases an easier mode to interact with virtual worlds. These interfaces require several linguistic modules such as speech recognition systems and speech synthesizers, syntactic parsers, semantic and pragmatic analyzers, and dialogue managers (Allen 1994). In a virtual environment, speech is only one mode of interaction in a multimodal context. Some adaptations must be made to classical dialogue architectures. Pointing devices must be integrated with speech.

Ulysse has been designed during a COST-14 program of the European Community on Computer Supported Cooperative Work (CoTech 1995). We have developed more particularly linguistic tools or spoken dialogue capabilities that could be embedded in virtual agents. *Ulysse*'s capabilities concern only navigation, which is certainly the most important capability a conversational agent can bring to help a user in virtual environments (Godéreaux et al. 1996:a; Godéreaux et al. 1996:b). *Ulysse* responds positively to motion commands and acts consequently. It transports the user embodiment within the virtual environment.

Ulysse

We collected a corpus of four dialogues in order to determine how spoken dialogue interface can help a user of virtual worlds. The design and the study of the first two dialogues allowed us to suppose that such an interface may help users in their navigation in virtual worlds. We've collected two others dialogues to examine the combination of speech and a deictic pointer (a mouse) by users in a situation of navigation.

From the corpora, we computed statistics on users' phraseology. It enabled us notably to build the lexicon and the application's grammar. It also served as a bootstrap to design and implement *Ulysse*.

Ulysse takes the form of a conversational agent that is incorporated within the user's embodiment. *Ulysse*'s overall structure is similar to that of many other interactive dialogue systems (Allen et al. 1994). It is inspired by a prototype we implemented previously to interactively dictate medical reports (Nugues 1993; Nugues 1994). It features speech recognition and speech synthesis devices, a syntactic parser, semantic, pragmatic, and dialogue modules. *Ulysse*'s architecture is also determined by the domain reasoner and the action manager.

We implemented *Ulysse* using the Distributed Interactive Virtual Environment (DIVE) (Andersson et al. 1994) from the Swedish Institute of Computer Science (SICS). DIVE enables us to build virtual worlds where users can connect from a remote location, move into, and meet other participants. Participants share the same geometric model of the world with a different point of view. Modifications of the world from user interactions are replicated to the other participant sites to keep the world consistent. Our navigation prototype in virtual universe (Godéreaux 1997) uses a modular architecture (Figure 4.1) similar to that of *TRAINS* project (Allen et al. 1994).

Speech recognition is carried out using the IBM's VoiceType device. VoiceType does isolated word recognition. The speaker must pause between words. It is primarily intended for report dictation.

Ulysse consists in seven modules. The main ones are **chart parser, semantic analyzer, pragmatic analyzer,** and **dialogue manager.**

The **parser** (El Guedj 1996) adopts a classical bottom-up chart algorithm with a dual syntactic formalism: It can operate using phrase-structure rules—Chomsky's grammars—and a dependency formalism. It uses a *lexicon* and a *grammar* written to accept all the 400 utterances of the corpus (Godéreaux 1997). The lexicon is using parts of speech similar to *MULTEXT* categories (Véronis and Khouri 1995). The chart parser is connected to the recognition device output, and it accepts the words. The chart detects and reconstitutes simple multiwords, such as « *au dessus de* » /above/ or « *à droite de* » /on the right hand side of/, defined in the lexicon. It computes zero, one, or more parse trees.

The **semantic analyzer** builds the semantic representation of an utterance according to lexical semantic data. It first performs a functional analysis, then computes lexical definitions of verbs. Complex verbs are defined in terms of simpler ones. The input of semantic analyzer is a parse tree, and the output is the corresponding semantic structure.

The **pragmatic analyzer** manages the assignation of referents. It links the objects of virtual worlds to the referential terms of utterances with the **reference resolution** module. It uses spatial knowledge derived by the **geometric reasoning.** It also processes the syntactic ambiguous utterances. It determines the semantic structures calculated by the semantic analyzer from different parse trees (note the double arrow between semantic and pragmatic analyzer in Figure 4.1). If irrelevant, it discards the parse tree and takes the next one. The pragmatic analyzer accepts parse trees from the semantic analyzer as well as the interactions deictic of user until pragmatic analysis succeeds. It identifies the speech acts and the corresponding virtual agent movements.

The **dialogue manager** is the core of architecture. According to speech acts, it generates and synthesizes the virtual agent utterances. It manages the turn taking between each speaker. It also manages conversational and linguistic ellipses by analyzing some lexical data with the chart parser and semantic and pragmatic analyzers. Then it transmits the movements calculated by the pragmatic analyzer to the **action manager.** These actions are executed in the VR system. Finally, the action manager signals the action's end to the dialogue manager.

In this article, we are going to present more precisely each of these principal modules by indicating their function in the processing of multiwords found in the corpus. The study of multiwords is primordial for a human-machine dialogue system.

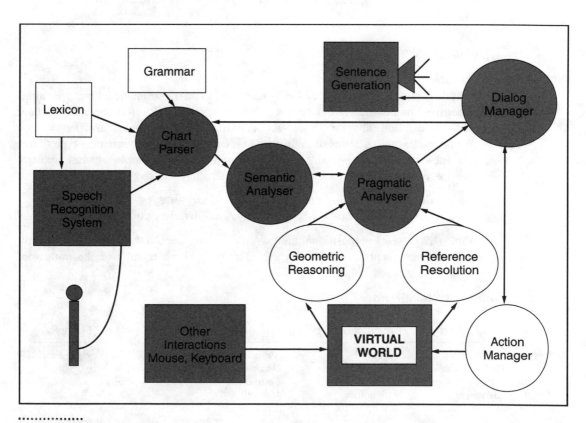

Figure 4.1 *Ulysse's* architecture.

Lexical Analysis

Introduction

International projects of lexicon normalization exist at the moment. *MULTEXT* (Veronis and Khouri 1995) groups together 24 European participants, universities and industries, and covers thirteen European languages. We have exploited the categories defined in the project *MULTEXT*. Thus, the lexicon uses inputs based on these parts of speech. A lexicon input associates a word and its parts of speech. When a word belongs to several different parts of speech, the lexicon will contain several inputs for this word. For example, the part of speech of DEVANT */ahead, in front of/* of utterance (1), is an adverb, whereas the one of utterance (2) is a preposition:

(1) OUI REGARDE DEVANT.   /YES LOOK OVER./

(2) VA VERS L'OBJET À GAUCHE   /GO TO THE OBJECT ON THE LEFT IN FRONT
    DEVANT TOI.   OF YOU./

In order to deal with these two possibilities, the lexicon contains the following two entries:

* DEVANT ADV
* DEVANT PREP

This definition of a lexical input is relative to a word. We have adapted it for group of words—a multiword.

Multiwords

Presentation

Table 4.1 lists multiword types (Rey 1993) according to their grammatical type. We empirically itemized multiwords in our corpus. We next developed different methods for their automatic recognition. These methods vary according to their grammatical type.

We have classified a series of multiwords according to their grammatical type (Table 4.2 and Table 4.3). We graphically represent a multiword by swapping spaces, apostrophes, and a hyphen between each word by score characters:

* we write **EN FACE DE** */in front of/* multiword as **EN_FACE_DE**
* we write **QU'EST-CE QUE** */mark of question/* multiword as **QU_EST_CE_QUE**

Table 4.2 presents prepositional multiwords we have kept in the corpus. It also indicates the derived prepositional multiwords. The derived multiwords of the multiword

Table 4.1 Multiwords

Multiword	Grammatical Type	Example	Example
Adverbial	Adverb	*À droite*	*On the right-hand side*
Conjunctive	Conjunction	*De sorte que*	*Such as*
Nominal	Noun	*Coup d'œil*	*Glance*
Sentence	Sentence	*Mon œil !*	*You bet !*
Prepositional	Preposition	*À côté de*	*Near*
Verbal	Verb phrase	*Prendre de la hauteur*	*To gain height*

AU_DESSUS_DE /*above*/ are **AU_DESSUS_D** (pour « *au-dessus d'un arbre* » /*above a tree*/), **AU_DESSUS_DES, AU_DESSUS_DU.** Most of the prepositional multiwords ending with the preposition *de* create three other multiwords of identical meaning and end respectively with the terms *d, du, des*.

Table 4.3 presents verbal multiwords of the corpus.

Analysis

We analyze these multiwords according to their lexical category. We have classified them in two categories according to an enumeration criteria and an ambiguity criteria:

• Ambiguous multiword: These multiwords are ambiguous because they may exist or not exist according to their context.

• Verbal multiword: These multiwords contain a verb that may be conjugated at any mode, any tense, any person. They may also contain additional adverbs. For example, the verbal multiword « *faire le tour* » may be used in « *je voudrais que tu fasses rapidement le tour de l'arbre* » /*I'd want you to walk fast round the tree*/. It is impossible to enumerate them.

• Adjective, adverbial, conjunctive, nominal, sentence, prepositional, proverbial, or pronominal nonambiguous multiword: These multiwords may be enumerated simply. For example, a prepositional multiword ended by preposition « *de* » creates four multiwords.

Table 4.2 **Prepositional Multiwords and Derived Ones from Corpus**

À_CÔTÉ_D	/NEAR/	À_L_INTÉRIEUR_D	/INTO/
À_CÔTÉ_DE		À_L_INTÉRIEUR_DE	
À_CÔTÉ_DES		À_L_INTÉRIEUR_DESS	
À_CÔTÉ_DU		À_L_INTÉRIEUR_DU	
À_DROITE_D	/ON THE RIGHT HAND SIDE OF/	AU_DESSOUS_D	/UNDER/
À_DROITE_DE		AU_DESSOUS_DE	
À_DROITE_DES		AU_DESSOUS_DES	
À_DROITE_DU		AU_DESSOUS_DU	
À_GAUCHE_D	/ON THE LEFT HAND SIDE OF/	AU_DESSUS_D	/ABOVE/
À_GAUCHE_DE		AU_DESSUS_DE	
À_GAUCHE_DES		AU_DESSUS_DES	
À_GAUCHE_DU		AU_DESSUS_DU	
À_L_EXTÉRIEUR_D	/OUT OF/	DU_CÔTÉ_D	/NEAR/
À_L_EXTÉRIEUR_DE		DU_CÔTÉ_DE	
À_L_EXTÉRIEUR_DES		DU_CÔTÉ_DES	
À_L_EXTÉRIEUR_DU		DU_CÔTÉ_DU	

Table 4.3 **Verbal Multiwords from Corpus**

FAIRE_DEMI_TOUR	/TO TURN BACK/
FAIRE_LE_TOUR	/TO WALK ROUND/
FAIRE_UN_TOUR	/TO GO FOR A WALK/
PRENDRE_DE_LA_HAUTEUR	/TO GAIN HEIGHT/
TOURNER_SUR_SOI_MÊME	/TO TURN ROUND/

More precisely, we classify the multiwords according to their enumeration and their ambiguity. If it's impossible to simply enumerate a type of multiword, such as verbal multiwords, or if the multiword is ambiguous, then we consider that a lexical analysis is insufficient to recognize them. The chart parser will detect them later with phrase-structure rules.

On the other hand, we have constructed a determinist algorithm analyzing the simple enumeration multiwords. Its principle is to take in input a list of words separated with spaces or quotes in order to detect in it the multiwords defined by a list of words separated with a « _ ». Beforehand, it holds an exhaustive list of simple enumeration multiwords sorted alphabetically. Here is an example:

EN_DESSOUS

EN_DESSOUS_D

EN_DESSOUS_DE

EN_DESSOUS_DES

EN_DESSOUS_DU

EN_FACE

EN_FACE_D

EN_FACE_DE

EN_FACE_DES

EN_FACE_DU

From this list, it analyzes the maximal list of words of the input corresponding to a multiword. For example, our algorithm will elaborate the following result:

- `input: REGARDE EN FACE, REGARDE LA VOITURE VERTE EN FACE DE TOI.`

  /LOOK OVER, LOOK THE GREEN CAR IN FRONT OF YOU/
- `output: REGARDE EN_FACE, REGARDE LA VOITURE VERTE EN_FACE_DE TOI.`

Then, we have created for all multiwords presented above additional inputs in the lexicon.

Syntactic Parsing

We have built a grammar to obtain a syntactic analysis of the user utterances of the corpus. It consists in parts of speech presented previously. We use phrase-structure rules and unification equations (El Guedj 1996).

Main Constituents

We summarize the syntactic analysis of an utterance: an utterance consists of words categorized with 40 parts of speech. We have elaborated the grammar, analyzing syntactically the utterances with these parts of speech. This grammar is composed of 68 constituents. The syntactic analysis allows us to represent an utterance with these 40+68 categories.

The resultant grammar is composed of

- 68 constituents
- 259 phrase-structure rules
- 552 unification equations

Table 4.4 **Principal Constituents**

1. a sentence
2. a proposition
3. a verb phrase
4. a noun sequence
5. a simple noun phrase
6. a determinant
7. a relation
8. a list of adverbs
9. a list of adjectives

We must define this multitude of categories in order to realize a precise syntactic analysis. However, too many categories will prejudice the analysis of the meaning of an utterance. That is why we have defined nine principal constituents (Table 4.4) based on the different constituents of our grammar.

Syntactic Analysis of Verbal Multiwords

Table 4.5 presents the analysis of five verbal multiwords of the corpus. We analyze them by creating a phrase-structure rule (constituent R_LOCVERB). For each of them (rules 1 to 5), each rule is composed of constituent R_VLOC, which analyzes a simple verb phrase in the imperative form (rule 6), conjugated form (rule 7), gerundive form (rule 8), or infinitive form (rule 8). The other constituents of R_LOCVERB correspond to parts of speech of other words of the multiword.

The unification equations

$$CATEG\_LEX \ MOT = ENTREE\_DU\_LEXIQUE$$

$$/POS \ WORD = LEXICAL\_ENTRY/$$

verify the spelling of the words.

Results

This grammar parses all the users' utterances of dialogue corpus. It parses utterances such as orders, questions, declarations (positives and negatives), conjunctive subordinate clauses, infinitive conjunctive subordinate clauses, gerundive conjunctive subordinate clauses, and relative subordinate clauses. These clauses may be juxtaposed or linked by a relational phrase.

Semantic Analysis

Semantic analysis is based on the addition of semantics information in lexical entries (Nugues et al. 1996). Previously, we constructed the functional representation of an utterance's parse tree. This construction consists in calculating some functions (subject, adverbial phrase of place, direct object phrase, attribute, and so forth) of main categories of utterance. For example, the following is the result of the functional analysis of an utterance.

```
---------------------------------
• VA ICI PUIS JE VOUDRAIS QUE TU AVANCES RAPIDEMENT VERS LA
  VOITURE À_DROITE_DE LA MAISON POUR ENSUITE ENTRER DANS
  CETTE PYRAMIDE . --> 1 PARSE TREE.
```

Table 4.5 Rules of Verbal Multiwords Analysis

Phrase-Structure Rules Groups	They Analyze:

1. `R_LOCVERB -> R_VLOC PREP ARTDEF N`
    `R_LOCVERB VINF = PRENDRE_DE_LA_HAUTEUR`
    `R_VLOC VINF = PRENDRE`
    `PREP MOT = DE`
    `ARTDEF MOT = LA`
    `N MOT = HAUTEUR`
	• multiword:
	prendre de la hauteur

2. `R_LOCVERB -> R_VLOC N`
    `R_LOCVERB VINF = FAIRE_DEMI_TOUR`
    `R_VLOC VINF = FAIRE`
    `N MOT = DEMI_TOUR`
	• multiword:
	faire demi-tour

3. `R_LOCVERB -> R_VLOC ARTDEF N`
    `R_LOCVERB VINF = FAIRE_LE_TOUR`
    `R_VLOC VINF = FAIRE`
    `ARTDEF MOT = LE`
    `N MOT = TOUR`
	• multiword:
	faire le tour

4. `R_LOCVERB -> R_VLOC PREP`
    `R_LOCVERB VINF = TOURNER_AUTOUR`
    `R_VLOC VINF = TOURNER`
    `ADV MOT = AUTOUR`
	• multiword:
	tourner autour

5. `R_LOCVERB -> R_VLOC PREP PROPERSDIS ADJINDEF`
    `R_LOCVERB VINF = TOURNER_SUR_SOI_MÊME`
    `R_VLOC VINF = TOURNER`
    `PREP MOT = SUR`
    `PROPERSDIS MOT = TOI`
    `ADJINDEF MOT - MÊME`
	• multiword:
	tourner sur soi-même

6. `R_VLOC R_VIMP0`
    `R_VLOC VINF = R_VIMP0 VINF`
    `R_VLOC RV_TYPE = R_VIMP`
	• a simple verb phrase in the imperative

7. `R_VLOC R_V`
    `R_VLOC VINF = R_V VINF`
    `R_VLOC RV_TYPE = R_VIMP`
	• a simple conjugated verb phrase

8. `R_VLOC R_V_P_I1`
    `R_VLOC VINF = R_V_P_I1 VINF`
    `R_VLOC RV_TYPE = R_V_P_I`
	• a simple verb phrase in the gerundive or infinitive form

▪ */GO HERE THEN I'D WANT YOU TO GO FAST TO THE CAR ON THE RIGHT HAND SIDE OF THE HOUSE THEN TO GET IN THIS PYRAMID./*

```
_____ proposition 1 : type=order
*PHRASE VERBAL :
  Verbe : vinf=ALLER intransitif -> sens=ALLER
  Champ :
    adv_lieu=EN_AVANT
  Adv : sens=ICI type=DÉICTIQUE

_____ proposition 2 : type=affirmation
*RELATION
  Connecteur=PUIS
*SUJET Groupe nominal :
  Nom : sens=JE type=PROPERSSUJET nombre=1.000000
*PHRASE VERBAL :
  Verbe : vinf=VOULOIR transitif -> sens=VOULOIR
  type=aux_mode

_____ proposition 3 : type=sub_conj
*RELATION
  Conjonction=QUE
*SUJET Groupe nominal :
  Nom : sens=TU type=PROPERSSUJET nombre=2.000000
*PHRASE VERBAL :
  Verbe : vinf=AVANCER intransitif -> sens=ALLER
  Champ :
    adv_lieu=EN_AVANT
  Adv : sens=VITESSE type=MANIÈRE quantite=2
*LIEU Groupe nominal :
  Prep : sens=VERS type=LIEU
  Det : sens=LA type=artdef genre=fem nombre=sing
  Nom : sens=VOITURE type=OBJET genre=fem nombre=sing
  texte=VERS LA VOITURE
 Groupe nominal :
  Prep : sens=À_DROITE_DE type=LIEU
  Det : sens=LA type=artdef genre=fem nombre=sing
  Nom : sens=MAISON type=OBJET genre=fem nombre=sing
  texte=À_DROITE_DE LA MAISON

_____ proposition 4 : type=sub_conj_inf
*RELATION
  Prep : sens=POUR type=?
*PHRASE VERBAL :
  Verbe : vinf=ENTRER intransitif -> sens=ALLER
  Champ :
    adv_lieu=DEDANS
  Adv : sens=ENSUITE type=TEMPS
*LIEU Groupe nominal :
  Prep : sens=DANS type=LIEU
  Det : sens=CETTE type=adjdem genre=fem nombre=sing
  Nom : sens=PYRAMIDE type=OBJET genre=fem nombre=sing
  texte=DANS CETTE PYRAMIDE

##### Transmission : 4 propositions.
```

This semantic structure is composed of four clauses: an order, a declaration, a conjunctive subordinate clause, and an infinitive subordinate clause.

Table 4.6 Negative Assessments Comments

PRÉPOSITION INCONNUE	*/UNKNOWN PREPOSITION/*
VERBE INCONNU	*/UNKNOWN VERB/*
OBJET INCONNU	*/UNKNOWN OBJECT/*
TERME INCONNU	*/UNKNOWN TERM/*
OBJET TROUVÉ SANS CET ADJECTIF	*/FINDING OBJECT WITHOUT THIS ADJECTIVE/*
OBJETS INVISIBLES	*/INVISIBLE OBJECTS/*
PLUSIEURS OBJETS POSSIBLES	*/SEVERAL POSSIBLE OBJECTS/*

Pragmatic Analysis

Introduction

The pragmatic analyzer links objects in the world to the referential terms. Then it identifies, in context, the speech acts, and calculates the resulting actions. Speech acts are sent to the dialogue manager, and actions are executed by the action manager. The pragmatic analyzer also analyzes the ambiguous utterances. For this, it uses the assessment of pragmatic analysis of an utterance.

Assessment of Pragmatic Analysis of a Sentence

The pragmatic analyzer calculates an assessment in the last part of pragmatic analysis. It consists in a positive or negative value, and a comment. During the pragmatic analyses of an utterance, the pragmatic analyzer may find multiple-failure cases. It associates a negative assessment to these failed analyses. Their comments are the reason for their failure. On the other hand, it associates a positive assessment for all successful pragmatic analyses. The comment will be a random message.

Table 4.6 indicates the list of failure causes of the pragmatic analyzer.

Ambiguities Resolution

Ambiguities may come from different possible syntactic analyses of an utterance. Particularly, if an utterance contains a verbal multiword, it will generally be ambiguous. We detect the presence of such a multiword with the elaboration of a phrase-structure rule specifically for this multiword (see rules from Table 4.5). This additional phrase-structure rule creates a first parse tree for the verbal multiword. Moreover, the classical rules of a clause create a second parse tree without multiword. For example, the analysis of the next utterance has two different parse trees:

PRENDS DE LA HAUTEUR. */GAIN HEIGHT/*

We present besides their respective semantic analyses:

```
------------------------------------
- PRENDS DE LA HAUTEUR . -> 2 arbres syntaxiques.

Analyse numéro 1:
_____

_____
_____ proposition 1 : type=ordre
```

```
*NOYAU VERBAL :
  Verbe : vinf=PRENDRE transitif ->  sens=PRENDRE
*DE Groupe nominal :
  Prep : sens=DE type=DE
  Det : sens=LA type=artdef genre=fem nombre=sing
  Nom : sens=HAUTEUR type=? genre=fem nombre=sing
  texte=DE LA HAUTEUR
```

```
Analyse numéro 2:
```

```
                                   proposition 1 : type=ordre
*NOYAU VERBAL :
  Verbe : vinf=PRENDRE_DE_LA_HAUTEUR transitif ->  sens=ALLER
  Champ :
    adv_lieu=EN_HAUT
```

In this example, the analyzer yields a parse tree using the classical phrase-structure rules of a clause. So, it recognizes the verb « *prendre* » /*take*/ and direct object phrase « *de la hauteur* » /*some height*/. The second parse corresponds to the phrase-structure rule analyzing the verbal multiword « *prendre_de_la_hauteur* ». It detects the multiword and calculates the meaning « *aller en haut* » /*go up*/. In the dialogue context of this utterance, the second semantic analysis represents the intention of the user.

However, it is possible that the order of these two analyses is inversed, because we can't control the order of phrase-structure rules firing. In this precise case, the first semantic analysis would correspond to the user's desire. It signifies that we should sometimes consider the first analysis and neglect the second one.

The pragmatic analyzer of a semantic representation of an utterance calculates a list of speech acts as well as the corresponding assessment: positive or negative. We use this specificity to start pragmatic analysis on the different semantic representations of a same utterance. For this, we have designed a pragmatic analysis algorithm that calculates the pragmatic representation of an ambiguous utterance (Figure 4.2).

```
1. Input: utterance E
2. Repeat
      3. Syntactic analysis for E to get its ith
         syntactic representation E_SYNi.
      4. Semantic analysis for E_SYNi to get its ith
         semantic representation E_SEMi.
      5. Pragmatic analysis for E_SEMi to get ith speech
         acts list and to calculate the corresponding
         assessment Bi.
6. Until pragmatic analysis calculates a positive
   assessment Bi or E_SYNi is the last syntactic
   representation for E.
```

Figure 4.2 Pragmatic analysis algorithm for an ambiguous utterance.

This algorithm can be illustrated with the example « *prendre de la hauteur* »: The user says to the virtual agent « *prends de la hauteur* ». The processing of this utterance creates two parse trees. The first one corresponds to the action of taking something, the second one to the action of going up (*aller en haut*). In our application, the agent doesn't know the verb *prendre /take/*. It indicates this in its first underlined message. It creates a negative assessment and continues in the processing of the second analysis. Then the agent accepts it and returns a positive assessment (second underlined message).

```
------------------------------------
- PRENDS DE LA HAUTEUR . --> 2 parse trees.
##### ANALYSE NUMÉRO 1 ######


_____

_____
_____ proposition 1 : type=order
*PHRASE VERBAL :
  Verbe : vinf=PRENDRE transitif ->   sens=PRENDRE
*DE Groupe nominal :
  Prep : sens=DE type=DE
  Det : sens=LA type=artdef genre=fem nombre=sing
  Nom : sens=HAUTEUR type=? genre=fem nombre=sing
  texte=DE LA HAUTEUR
_____

_____
##### Transmission : 1 proposition.
##### Réponse de l'agent virtuel en 0.3 second : Bilan négatif.
     Comment : VERBE INCONNU



##### ANALYSE NUMÉRO 2 ######
_____
_____ proposition 1 : type=order
*PHRASE VERBAL :
  Verbe : vinf=PRENDRE_DE_LA_HAUTEUR transitif -> sens=ALLER
  Champ :
    adv_lieu=EN_HAUT
_____
##### Transmission : 1 proposition.
##### Réponse de l'agent virtuel en 0.4 second : Bilan positif.
     Comment : OK .
```

Dialogue Manager

Dialogue manager carries out the interpretation of speech acts, as well as the following:

- Sentence generation of the agent
- Linguistic and conversational ellipses management
- Turn taking management

It uses the output of the pragmatic analyzer to generate the sentence of the agent. The natural language generator uses template messages and possibly selects a random one. It also uses the dialogue history, which dynamically records the conversation structure. The history allowed the management of some ellipses.

Utterances and Movements of the User	*Utterances of the Agent*	*Snapshots Presenting the Movements of the Agent* (*See color plates on pages 308–312.*)
	BONJOUR CHRISTOPHE BIENVENUE DANS LE MONDE ITHAQUES.	
BONJOUR JE VOUDRAIS ALLER DEVANT LA MAISON.		
	C'EST PARTI.	
	J'AI FINI.	
EST-CE QUE TU PEUX REGARDER À DROITE?		
	ET VOILÀ.	
J'AIMERAIS BIEN MONTER SUR LA VOITURE.		
	OUI.	
	J'AI FINI.	
RETOURNE DEVANT LA VOITURE.		
	D'ACCORD.	

(continued)

Utterances and Movements of the User	*Utterances of the Agent*	*Snapshots Presenting the Movements of the Agent*
	J'AI FINI.	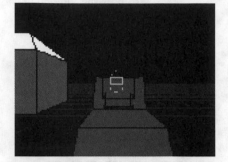
NON DEVANT LA MAISON.	BIEN.	
	ÇA Y EST.	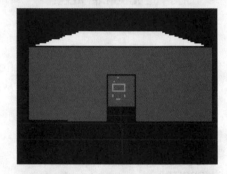
REGARDE À DROITE.		
EST-CE QUE TU PEUX TE RENDRE À CÔTÉ DU GRAND CUBE VIOLET NON JAUNE?	OUI.	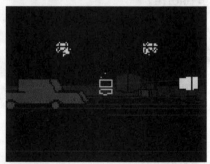
	ÇA Y EST.	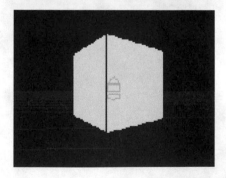
JE VOUDRAIS MAINTENANT QUE TU REVIENNES VERS LA MAISON S'IL TE PLAÎT EN PASSANT SOUS LE DRAPEAU.	BIEN.	

Utterances and Movements of the User	*Utterances of the Agent*	*Snapshots Presenting the Movements of the Agent*

J'AI FINI.

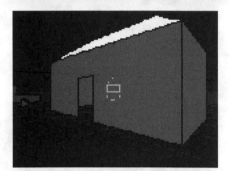

DIRIGE TOI ENSUITE VERS UNE TABLE.

BIEN.

EST-CE QUE TU VEUX BIEN TE PLACER DERRIÈRE LA CHAISE?

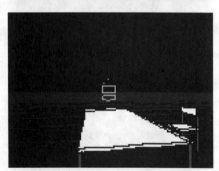

OUI.

VOILÀ.

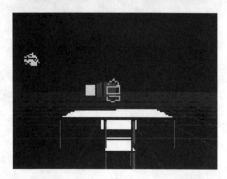

JE VOUDRAIS QUE TU LA CONTOURNES.

D'ACCORD.

(continued)

Utterances and Movements of the User	*Utterances of the Agent*	*Snapshots Presenting the Movements of the Agent*
	ÇA Y EST.	
EST-CE QUE TU PEUX LA PRENDRE?		
	JE NE SAIS PAS PRENDRE.	
MAIS PEUX TU PRENDRE DE LA HAUTEUR.	OUI.	
STOP.		
REDESCENDS DEVANT LA MAISON.	IL Y EN A PLUSIEURS.	
CELLE CI. *(L'utilisateur clique on la maison de gauche)*	OK.	
	VOILÀ.	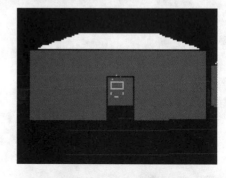
REGARDE À DROITE.		
VA DERRIÈRE UNE VOITURE.		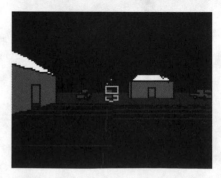
	OK.	

Utterances and Movements of the User	*Utterances of the Agent*	*Snapshots Presenting the Movements of the Agent*
AU REVOIR.		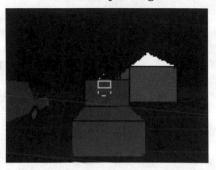
	AU REVOIR CHRISTOPHE.	

Conclusions

We have implemented an interactive spoken dialogue interface to navigate into virtual reality environment. *Ulysse* consists of a speech recognition device, together with a speech synthesis circuit. It relies on a modular architecture embedded within a virtual environment. The entities of the prototype are to process syntax, semantics, and pragmatics together with dialogue. We have adapted the agent to the spoken manipulation of brains reconstructed from MRI images.

We think that this kind of prototype can help discover and master new virtual workspaces in Computer Supported Cooperative Work (CSCW). More generally, it offers a unique experimentation framework for dialogue, computational linguistics, or geometric reasoning. The spoken dialogue brings a new dimension to virtual environments, and vice versa—virtual environments present a fruitful framework for interactive spoken dialogue.

References

J.F. Allen, *Natural Language Understanding,* 2d ed., Benjamin Cummings, Redwood City, Calif., 1994.

J.F. Allen et al., *The TRAINS Project: A Case Study in Building a Conversational Planning Agent,* TRAINS Tech. Note 94-3, Univ. of Rochester, New York, Sept. 1994.

M. Andersson et al., *DIVE, The Distributed Interactive Virtual Environment, Technical Reference,* Swedish Inst. of Computer Science, Mar. 1994.

G. Ball et al., Likelike Computer Characters: The Persona Project at Microsoft Research," *Software Agents,* J. Bradshaw ed., MIT Press, 1995.

R.A. Bolt, "Put That There: Voice and Gesture at the Graphic Interface," *Computer Graphics,* Vol. 14, No. 3, 1980.

CoTech, "Virtual and Augmented Environments for CSCW," *Minutes of the COTECH Workgroup,* Dept. Computer Science, Univ. of Nottingham, Angleterre, 1995.

P.O. El Guedj, *Un analyseur syntaxique combinant grammaire syntagmatique et de dépendance,* Thèse de l'Université de Caen, spécialité Sciences, 1996.

S. Everett, K. Wauchoppe, M. A. Pérez, *A Natural Language Interface for Virtual Reality Systems,* ACM/SIGCHI Conference, 1995.

C. Godéreaux, Un modèle d'agent conversationnel pour naviguer dans un monde virtuel, thèse de doctorat, Université de Caen, spécialité Informatique, janvier 1997.

C. Godéreaux et al., Interactive Spoken Dialogue Interface in Virtual Worlds, *Linguistic Concepts and Methods in Computer-Supported Co-operative Work,* Springer-Verlag, London, 1996a, pp. 177–200.

C. Godéreaux et al., *Un agent conversationnel pour naviguer dans les mondes virtuels,* GRKG Humankybernetik, Band 37 Heft 1, März 1996b, pp. 39–51.

J. Karlgren et al., "Interaction Models, Reference, and Interactivity in Speech Interfaces to Virtual Environments," *Eurographics Workshop,* 1995.

P. Nugues et al., "A conversational agent to navigate in virtual worlds," *Twenty Workshop on Language Technology TWLT 11, Dialogue Management in Natural Language Systems,* S. Luperfroy, A. Nijholt, and G. Veldhuijzen van Zanten, eds., June 1996, pp. 23–33.

P. Nugues et al., "Système de dialogue Homme-Machine pour la génération de comptes rendus médicaux," *Innovation et Technologie en Biologie et Médecine,* Vol. 14, No. 4, Sept. 1993, pp. 469–480.

P. Nugues et al., Un système de dialogue oral guidé pour la génération de comptes rendus médicaux, *9ème congrès de l'AFCET-INRIA, Reconnaissance de Formes et Intelligence artificielle,* Paris, Vol. 2, janvier 1993, pp. 79–88.

A. Rey, "Dictionnaire des expressions et locutions, Les usuels du Robert, *Dictionnaires Le Robert,* Paris, 1993.

J. Véronis and L. Khouri, "Ètiquetage grammatical multilingue: le projet MULTEXT," *Traitement Automatique des Langues,* Vol. 36 No. 1–2, 1995, pp. 233–248.

....

Chapter 5

Information Drill-Down Using Web Tools

Mikael Jern

Vice President Technology
AVS/UNIRAS

• •

Abstract

The paper reviews the information visualization and interaction techniques needed to add another dimension to surfing the Web, *information drilling* and *interactive data querying,* sometimes also referred to as *Visual Data Mining.* Information visualization can be used to explore relationships by "drilling down" and retrieving more data within a region of interest in the visualized data, combining data mining, direct manipulation, and data visualization with 3-D Web tools. It is now possible to create desktop visualization applications that let users interact with databases with larger datasets over the network using both 2-D and 3-D interaction metaphors. The VRML standard allows users to view and navigate through 3-D information data worlds and hyperlink to new worlds. Information drilling based on HTML's Image Map, VRML's anchor node, and multiple predefined viewpoints will be explained and demonstrated. The image map in 2-D and 3-D graphics objects (glyphs, etc.) will represent the Visual User Interface to the information stored in the database. Also, the advantages of using distributed component techniques based on plug-ins, Java Beans, and ActiveX, providing client-side data manipulation, will be reviewed and illustrated. Over the next couple of years, we shall see 3-D visualization evolve in giant steps into interactive data drilling on the Web, providing visualization technology closely integrated with the data warehouse and multidimensional abstract and geospatial data models.

Key Terms

information visualization	interaction	Java
data drill-down	collaboration	VRML

Information Visualization

Information visualization can be used to explore and analyze relationships in multidimensional, large datasets by "drilling down" layer after layer and retrieving more information within a region of interest. The graphics in each layer can distinguish types of information by color, height, size, pattern, outline, texture, arrows, or shapes to name but a few.

It is now possible to create desktop visualization applications that let users interact with large datasets on the Web using fully 3-D interaction metaphors. This is also the foundation for "Collaborative Visualization—Share my data and design." This paper considers the potential for information visualization to bridge the gap between the abstract, analytical world of the digital computer and the human world. Concepts such as publishing, interactive 3-D Web browsing, interactive data querying, data surfing, drill-down, and so forth will be explained.

"Thin" Versus "Fat" Visualization Clients

The widespread popularity of Web technology has created a new information visualization technology model in which browsers enable the widespread distribution of information using standard HTML techniques. The explosive growth of the Web has changed user expectations concerning the delivery of information to a client. The traditional information visualization approach allowed any client to communicate as a peer to any available server. The GUI model was either written in Motif for UNIX or Windows for the PC desktop. The Web introduces a new model in which the client GUI, based on HTML, is less functional and relies upon the data or application servers for visualization traditionally executed on the client.

In the Web-enabled world, the client is effectively reduced to a browser (viewer) of information supported by a server. A true Web client is not capable of program execution unless the executables are downloaded to the client as either Plug-ins or Components. This client is normally referred to as the "thin" client. A thin client, by definition, has minimal software requirements necessary to function as a user interface front end for a Web-enabled application.

Local data manipulation, information drill-down technique, context sensitive menus, object picking, and other interactive user interface functions that traditionally have been available on the client are now controlled by the visualization server. In the thin client model, nearly all functionality is delivered from the server side of the visualization engine while the client performs very simple display and querying functions.

The most appealing aspect of the "thin" visualization client to information visualization users is that the overall cost of software and maintenance can be dramatically reduced. The "thin" client allows the application developers to eliminate the notion of software distribution at the client level (no license issue!), eliminate the notion of maintaining local software, and supporting multiple operating systems on remote clients.

The concept of a "thin" client, however, raises the issue of client versus server data visualization rendering (Figure 5.1). The standard Web browsers are "static" and do not permit any visual data manipulation at the client side. The user interaction is dependent upon the network bandwidth. Partitioning the visualization process between clients and servers is an effective way to distribute the computing resources. The most flexible visualization system allows the application developers to control the visualization partitioning.

Java is now being used to overcome some of the limitations. Java allows the creation of components "applets" or "JavaBeans," which are automatically downloaded and executed

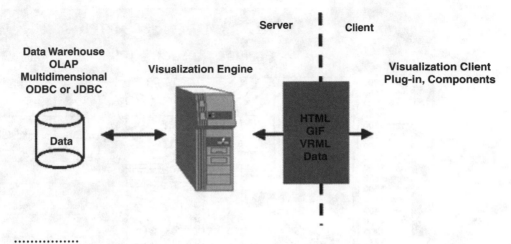

Figure 5.1   Client vs. Data Rendering.

on the local client. These components can significantly increase the data interaction between the client application and user and allow tasks to be executed on the client. Java applets, for example, are interpreted on the client by the Java Virtual Machine, which is usually embedded in the Java-enabled browser, such as Netscape's Navigator or Microsoft's Explorer.

These Java applets that deliver locally available executables are, however, still dependent on the network bandwidth. Depending on the scope and application, Java applets and its datasets must be downloaded. Java applets are only resident during execution and are therefore removed from the local disk after the completion of the task. As the demand for larger applets and datasets grows, significant download time could be incurred, and the network becomes the bottleneck. Keeping commonly used applets resident on the client would significantly reduce download time, although this practice is counter to the Java applet architecture.

The most compelling reason for the use of an "intelligent" client, such as a Plug-in or Component in a visualization application is the need for sophisticated user interface and data manipulation. Ease-of-use is often the primary factor for considering a Web-based solution. However, the current limitations with HTML and Java class libraries make the implementation of complex visual interface front ends very difficult. An intelligent client offers the opportunity to deliver highly graphical, highly interactive user interfaces that provide point-and-click navigation through multidimensional data models, such as exploring complex Data Mining trends and subsetting dimensions in an OLAP environment.

An "intelligent" visualization client provides local functionality through Plug-ins or Web components (ActiveX or JavaBean). Visual data manipulation is provided at the client side through locally stored components. Highly interactive user interface tasks are delivered that provide point-and-click navigation through multidimensional data structures. Visual data interfaces such as information drilling, moving a cutting plane through a volume data set, and so forth are supported. Clearly, a full-featured visual data manipulation has many advantages over the rudimentary offerings of Java applets and HTML query forms.

Figure 5.2 shows an example of an intelligent visualization client produced with AVS' GSHARP Web Edition. The 2-D contour map, color legend, and the two charts to the left were produced at the server side with AVS' GSHARP using Java2-D graphics imbedded in a Java applet. The special GSHARP Java Profile applet was transferred from the server and executed at the client side doing the profile calculation and drawing of the horizontal and vertical profiles.

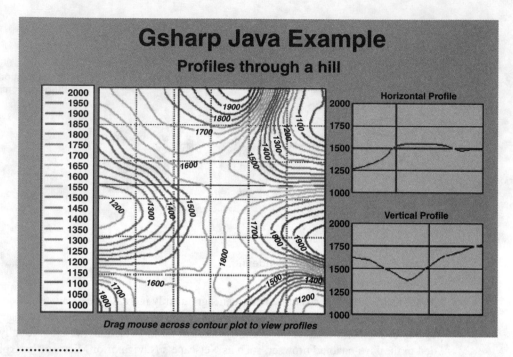

Figure 5.2 Example of an "intelligent" visualization client produced with AVS' GSHARP Web Edition. The underlying 2-D graphics were produced with AVS' GSHARP on the server using Java2-D graphics in an applet, while the profile Java code was transferred to be executed on the Visualization client side. (*See color plate on page 312.*)

Information Drill-Down on the Web

It is now possible to create desktop visualization applications that let users interact with databases with larger datasets over the network using both 2-D and 3-D interaction metaphors. The following reviews the information visualization and Web interaction techniques needed to add another dimension to surfing the Web: information drilling and interactive data querying, sometimes also referred to as Visual Data Mining. Web-based information visualization can be used to explore relationships by "drilling down" and retrieving more data within a region of interest in the visualized data, combining data mining, direct manipulation, and data visualization with either 2-D or 3-D Web tools.

3-D interactive graphics on the network require a 3-D interactive format and a navigation system that combines the 3-D input and high performance rendering capabilities. Virtual Reality Modeling Language, VRML, is the language for describing multiuser interactive simulations—virtual worlds networked via the global Internet and hyperlinked within the Web.

VRML is an open, platform-independent, standard file format for 3-D graphics that grew out of Silicon Graphics' object-oriented Open Inventor 3-D tool kit in 1994. By defining a new file format to represent 3-D scenes and by creating stand-alone "client" viewing programs for that file format, today's Web browsers can also handle 3-D scenes on the PC desktop platforms. VRML, introduced in early 1995, is now driven by an "open," growing consortium, which is quickly broadening its horizons and future development:

VRML 1.0   Geometry   3-D object representation
    1995

VRML 2.0   Behavior   Put the static geometry into motion.
    1996

VRML 3.0   Sociality   Interface for multiuser interactivity in virtual spaces.
    1997

VRML's anchor node and multiple predefined viewpoints allow the users to view and navigate through 3-D information data worlds and hyperlink to new worlds. Information drilling can also be implemented in 2-D based on HTML's Image Map. The image map in 2-D and 3-D graphics objects (glyphs, etc.) will represent the Visual User Interface to the information stored in the database.

Image Map Is Used to Implement Information Drill-Down

One of the most powerful uses of 2-D graphics found on the Web is the Image map in HTML. Image maps are regions of your screen assigned to links. Clicking on one area of the image will take the user to one location, while clicking on another will take him somewhere else. The graphics that are displayed are just ordinary GIF or JPEG graphics. However, an additional file is kept with the image called a map definition file. This is an ordinary ASCII file that contains the definition of where the active "clickable" areas on the image are located. This information is stored as coordinate pixel locations with corresponding links. These areas can be defined as rectangles, circles, and even arbitrary polygons.

In addition to the image and map definition file, an image map needs a Common Gateway Interface (CGI) script. This script is a special program that acts as a middleman between the browser and the map definition file. When the user clicks on the image map, the CGI script looks in the map definition file to see what to do. It then points the browser in the right direction, and the user is transferred to another site.

A typical image map definition file contains several pieces of information for the server's CGI image map routine. The contents include lines starting with words such as rect, line, and polygon, which represent the type of areas being defined.

```
<IMG SRC="graph.gif USEMAP=#map1>
<MAP NAME="map1">
  <AREA SHAPE=RECT COORDS= "0,0 100,100"
   HREF="/cgi-bin/map.cgi?pick=upperLeft>

  <AREA SHAPE=POLYGON COORDS="
   262,231
   263,231
   ...
   203,169"
   HREF="/cgi-bin/map.cgi?pick=DEN>
</MAP>
```

GSHARP Web Edition from AVS supports Information Drill-down based on Image maps. GSHARP's object-oriented architecture provides the framework for automatically generated Image maps. The countries in Figure 5.3 were generated by GSHARP and are "clickable" areas with hyperlinks attached to them.

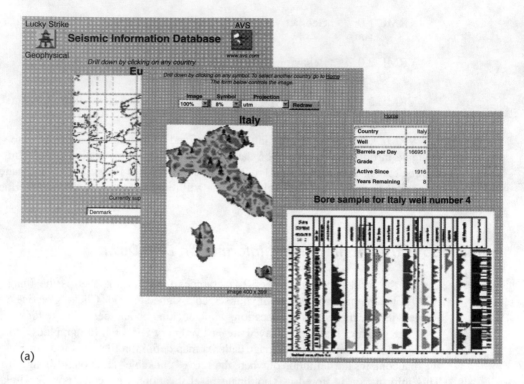

(a)

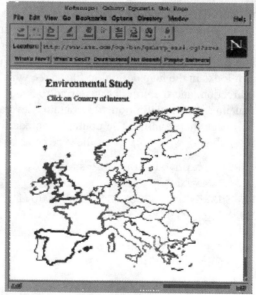

(b)

....

Figure 5.3   (a) Image maps (countries) generated by GSHARP's Web Edition. (b) Information drill-down generated automatically with AVS' GSHARP Web Edition. The user drills down one step at a time. 1. Select country. 2. Select Oil Well 3. Study Well log. (*See color plate on page 313.*)

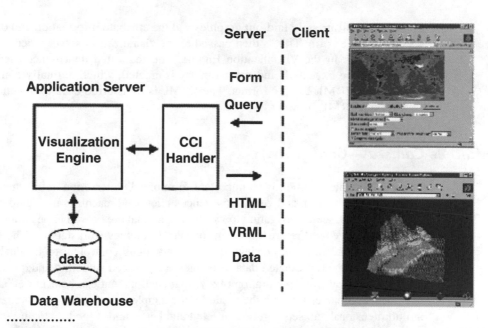

Figure 5.4   A diagram of a dynamic Web VRML application. (*See color plate on page 314.*)

Dynamically Created Visualization

The VRML files can be either static or created dynamically.

The user can view a database of already existing static VRML scenarios. The user interacts with a single VRML file at a time in a 3-D browser. Any such selection results from a server being contacted to deliver the appropriate VRML file. The user sends a request (with a HTML generated "form") to the server and receives the selected static VRML scene from a database.

Virtual VRML scenarios defined on the basis of a simulation or any analytical expression are generated interactively with CGI scripts. The HTML user interface form permits the user to control the visualization method and its attributes dynamically.

The CGI is a powerful mechanism for transferring images and data that has been produced interactively over the Web. The architecture of an Application Visualization Server includes a number of parts:

User Interface implemented with HTML forms and a Web browser

Application Server, which utilizes a Visualization Engine for batch graph generation

CGI Handler Program controlling the dynamics of the Web application

VRML browser for viewing the resulting visualization

The interaction between the Visualization Server software and the client "Web browser" is described in the diagram in Figure 5.4. The visualization Web interface is created with standard HTML, which provides limited tools for designing a user interface called a "form." This form contains a number of fields to be set by the user, which controls not only the data to be visualized, but also the visualization attributes. The user accesses the application server through this HTML page in a Web browser.

On submission of the form, a CGI script is executed on the Web server machine. The script contains the sophistication necessary to guide the visualization software in producing appropriately laid-out graphics and the attribute information and data request specified in the form. This is then passed to the visualization server where it is used to set parameters in the Visualization Engine. The requested data is accessed, the map instructions are executed, and the geometry is created, which is finally converted into the standard VRML 2.0 file format. The VRML is transferred back to the client and the appropriate VRML browser.

Multiple Cameras—Guided Tours

The virtual camera motion is an important form of 3-D interaction in information visualization. The navigation and visual experience on low-end machines use "guided-tour" and "point-and-click seek" navigation tools, techniques that are proven to be almost as appealing and effective as "free roaming" on powerful graphics workstations. The visual cues provided by interactive 3-D viewing with continuous control offer invaluable help in understanding the represented data. If images are rendered smoothly and quickly enough, an illusion of real-time exploration of a virtual environment can be achieved as the simulated observer moves through the model. For example, in information visualization, large multidimensional datasets can be inspected and better understood by the user by walking through the 3-D virtual data projections.

An example of these navigation tools is the "multiple virtual camera" feature in a VRML file. A number of cameras can be specified in the VRML file and are listed inside a Switch node. This allows the browser to give the user a selection of viewpoints. The user clicks on the "Viewpoints" menu item, and a list of the available camera positions will pop up. The camera node can also be assigned a name, which can then be referenced as an "entry point" of the virtual world. Cameras can be placed in the object hierarchy just like any object or light and are therefore affected by the transformation nodes.

The multiple-cameras feature is supported in a very professional way in most 3-D browsers (Figure 5.5 shows a guided tour with the WebSpace 3-D browser). By selecting a new viewpoint, the user is taken on an animated tour from the existing viewpoint through the landscape to the new selected viewpoint. Such a "guided tour" is useful, to guide the user through a complex scene and highlight special points of interest. The camera positions, affected by the transformation nodes, allow the modeler to zoom in at selected

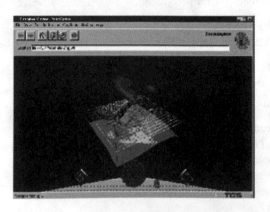 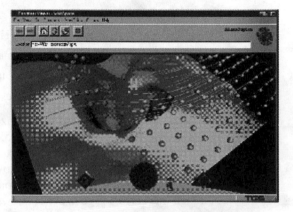

Figure 5.5   Example of a guided tour in the WebSpace Browser. (*See color plate on page 314.*)

objects. These predefined camera positions are interactively specified by the modeler and stored in the VRML file.

Attaching Data Attributes to Graphical Objects

The effectiveness of interactive 3-D viewers for communicating information about 3-D environments can be dramatically enhanced by attaching annotations to the 3-D scenes. Links in VRML work in precisely the same way as they do within HTML, thus, pointing to an object with a link will first highlight the object "visual cue" and if demanded, bring up an application or data attribute that is designed for the selected object. These links can be used to develop information drilling in a 3-D space. The WWWAnchor node in VRML provides the framework to have links to other worlds, animations, sound, and documents.

Two ways of attaching attributes to a graphics object:

1. The WWWInline node represents one of the most powerful features of VRML and is used to load additional VRML files from elsewhere on the Web into the scene. With this feature, large and complex scenes can be composed from a repository of smaller objects.
2. The WWWAnchor node is the equivalent to the anchor tag in HTML. Hence, it represents the Web hyperlink. You can create an anchor to anything in the Web, to a Web text, a movie, or another VRML world (Figure 5.6).

Five Dimension Display—"Glyph" Visualization on the Web

A 3-D scatterplot display uses "glyphs" to represent multivariate data, where each characteristic is determined by the data. Each data point is represented by a sphere, or "bubble." The position ($x, y,$ and z), size, shape, and color of the sphere each can be used to represent a variable in the data, an example of a multivariate display in five dimensions. The user can also view attached additional abstract information of a selected sphere.

The glyph visualization helps the user to see trends in very large datasets. However, the user is not limited to an overview of the data but can retrieve full and rich descriptions

Figure 5.6   Interactive 3-D information visualization on the Web using the WWWAnchor Node:"http://www.tel.com/call.htm"
    description
      "AREA CODE: 617; Day of Month: 29; Avg Call; 67; NUM ....."
(See color plate on page 314.)

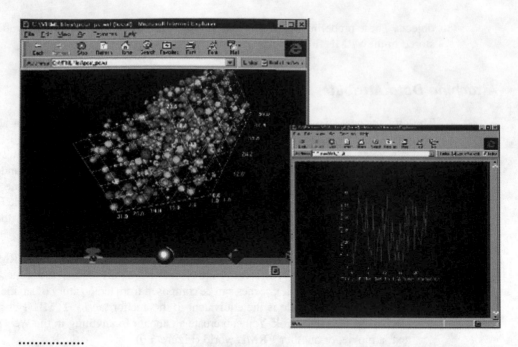

....
Figure 5.7 Example of "Visual Data Mining" on the Web with a large volume of telecommunication data stored in an Oracle database. Dynamic queries are formulated through direct manipulation of graphical widgets, such as icons, buttons, and sliders. The result of the query is a 3-D glyph display, representing selected customers in a market search. A 3-D VRML browser allows the user to walk through the data space in real time. The user can view attached abstract information by selecting objects, "bubbles" of special interest. The scene also highlights the utility of 3-D hyperlink facility in VRML, since every glyph is hyperlinked to some metadata. (*See color plate on page 315.*)

of underlying database attributes. Only showing details when they are requested is vital for the concept of dynamic queries. The user applies "details-on-demand" by pointing at a glyph object and clicking the left mouse button. The object is highlighted to indicate that it was selected. The client creates a query that is sent to a server.

The commercial application in Figure 5.7 (produced by a prototype application developed with AVS/Express software) uses the "glyph" display technique. The example shows the possibility to construct a dynamic query system "information drill-down" to a database using HTML forms, VRML, and a CGI script. The application provides controls to choose a dataset, select a variable of interest, and navigate among the dimensions of the variable.

An SQL query, given in an HTML form, is sent to the Web server, and the user receives a 3-D glyph display in VRML format in return. The user then interacts in 3-D space by pointing directly to a glyph (sphere in this case). Each glyph is represented as a WWWAnchor Node, which hyperlinks the object to additional information about the selected data. This drill-down technique provides an immediate graphical representation of data and its attributes, without the intervention of agents like 2-D controls (scroll bars and other traditional direct manipulation tools) or 3-D widgets that are separate from the data representation. By allowing users to interactively recall and view the attached information by selecting objects of interest during navigation, the interactive 3-D viewer becomes a natural front end for dynamic querying of information.

Viewing Versus Client Application Plug-Ins

Despite new types of information, developers and information sources still cling to the standard HTML, GIF, and VRML formats for a simple reason—it's practically guaranteed that the widest Internet audience possible can view the information. Real-time visual data manipulation, however, doesn't translate well into these standards. While the VRML file format allows distribution of visualization scenes to the Web, the user has no interactive control of the actual underlying data. The mapping of numerical data into geometry format (VRML) takes place on the server side. The client 3-D browser can navigate only in the 3-D world. Manipulations of the data and dynamic control of the visualization attributes that have been used in traditional data visualization systems take place on the Client side.

For example, the user wants to interactively slice through a 3-D volume of data. The Visualization Engine at the Server side must generate a new VRML file for every new data slice. The user on the Client side clicks on the browser's reload button every time to get a new updated VRML. Unfortunately, with low network bandwidth and maybe several hundred people clicking their reload button every minute, the Web server would become overloaded and prevent anyone from getting images or data. Clearly, in some situations the wide acceptance of HTML can't offset the inherent limitations of the format. In these cases, another option is needed.

If you need to interact directly with your data in real time, your information just can't survive a translation into HTML and VRML. The solution is to move part of the actual data-rendering process from the Server to be imbedded in your Web browser at the Client side. Any of the following techniques are available: "Visualization Plug-ins," "Helper Applications," Java applets, or ActiveX controls.

The most compelling reason for the use of local tasks is the need for sophisticated user interfaces. Ease of use is the primary factor for considering a Web-based solution. However, the limitation of HTML makes the implementation of complex front ends very difficult. A more sophisticated Client can deliver highly graphical, highly interactive user interfaces that provide point-and-click navigation through complex data models and data drilling.

Java applets are still dependent on network bandwidth to deliver the executables. Depending on the scope and capability of the applets, large numbers of executables may need to be downloaded in order to accomplish the task. Executables are only resident during execution and are removed from the local disk after the completion of the task. As the demand for larger applets grows, significant download time could be incurred.

Plug-in modules are programs specifically written to run "embedded" within a particular Web browser. Netscape Plug-ins are the most popular format and are now emerging as the standard. Visualization Plug-ins can be used to let the user (Client) read a script language, which controls the visualization type, attributes, and the data to be visualized. The Web browser knows about the Visualization Plug-in and will automatically launch it and load the Plug-in into it once the data transfer finishes. The Visualization Plug-in will perform the data manipulation and rendering locally at the client side (Figure 5.8).

There are, however, always trade-offs. The application publisher scenario is not suitable for transferring very large datasets over the Internet. Here the VRML scenario would be a more appropriate Web visualization method.

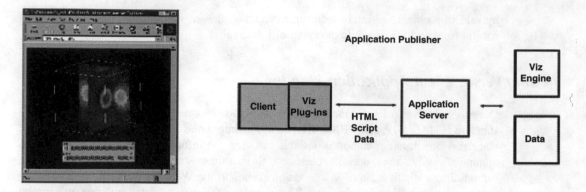

Figure 5.8   The Application Plug-in scenario. The mapping of data into geometry and rendering is performed at the Client side. The user can interactively manipulate the data. A script language can be transferred together with the data to set up the appropriate visualization method. This special visualization Plug-in performs "data slicing" through a volume dataset. (*See color plate on page 315.*)

Client-Side Behavior with Visualization Components

The ultimate Web-based visualization capabilities will be delivered through Web components. Visualization on the Web will become even more active and dynamic when JavaSoft's JavaBeans and Microsoft's ActiveX visualization components begin streaming down the Internet to the Web browsers.

In many applications, sophisticated components are increasingly becoming the main feature, providing capabilities now that would take a considerable amount of development time and expertise to code. Developing some of the intricate components that are needed in interactive Web applications on the market today can be prohibitively expensive and require deep knowledge.

ActiveX essentially extends Microsoft's existing and proven Object Linking and Embedding (OLE) and Component Object Model (COM) technologies to the Web and is optimized for the Windows environment. Microsoft's Internet Explorer (IE) 3.0 is the first major application to exploit the ActiveX technology. ActiveX components can be used as "plug-ins" or you can create appletlike programs (programmed in Java!) embedded inside HTML pages that use <OBJECT> tag to refer to components "ActiveX controls." In Figure 5.9, a VRML browser is embedded in a Word document as an ActiveX component.

AVS/Express is a visual component development framework in which developers design, build, and customize visualization components using a Visual Programming technique. These components are either C++ classes, ActiveX, or Plug-ins that can be combined to create Web applications. Such sophisticated components do everything from displaying information in sophisticated 3-D and 4-D scientific visualization techniques, to providing real-time graphs and images on everything from medical MRI images to sensitive financial information.

Conclusions and Future Trends

In this exploding world of abstract data, there is great potential for information visualization to increase the bandwidth between us, and an ever growing and ever changing world of data.

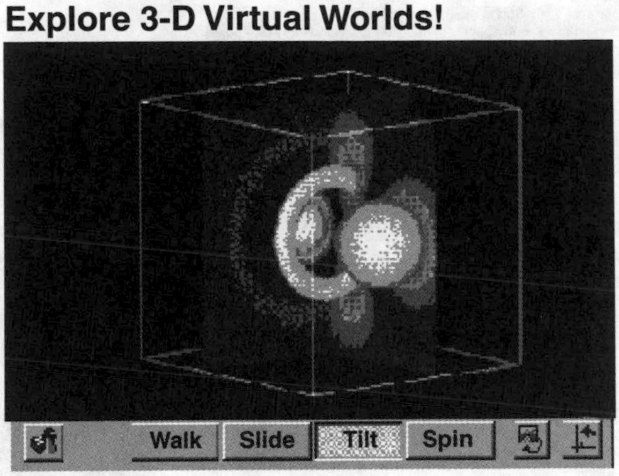

Explore 3-D Virtual Worlds!

Figure 5.9   VRML browser embedded in a Word document as an ActiveX component. The VRML file was produced with AVS/Express advanced 3-D visualization software. (*See color plate on page 316.*)

The future trends and improvements in information visualization for the Web can be summarized as follows:

Information drilling on the Web

3-D visualization on the PC desktop

Nonimmersive VR navigation using "Visual" User Interface technology

Real-time visualization of very large datasets

Components—JavaBeans and ActiveX

Collaborative multiuser visual environment—Guided analysis

Web database visualization—Visual Data Mining

Over the next couple of years, we will see VRML evolve in giant steps into interactivity and multiuser participation based on the new emerging standard VRML 2.0 and the future evolving more collaborative VRML 3.0. Visualization will develop into interactive data drilling on the Web, providing visualization technology closely integrated with the database.

As the technology of information visualization becomes more important to the decision-making process, it will have a natural tendency to migrate to the Web so that it can be made available to the largest number of users. The VRML virtual camera motion feature provides users with an animated tour of the data. Glyphs can be used to construct dynamic query systems that operate on the Web. Finally, plug-ins, Java applets, and ActiveX controls move the data-rendering process to the client side to overcome bandwidth limitations (Figure 5.10). Each of these technologies is contributing to the rapid transfer of information visualization on the Web.

References

R. Abraham, F. Jas, and W. Russell, *The Web Empowerment Book,* Springer, 1995.

J. Brown, et al., *Visualization, Using Computer Graphics to Explore Data and Present Information,* John Wiley & Sons, New York, 1995.

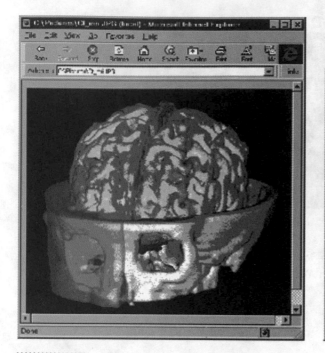
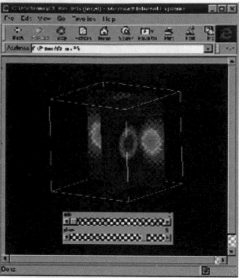

Figure 5.10   Example of two Application Plug-ins prototyped by AVS/Express to perform Volume Rendering and Data Slicing at the Client side. Compared to general purpose VRML and http, the Application Plug-ins allow more sophisticated interaction between the client application (the Web browser plus plug-in) and the visualization server. It supports direct manipulation of both data and the visualization parameters. For example, dynamic specification of data/color mapping is accomplished using specialized interaction tools. (*See color plate on page 316.*)

R.A. Earnshaw and J.A. Vince, *Computer Graphics, Developments in Virtual Environments,* Academic Press, 1995.

A. Ford, *Spinning the Web, How to Provide Information on the Internet,* Thomson Computer Press, 1995.

M. Gobel, ed., *Virtual Environments '95,* Springer Wien, 1995.

M. Gobel, H. Miller, and B. Urban, eds., *Visualization in Scientific Computing,* Springer Wien, 1995.

M. Grave, Y. Le Lous, and W.T. Hewitt, eds., *Visualization in Scientific Computing,* Springer Wien, 1994.

M. Gross, *Visual Computing the Integration of Computer Graphics, Visual Perception and Imaging,* Springer, 1994.

W. Herzner and F. Kappe, eds., *Multimedia/Hypermedia in Open Distributed Environments,* Springer Wien, 1994.

B. Pfaffenberger, *Publish It on the WEB,* Academic Press, 1995.

R. Scateni, J. van Wijk, and P. Zanarini, eds., *Visualization in Scientific Computing '95,* Springer Wien, 1995.

E.R. Tufte, *Envisioning Information,* Graphics Press, Cheshire, Conn., 1990; J. and N. Randall, December 1994.

R.C. Veltkamp and E.H. Blake, eds., *Programming Paradigms in Graphics '95,* Springer Wien, 1995.

D. Woo, *The World Wide Web Book,* Springer, 1995.

Chapter 6

Generic Uses of Real World Data in Virtual Environments

M.W. Wright, G.C. Watson, and R.L. Middleton
Edinburgh Virtual Environment Centre
University of Edinburgh

●●●

Abstract

Our goal is the delivery of a tool kit for the development of virtual environment applications for education and research. If the tool kit is to be useful over a wide range of academic subjects, then it must cater to generic educational processes rather than be tailored to specific ad hoc applications. We have identified two generic processes involving the use of real world data for education: visualization and conceptualization.

Visualization concerns the comprehension of datasets as a whole. The emphasis is on intuitive understanding, summary, and description. Conceptualization concerns the application of theoretical models to datasets. The emphasis is on explanation, simplification, and theoretical modelling. We identify and discuss the generic technical issues, such as delivery and data reduction, which arise when attempting to support visualization and conceptualization in a virtual environment.

From this analysis we note that many of the requirements of educators are solved by standard methods and all that remains is to package them effectively. However, some of the requirements of education using virtual environments are poorly catered to, and it is clear that new techniques must be developed. In particular we note that conceptualization has few solutions at present. We give examples of adaption and development of suitable techniques on our project.

Introduction

Education is an important application of virtual environments. As part of the VLDTK project (Middleton 1996), researchers at the Edinburgh Virtual Environment Centre (EDVEC) are creating a tool kit for educators to construct virtual environment teaching applications. A particular focus of the project is the use of real world data from the natural sciences, and this paper addresses some technical issues in their use. A major focus of the project is the completion of four pilot studies chosen from diverse academic disciplines. These include the following:

- Geology—Virtual geology field course
- Genetics—Mouse embryo development and gene expression
- Meteorology—Spatial and temporal appreciation of complex atmospheric phenomena
- Veterinary Science—Cerebral and neural anatomy

The principal approach of the project is to identify *generic* processes and problems arising from these pilot studies that are common to many or all applications of virtual environments in education. Once these generic processes and problems are identified, a decision can be made as to which are adequately addressed by existing approaches and which need additional work. The structure of this paper is as follows: First, we describe two generic processes common to all the pilots. Second, we discuss the technical issues that arise in the implementation of these processes. Third, we describe examples of two of these technical issues as they have been studied on the VLDTK project.

Generic Features of Our Virtual Environment Applications

In identifying generic features of virtual environments there are many ways in which we can categorize them. We have identified that there are two generic uses of VEs that occur in all applications; these are the complementary processes of visualization and conceptualization.

Visualization

Visualization concerns the comprehension of datasets as a whole. The emphasis is on intuitive understanding, summary, and description. The goal of visualization is to allow the user to understand the fundamental forms, correlations, and associations within the data by visual means.

The key underlying process of visualization is exploration. Exploration involves real-time interactive viewing of the data. The primary variable is the viewpoint, but many other variables can be adjusted that affect the appearance of the data, such as image-processing parameters.

We now show two examples of visualization typical of those used in the subject areas of our four pilot studies. Figure 6.1 shows a volume rendering of a section through a horse's head from an MRI scan. This data appears courtesy of Liverpool University Veterinary School. A planar clipping boundary can be moved to view the interior of the volumetric data, but it is not possible to isolate separate organs using such a method. Figure 6.2 shows a visualization of a mouse embryo in the genetics pilot study (Baldock et al. 1992). Volume rendering was done using "volren," an interactive renderer using texture mapping that runs on Silicon Graphics platforms. In each of these examples there is little need to transform the data from their raw state other than to aid their delivery to the user. Any transformations made are typically information preserving.

Conceptualization

Conceptualization concerns the application of theoretical models to datasets. The emphasis is on explanation, simplification, and theoretical modelling. Conceptualization involves

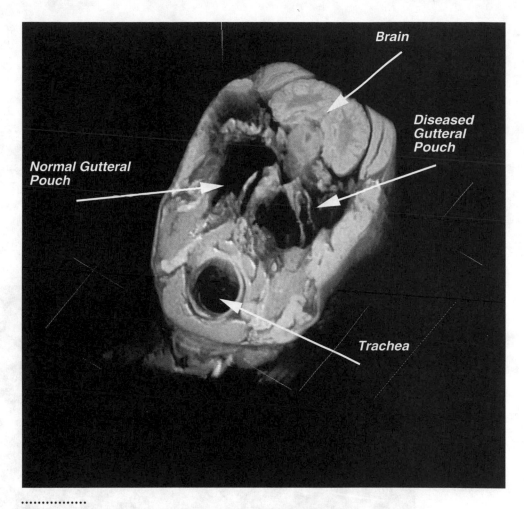

Figure 6.1   Volume rendering of an abnormal horse's head from MRI data.
Data courtesy of the MRI Resear Centre, Faculty of Veterinary Science, University of Liverpool.

the imposition of a particular world view or theory on a dataset. The goal of conceptualization is to convey to the user how a particular theory applies to or explains a dataset.

The key underlying process of conceptualization is *segmentation*. Segmentation is a process that takes a dataset as its input and defines spatial subsets of the data such as points, lines, areas, and volumes, which correspond to theoretical meaningful entities.

We now show examples of conceptualization typical of those used for education in the subject areas of our pilot study. Figure 6.3 shows an example of conceptualization of the mouse data of Figure 6.2. A human genetics expert has taken each slice and manually segmented out the neural tube of the mouse (Baldock et al. 1992). These separate segmentations are combined to form the surface geometry of the entire tube in the lower right of the figure. This surface constitutes a conceptual model that is "close" to the dataset in the sense that there is little abstraction.

Figure 6.4 shows an example of visualization followed by conceptualization in meteorology (Watson 1996). The figure illustrates circulation of the middle atmosphere. The flow pattern visualized in the upper figure is from a comprehensive atmospheric

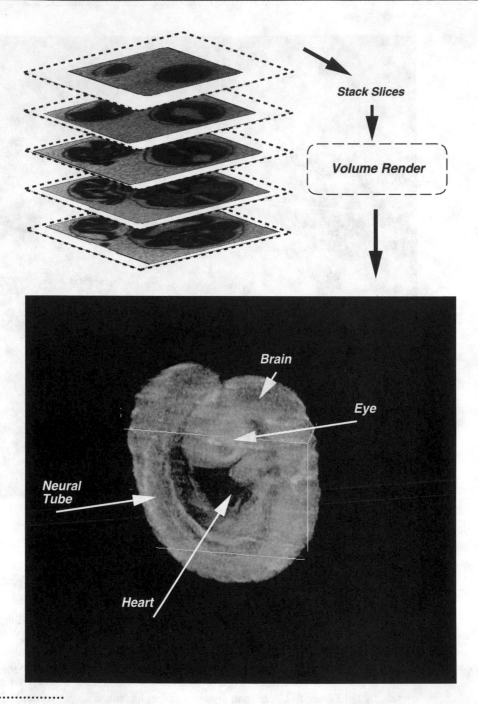

Figure 6.2 Visualization of the mouse embryo.

simulation of the middle atmosphere. A cold vortex forms around the pole, which is shown as an isosurface of a quantity known as potential vorticity, which is a measure of spin. Flow is visualized by the use of massless particles injected in the top of the model. In early winter these can be seen in the tropics at upper levels. During the course of the winter these can be seen to descend and scatter throughout the lower regions of the

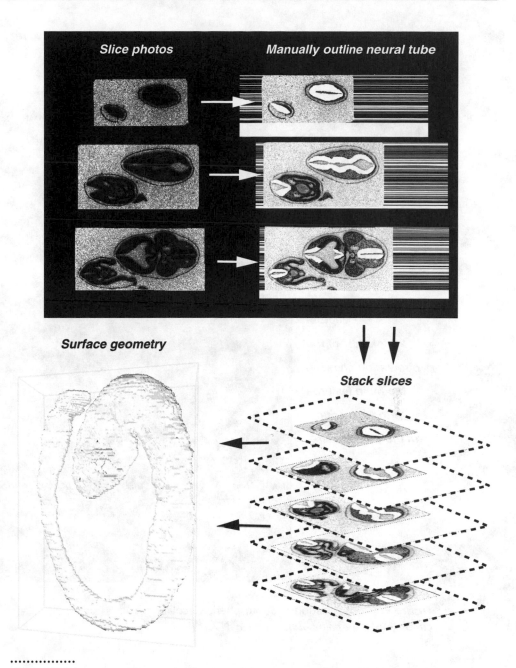

·················
Figure 6.3   Stages in Conceptualization of the mouse embryo. (*See color plate on page 317.*)

model. This circulation in the middle atmosphere is known as the Brewer-Dobson circulation. The upper figures are visualizations of the circulation; the lower figure is a conceptual representation of the flow. This is a more abstract example of conceptualization than the mouse neural tube of Figure 6.3.

In each of these examples there is a clear progression starting with the raw or near raw data of visualizations and proceeding to the more abstract representations of conceptualization.

Three frames from a visualization using an isosurface of vorticity to represent the edge of the vortex, and particles to trace the flow.

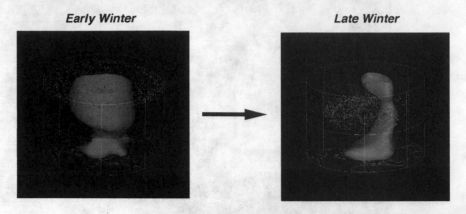

Early Winter   *Late Winter*

A Conceptual 2D 'sketch' of the circulation showing the main features of the winter flow in the stratosphere.

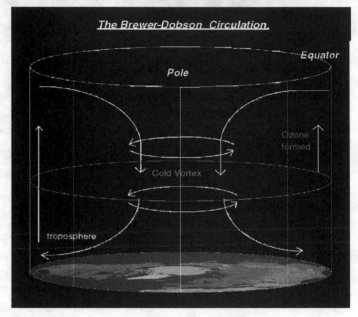

Figure 6.4 Visualization of Stratospheric Circulation, and a Conceptual 'sketch' of the circulation. (*See color plate on page 317.*)

What Are the Generic Technical Issues?

We have identified visualization and conceptualization as generic features of any educational application of virtual environments. We have also found that the key processes underlying these features are exploration and segmentation. The next step is to identify the generic technical issues involved in these processes:

- Segmentation techniques
- Conceptual model building

- Visualization techniques
- Data reduction

Segmentation Techniques

There is extensive literature on segmentation, including the disciplines of image processing (Gonzalez and Wintz 1977), computer vision (Marr 1982; Faugeras 1993; Blake and Yuille 1992), and medical imaging. Despite this, it is true to say that the general automatic segmentation of data remains an unsolved problem. It is interesting to characterize this problem in terms of the relationship between segmentation and conceptualization. From this perspective, the central issue is that it is difficult to get the output of an automatic segmentation process to correspond directly to a conceptual model. In medical imaging, for example, CT data (Herman and Liu 1979) can easily be thresholded to segment (i.e., conceptualize) bone, whereas soft tissue segments poorly. MRI, on the other hand, provides high definition of soft tissue but discrimination of conceptual entities on the basis of water content works poorly except for certain cases such as tumors.

Another crucial task is surface extraction and surface tessellation of raw data points. The Marching Cubes algorithm (Lorensen and Cline 1987) has provided a solution to this problem although further data reduction is often required (further discussion to follow). A more recent algorithm is that of Hoppe and others (1992).

Progress has been made regarding the development of new segmentation techniques such as edge detectors, morphological filters (Serra 1982), and entropy-based methods. However, it is doubtful whether a general solution to the segmentation problem exists as success depends largely on specific application parameters such as the source of the data, sensing modality, and the required final conceptual models. At present, the best general solution is to provide the user with a variety of segmentation tools in the form of an "image-processing package" (Khoros Group 1992) from which segmentation solutions can be constructed that are tailored to particular applications.

We believe that the advent of virtual environments has introduced a need for an extension to such packages. The most obvious is that the segmentation tools must work for 3-D as well as 2-D data. There is also a requirement to support manual and semiautonomous segmentation. Users can define surfaces or volumes where segmentation tools are to be localized. Model-based segmentation and tracking (Blake and Yuille 1992; Terzopoul, Witkin, and Kass 1988) has developed as a powerful technique in the computer vision research community. In this paradigm, templates such as splines and meshes can be deformed using physics-based modelling principals for a variety of applications, such as tracking of organ shape deformation in medical imaging. There is a requirement for these methods to be used in segmentation packages, particularly in 3-D virtual environment applications where they are particularly suitable.

Conceptual Model Building

There is a clear need for tools to build conceptual models from data. Very little work has been done in this area regarding virtual environments although prerequisite tools do exist. Drawing and modelling packages exist that could in principle be used for 3-D and 2-D conceptual modelling. Animation is a key process in conceptual modelling; tracking, fitting, and key frame animation techniques have been developed that can be used to aid this process. An interesting outstanding problem regarding animation is the interpolation of evolving surfaces where a change in topology is involved. Our genetics pilot study provides an example of this problem. As the mouse embryo develops, separate islands of tissue appear and disappear as slices of the dataset are traversed. Even within the 3-D data as a whole, holes and loops appear with time thus changing the topology of the embryo.

Visualization Techniques

Visualization is a rapidly evolving area. Many sophisticated techniques exist for the visualization of data. There are many outstanding problems with regards to the use of visualization in virtual environments, and we now describe some of them.

An interesting emerging research and application area in visualization is the depiction of real locations using image-based VR (Chen 1995). This is a form of visualization where the data typically corresponds to a real landscape or indoor environment. Viewpoints of these real environments are captured at a small finite number of locations. A navigation problem arises on switching from one location to another. The user is effectively instantaneously "teleported" from one location to the next. The temporal and spatial discontinuity thus generated inevitably induces a sense of disorientation in the user. We have devised a simple real-time solution to this, which we describe later in the chapter.

Volume rendering (Herman and Liu 1979) is an important visualization tool well suited to the 3-D nature of virtual environments. Typically, planar clipping boundaries are defined to view the interior of a data volume. These clipping planes cut across the data indiscriminately without any conceptualization. We are addressing the problem of bridging the gap between visualization and conceptualization in volume rendering by investigating the use of nonplanar clipping planes, which lock onto surfaces such as tissue boundaries.

Data registration is another generic technical issue for visualization. Datasets of objects are acquired either through the same sensor at different times or from different types of sensors. CT image slices of a brain may be taken over several weeks and then need to be registered to assess the course of a disease or effectiveness of treatment. Satellite images may be registered with other datasets, such as geographic and geological maps. The main problem is not the application of transformations to the data for registration but the identification of reliable landmarks that themselves could be evolving over time. Our genetics pilot study involves registration of slices of mouse embryo, segmenting the anatomy and mapping the gene expression.

Data Reduction

The requirement for real-time interactive exploration raises data reduction as an important issue. Data reduction is fundamental to the delivery of both visualizations and conceptual models. Low-end workstations are unable to support real-time interaction with very large datasets. There is a generic problem therefore to reduce the amount of data while maintaining perceptual fidelity.

Data reduction methods can be divided into two broad categories of geometric model-based or image/texture-based approaches. An important example of geometric model data reduction is mesh decimation; see Erikson (1996) for an overview. We have evaluated mesh decimation schemes for natural science data and will present some results in a following section.

One example of texture-based data reduction comes from volume rendering mentioned previously. Volume rendering of complex datasets can lead to extremely large and complex meshes being produced. It has been shown that human perception of shape is strongly influenced by texture, and this fact can be utilized to replace areas of complex and finely detailed geometry with textures (Aliaga 1996). For this technique to work successfully with shaded models, the object has to be pre-lit and have shading taken into account.

The textured polygons can be made partially transparent, thus replacing complex planar shapes or a number of parallel textured polygons can be blended together to give the impression of a shape with depth. Texture blending can be taken one stage further to implement an entire volume renderer by creating a large number of parallel polygons that are textured with the volume data at the current location (Cabral, Cam, and Foran 1994). These are then blended together to produce the image.

Hierarchical methods are an additional approach to data reduction that can be applied both to geometric models (Hoppe 1996) and textures (Haley and Blake). In a hierarchical method, data representations of objects are created at a number of distinct spatial scales. Large, fine scale representations are defined for rendering of objects appearing close to the viewer, but data and time savings are made by rendering small, low-detail, coarse scale representations when the object is far away from the viewer.

In summary, the central problem is that there is a lack of generic tools for the above. Our project attempts to address some of these problems by providing these generic tools.

Technical Issues Addressed in VLDTK

We now describe work on two of the generic technical issues addressed as part of the VLDTK project. These issues are mesh decimation and directional awareness in navigation.

Mesh Decimation

Mesh decimation (Erikson 1996; Schroder, Zarge, and Lorensen 1992) is a solution to the problem of data reduction, which arises due to the real-time requirements of visualization and the need to render large and complex surface models in conceptualization. As good mesh decimation algorithms exist already, our goal is not to invent new ones. Instead, we evaluate their suitability for our purposes and then present them as part of a tool kit that is more accessible to the general user of virtual environments.

Figure 6.5 shows data of a human head that has been segmented into surfaces using the Marching Cubes algorithm (Lorenson and Cline 1987). It can be seen that the surface is vastly overtessellated with a very high polygon count. The data are therefore not suitable to manipulate for visualization on a low-end computer.

Surface data from the head scan can be seen in Figure 6.6. The full resolution model is shown to the left of the upper figure and at the bottom of the lower figure. This model consists of 278,000 polygons—clearly impractical to render on a low-end machine. The two middle heads represent mesh decimation by linear data reduction. This is simply achieved in the Marching Cubes algorithm by reducing the resolution of the source volumetric data. The lower resolution of these two heads exhibits topological degeneration, but at 16,000 polygons is still a large model to manipulate on a low-end machine. On the right can be seen results from an intelligent decimation algorithm (Cohen et al. 1996). This algorithm utilizes two envelopes closely spaced around the surface. By ensuring the decimated surface does not intersect either of these surfaces, we guarantee preservation of model topology. The figure represents a maximum error of 1 percent from the original surface, and at 10,000 polygons represents a manageable model.

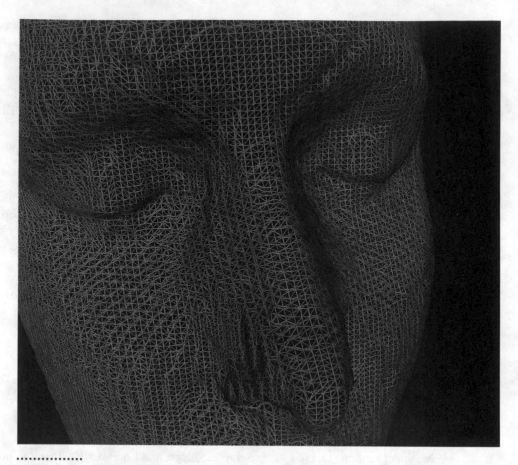

Figure 6.5 Overtessellated mesh produced by the Marching Cubes algorithm.

Figure 6.7 shows decimation of digital terrain model data for the Isle of Skye. Terrain models represent a single surface topology, which can be decimated more easily than general models. The upper figure shows the original data for the island. These data are dense at 600,000 polygons for the 100-km by 200-km region. The third figure shows the decimated model reduced, using the method of Garland and Heckbert (1995) to 2,000 polygons. This reduction ensures real-time interaction. As can be seen, addition of textures in the second and fourth figures mitigates the perceptual effects of geometric reduction.

Figure 6.8 illustrates a useful feature of the mesh decimation algorithm just introduced that improves the perceptual quality of the output. The upper figures show the decimation method of Garland and Heckbert. Points are chosen in a pseudorandom manner guided by the topographic gradients in the data. This error diffusion method weights the point density according to the topographic variance. Delaunay triangulation is used to produce a mesh from the point set. The lower figure illustrates how a texture map can be used to weight the point distribution in order to maintain detail in perceptually important areas. In this case, a texture is defined such that sharp gradients occur along the coastlines, which biases the algorithm to maintain topographic detail along the coasts. Equally, this texture map could be a geological or demographic map, preventing the algorithm from tessellating across important conceptual features.

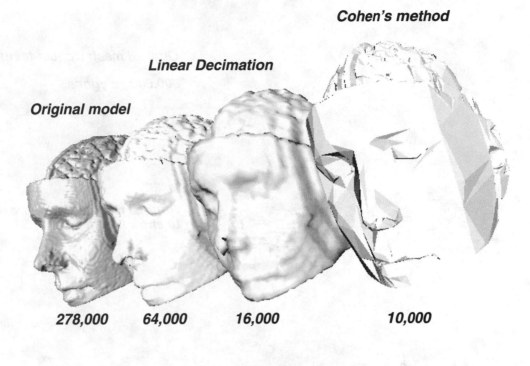

Cohen's method

Linear Decimation

Original model

278,000   64,000   16,000   10,000

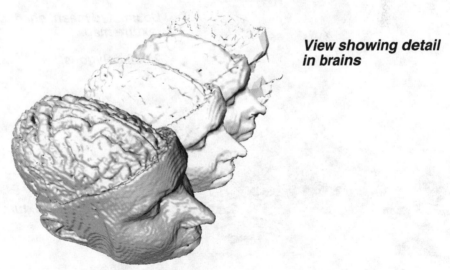

View showing detail in brains

Figure 6.6   Decimation of the MRI surface data. (*See color plate on page 318.*)

Directional Awareness in Navigation

In the display and navigation of real locations, a navigation problem arises on switching from one location to another. The user is effectively instantaneously "teleported" from one location to the next. The temporal and spatial discontinuity thus generated inevitably induces a sense of disorientation in the user. We have devised a simple real-time real location. Panoramas can be created to give 360° views from distinct locations within an environment.

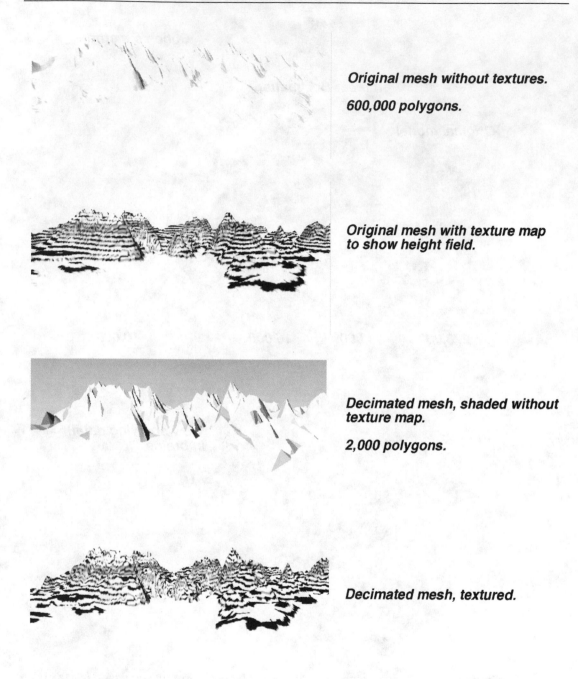

Original mesh without textures.

600,000 polygons.

Original mesh with texture map
to show height field.

Decimated mesh, shaded without
texture map.

2,000 polygons.

Decimated mesh, textured.

Figure 6.7   Mesh decimation of height fields: The Isle of Skye. (*See color plate on page 318.*)

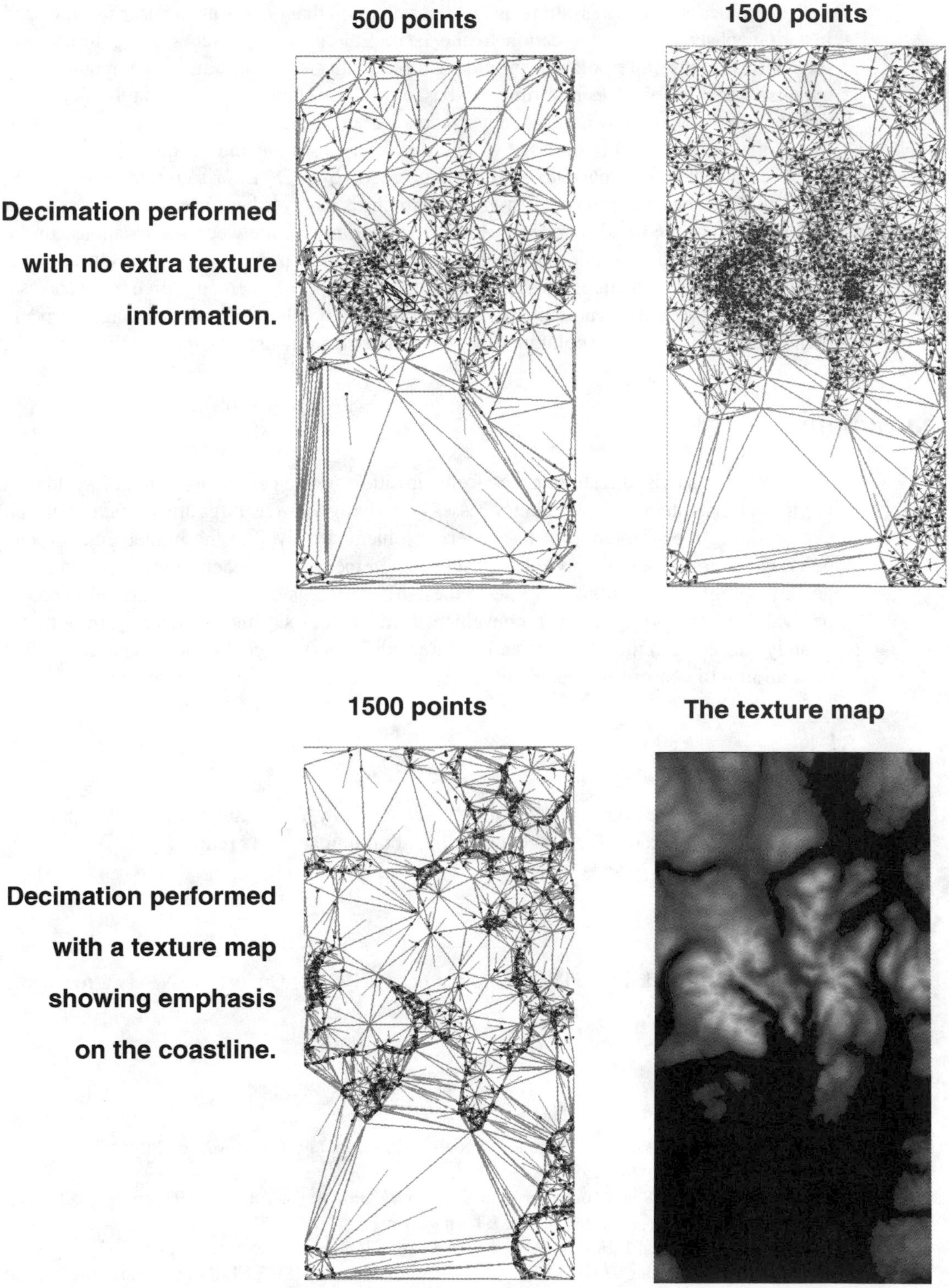

500 points

1500 points

Decimation performed with no extra texture information.

1500 points

The texture map

Decimation performed with a texture map showing emphasis on the coastline.

Figure 6.8   Decimation of the Skye terrain data.

A problem with such systems is the disorientation that occurs as the user is teleported instantaneously from one location to the next. Emerging approaches using image and model-based view interpolation provide a high-quality solution but are computationally expensive and unsuited to real-time PC-based implementations. As the main objective is maintenance of spatial awareness rather than accurate scene interpolation, we suggest that image manipulation using minimal geometries offers a simple and adequate solution.

We have therefore introduced a technique called SMUDGE, or stereo morphing using differential geometric epiboles. This is a morphing strategy where images at the adjacent locations are warped and blended to imitate smooth motion between the locations. In the warping process, the evolution of mesh node positions are confined to epipolar lines, which arise as geometric constraints in stereo imaging of rigid scene geometry. Figure 6.9 shows a SMUDGE between two panoramas in a corridor. The left column shows the image warp and blending process. The right column shows the geometric mesh used for the morphing process.

Conclusions

We conclude that visualization and conceptualization are the generic uses of real world data in virtual environments. The main processes underlying these features are segmentation and exploration. An examination of the generic problems involved in the implementation and delivery of these processes provides a list of requirements for a generic virtual environment designers tool kit. Solutions to some of these requirements exist already and only need to be evaluated and presented in a convenient form for users. Other problems are not adequately addressed in the literature and require additional research. We note in particular that the building of conceptual models is poorly understood and supported at present.

Future Work

Future projects have been specified to address other key technical issues. Wc will pay particular attention to tools for conceptual model building and data reduction.

Acknowledgments

This work was funded by JTAP grant number 92/102/01. QuickTime VR is a registered trademark.

References

D.G. Aliaga, "Visualization of Complex Models Using Dynamic Texture-Based Simplification," IEEE Visualization '96, 1996, pp. 99–108.

R. Baldock et al., "A Real Mouse for Your Computer," BioEssays, Vol. 14, 1992, pp. 501–503.

A. Blake and A. Yuille, Active Vision, MIT Press, 1992.

B. Cabral, N. Cam, and J. Foran, "Accelerated Volume Rendering and Tomographic Reconstruction Using Texture Mapping Hardware," Proc. ACM/IEEE Symposium on Volume Visualization, 1994.

S.E. Chen, "QTVR—An Image-Based Approach to Virtual Environment Navigation," Computer Graphics SIGGRAPH '95 Proceedings, 1995, pp. 29–37.

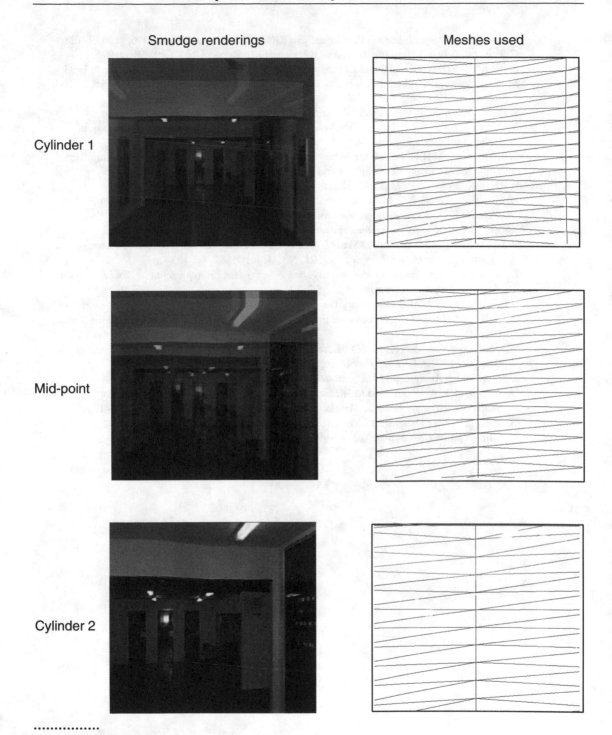

Figure 6.9 Smudge.

J. Cohen et al., "Simplification Envelopes," SIGGRAPH '96 Proceedings, 1996, pp. 119–128.

C. Erikson, Polygon Simplification: An Overview, Tech. Report UNC TR96-016, 1996.

O. Faugeras, Three-Dimensional Computer Vision: A Geometric Viewpoint, MIT Press, 1993.

M. Garland and P. Heckbert, Fast Polygonal Approximation of Terrains and Height Fields, Tech. Report, CMU-CS-95-181, 1995.

R.C. Gonzalez and P. Wintz, Digital Image Processing, Addison-Wesley, 1977.

M. Haley and E. Blake, "Incremental Volume Rendering Using Hierarchical Compression," Computer Graphics Forum, Vol. 15, No. 3, pp. C-45–C-55.

G.T. Herman and H.K. Liu, "Three-Dimensional Display of Human Organs from Computed Tomograms." Computer Graphics and Image Processing, Vol. 9, No. 1, Jan. 1979.

H. Hoppe, "Progressive Meshes," SIGGRAPH '96 Proceedings, pp. 99–108.

H. Hoppe et al., "Surface Reconstruction from Unorganised Points," *Computer Graphics SIGGRAPH '92 Proceedings,* Vol. 26, No. 2, July 1992, pp. 71–78.

Khoros Group, *Khoros Programmers Manual,* University of New Mexico, 1992.

W.E. Lorensen and H.E. Cline, "Marching Cubes: A High Resolution 3D Surface Construction Algorithm," *Computer Graphics,* Vol. 21, No.. 4, July 1987, pp. 163–169.

D. Marr, *Vision: A Computational Investigation into the Human Representation and Processing of Visual Information.* Freeman, 1982.

R.L. Middleton, "Virtual Laboratory Developers Toolkit (VLDTK)," *EdVEC Tech. Report 96001.* Delivered at the AGOCG computer graphics and visualization supporters event, Manchester, UK, 1996.

W.J. Schroder, J.A. Zarge, and W.E. Lorensen, "Decimation of Triangle Meshes, *Computer Graphics SIGGRAPH '92 Proceedings,* Vol. 26, 1992, pp. 65–67.

J. Serra, *Image Analysis and Mathematical Morphology,* Academic Press, 1982.

D. Terzopoulos, A. Witkin, and M. Kass, "Constraints of Deformable Models: Recovering 3D Shape and Nonrigid Motion," *Artificial Intelligence,* Vol. 36, No. 1, 1988, pp. 91–123.

G. Watson, "Three Dimensional Visualization of Winter Polar Stratospheric Flows," *Proc. Eurographics UK Chapter,* Vol. 1, 1996, pp. 141–159.

## Chapter 7

A Generic Functional Architecture for the Development of Multiuser 3-D Environments

Tao Lin and Kevin Smith

CSIRO Mathematical and Information Sciences

Abstract

The development of multiuser 3-D environments for modelling and simulation is a complicated task for four reasons: (1) the need to often integrate existing 3-D modelling and simulation systems, which on the whole have not been designed for integration; (2) the communication protocols of the shared context between the collaborative systems are difficult to define; (3) a multiuser 3-D environment should be able to support dynamic reconfiguration; and (4) a multiuser 3-D environment should support changes of collaborative scenarios at runtime. To solve these problems, this paper presents a generic functional architecture that explicitly decomposes the functionality of a multiuser 3-D modelling system into functional tools and specifies the interfaces between them. This generic functional architecture also defines the possible collaborative contexts and specifies the communication protocols between the integrated systems. It can also support environment re-configuration and dynamic collaborative scenario changes.

Key Terms

collaborative environment   functional architecture   3-D graphics
CSCW   multiuser environment
framework   system integration

Introduction

Many application areas such as mining, medicine, manufacturing, and even information processing require the integration of 3-D modelling systems and other systems with specific functions to provide the necessary combined functionality. For example, an underground mining vehicle management environment including facilities for simulation, training, and teleoperation requires a number of 3-D modelling systems:

- 3-D modelling of the geology including ore grade distribution and rock quality
- 3-D modelling of the mine infrastructure (e.g., drives, stopes, and shafts)
- 3-D/4-D kinematic modelling of the vehicle
- 4-D mechanico-physical modelling of machine/rock interaction

Also needed is a system to drive the vehicle operator interface, the cost management system, risk prediction system, and personnel management system. Only by combining these systems can the desired environment be built.

In the past, the users of each individual system have worked together to produce a coherent, combined, and consistent model. We now need to integrate these systems into a coherent multiuser 3-D environment, and mechanisms need to be provided for collaboration, communication, and coordination.

Often these various 3-D modelling systems were developed independently with little or no consideration given to the future need to integrate them into a more complex system. In an integrated environment, data transformation between these individual systems is an essential task. Given the diverse functionality of 3-D modelling systems and the other systems in an integrated environment, there could be numerous ways to integrate them. It is very difficult to identify in advance the exact functions expected from each individual system. For complex applications there is a large diversity of possible data types to be transformed, the task of developing the data transferring modules to convert data between the individual systems is considerable.

As a collaborative environment, a multiuser 3-D environment needs to transform not only the model data between the individual systems, but the information on the working environment as well (e.g., the viewing point of a scene). For a particular system, the model data and the working environment information to be received and sent out together form its *collaborative configuration,* and the collaborative configuration of each individual system within the environment forms the *collaborative scenario* of the collaborative environment. Given the diversity of applications and numerous possible collaborative scenarios, the communication between the integrated systems in a multiuser environment becomes a very difficult task.

Another problem that environment developers face is dynamic reconfiguration. In a flexible multiuser environment there is the need to access the services of different but related systems: this requires either rapid or dynamic reconfiguration of the environment (i.e., adding or removing individual systems). Additionally, when a work task changes, the collaborative configurations of each integrated system need to be adjusted. Some changes require the reconfiguration of the environment, which is a system integration problem, and some can be done dynamically.

In this paper, we propose to use a generic functional architecture of a 3-D modelling system with full functionality as the backbone for integrating individual systems and enabling multiuser 3-D environments. We are using this generic functional architecture as the *reference model* for the development of multiuser 3-D environments. This generic functional architecture explicitly decomposes the functional components of a 3-D modelling system. The links between the functional components, the interfaces to each functional component, and the communication protocols between the functional components are well defined in this generic architecture.

The four problems discussed previously can be solved based on this generic functional architecture:

• The developer needs to identify only the equivalent functional components of the generic functional architecture within each integrated system. Thus the integration can be much simpler.

• The communication protocols between the functional components of the generic functional architecture can be used as the communication protocols for the communication between the equivalent integrated systems.

• Since the interfaces between the functional components in the generic functional architecture are fixed, dynamic reconfiguration can be supported by replacing the integrated system with another having the same functional interface.

• By considering the integrated systems as the functional components in a multiuser 3-D collaborative environment, the potential collaborative scenarios can be defined, and the required facilities to enable changes to the collaborative scenarios can be provided.

The rest of this paper is organized as follows. The next section gives the background. Then a generic functional architecture for a 3-D modelling system with full functionality is described. Then we discuss the extension of this generic functional architecture for system integration and describe the extension of this generic functional architecture for the development of multiuser 3-D environments. Conclusions can be found in the final section.

Background

3-D Modelling

A 3-D modelling process takes *input models* as the observed data to generate an *output model*. In a geological modelling process, the input model can be aeromagnetic, gravity, or drilling data, and the output model is a geological block model (Lin et al. 1995). In a medical modelling process, the input model can be X-ray images and a generic anatomical model, and the output model is a patient specific anatomical model (Wilson et al. 1996).

Figure 7.1 illustrates the architecture of a 3-D modelling system. The local dataset contains the input models and partially generated output model. The local dataset is visualized in pictures that can be observed by the users. Users can invoke a modelling operation from the modelling and analysis library (such as a triangular surface generation operation) and interactively specify the parameters (such as choosing the nodes that can be used by the triangular surface generation operation). After users generate portions of the output model, they can use the analysis operations to analyze the results. During this stage, the experience of the users is critical to making the right decisions since there is often no unique solution. The users can also apply some of the modelling operations to modify the output model. The modelling and analysis processes can be iterated until the users generate the final output model. The modelling process may last several months in the case of geological modelling, and an output model may need to be modified again in the future. Therefore, the management of the output models is also important.

The analysis operations can include topological and geometrical consistency checking, specific domain context consistency checking (such as structural geology, which is

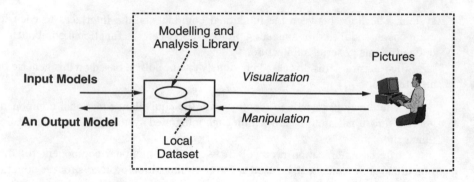

Figure 7.1   The architecture of a computer-aided 3-D modelling system.

historically consistent), and property measurement (such as volumes of regions of similar ore grade). It is not necessary to support data consistency on the local dataset at all times during the modelling process, but it is required that users should be able to invoke data consistency checks at any time. As there are many possible solutions in each stage of a modelling process, it may be necessary to create multiple versions of the output model from the same input model.

Many generic visualization systems (Rasure and Williams 1991; Unson et al. 1989) can support a wide range of visualization functionality. A 3-D modelling system needs to support the modelling for specific domains (Lin et al. 1995; Mallet 1992; Power et al. 1995), and the visualization operations should dynamically reflect the modelling context and be coherently integrated with modelling operations. It is very difficult to extend a generic visualization system to meet these requirements. Note that a 3-D modelling system means a system that is usually application domain specific, and, furthermore, the visualization operations supported in such a system can affect only that domain and are coherently integrated with domain-specific modelling operations. Unlike a generic visualization system, all the visualization operations supported by a 3-D modelling system are required at sometime in the modelling process.

Data Management

Data management is also very critical for 3-D modelling. A 3-D spatial domain can be divided into cells. Associated with a single cell, there could be (illustrated in Figure 7.2).

- Different model types (Sriam et al. 1989) (such as geophysical, geotechnical, geochemical, and lithological in geological modelling)
- Multiple model versions (ibid.) (in geological modelling, these could be several versions of the geological block model)
- Multiple systems creating the models associated with the same cell
- Multiple data storage devices storing the models associated with the same cell

Two functions are required:

- *Maintaining consistency of data model.* It is necessary to generate a consistent data model that is topologically and geometrically consistent for all data types. Also, it is desirable to achieve semantic consistency.

- *Data transformation.* As there are different data formats for different model types and modelling systems, data transformation between different data formats is required.

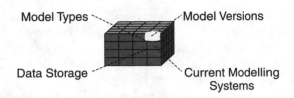

Figure 7.2   Data management issues.

Given the wide variety of possible data formats, a universal data format should be used as the intermediate data format (Power et al. 1995; Sriam et al. 1989) for data exchange between different data formats and for checking the consistency of the data in different data formats for the same spatial region.

The traditional 3-D modelling systems do not support these data management capabilities. We call a 3-D modelling system with data management functionality, as described previously, a *3-D modelling system with full functionality* and, a traditional 3-D modelling system, a *simple 3-D modelling system.*

System Integration

As described earlier, an underground mining vehicle management system requires the integration of various complex 3-D modelling and nonmodelling systems. A similar analysis of a minimal access surgical modelling and simulation environment yields the following models:

- 3-D anatomical model that includes geometric structure, shape, tissue properties, relationships between organs for a wide range of patient types (e.g., age and sex)
- 2-D/3-D models of the view through an endo- or laparoscope
- 2-D/3-D models of what a trainee surgeon would see using various imaging modalities such as ultrasound
- 3-D/4-D mechanico-physical model of the surgical instrument/tissue interactions

The models also include the systems for patient record management, cost estimation, and risk analysis.

Thus, the scope of some 3-D applications is so broad that no single 3-D modelling system can meet all needs. It is necessary to integrate various systems to provide the required functionality. The most critical issues to be faced in integrating the various modelling systems are to specify

- The links between the integrated systems
- The interface to each integrated system
- The protocols for data transformation between the integrated systems

These are very difficult to specify because of the diversity of functions in each of the integrated systems.

Collaborative Processing

In the examples discussed previously, it is often necessary for the 3-D modelling systems to support multiple users. The advantages of a multiuser 3-D environment are the following:

- *Sharing of the modelling workload.* As the scale of a modelling task increases, it can be completed only in a reasonable time by multiple users sharing the task.

• *Increased scale of the operation.* With a distributed 3-D collaborative environment, the collaborators can access remote locations through the environment. For example, specialist doctors in the city can work with doctors in rural areas to improve patient care, and consultant geologists in the city can construct complex 3-D geological models given observation data from mine-site geologists.

• *Sharing of expertise.* A multiuser 3-D environment supports the sharing of both human and software system resources. In the example discussed in the previous item, the expertise of specialized users is shared. In a collaborative modelling environment composed of different 3-D modelling systems, there is access to greater functionality.

A multiuser 3-D environment is a groupware system. Much generic research on groupware and CSCW has been done (Benford and Fajlen 1993; Dourish 1995; Ellis et al. 1991; Mandviwalla and Olfman 1994), and numerous systems (Bentley et al. 1994; Hill et al. 1994; Reinhard et al. 1994) have been developed. Currently, the emphasis is still on text and multimedia-based environments. There are only a few systems being developed for handling graphics, and they are still in the research stage (Greenberg et al. 1995; Li and Flowers 1992). Recently, with the availability of virtual reality systems, many distributed multiuser virtual environments (Calvin et al. 1993; Carlsson and Hagsand 193; Codella et al. 1993; Shaw et al. 1993; Singh et al. 1995) have been developed. However, most graphics-based multiuser environments focus only on the modelling of graphic objects. The development of multiuser environments is still done in an *ad hoc* manner. In the 3-D modelling area, there is a lack of understanding of the relationships between the following factors:

• The possible shared context
• The implementations of the components
• The physical constraints (such as network bandwidth, latency, computational systems, frequency of use, display, and input devices)

A groupware system requires three key technologies: collaboration, communication, and coordination. Some conceptual models (Ellis and Wainer 1994; Hawryszkiewycz 1994) have been developed to provide a fusion of these technologies in groupware systems. In these models, the components in a groupware system can be divided into three categories:

• A description of the objects and operations on these objects available in the system
• A description of the activities
• A description of the interface of users with the system

These models provide a very good guide for the development of groupware systems. Unfortunately, these models are too generic to describe the relationships between the factors mentioned previously. In this paper, we intend to address the relationship between these factors for multiuser 3-D environments using a single fully integrated reference model.

Other Development Requirements

The main aim of this paper is to solve the problems mentioned in the first section. However, we also intend to give solutions to some other development requirements:

• *A basis for analyzing the role of a technology.* A multiuser 3-D environment utilizes a wide range of capabilities. Some of these capabilities are changing very quickly in recent years, such as 3-D display facilities. Relevant new technology, new devices, new facilities, and new software systems come out almost every day. To cope with this situation, we need a basis from which to analyze the role of these new capabilities in multiuser 3-D environments.

• *A basis for producing a balanced design.* There could be many possible implementations of a multiuser 3-D environment. Designers need to make decisions based on the functional requirements, budget, and physical constraints. Designers need to be able to compare different implementations of each functional component of the conceptual architecture and understand the potential dependencies of these implementations.

• *A basis for the management of the repository of implemented components.* In our situation, it is already possible to develop several multiuser 3-D environments. Although the purpose of each multiuser 3-D environment differs from the others, many implemented components can be reused across different environments to reduce the development cost. To achieve this aim, a generic conceptual architecture that can be applied to a wide range of the possible multiuser 3-D environments is needed.

Functional Architecture-Based Software Development

Currently, object-oriented frameworks have become one of the most active research areas in object-oriented technology (Kochhar and Hall 1996; Lin et al. 1995). An object-oriented framework consists of two parts: a functional architecture and a set of class libraries. A framework is used to develop a set of applications with a similar architecture. Figure 7.3 illustrates the concept of an object-oriented framework. The functional architecture is used as the base architecture for all possible applications. An application might use only a portion of the functional architecture. The components in the functional architecture are called *functional components,* and the implemented components in the applications are called *implementation components.* All implementation components have corresponding functional components. The implementation components form the libraries. To develop an application with an object-oriented framework, a developer bases it on the functional architecture and selects the implementation components with a particular behavior.

An application might implement only a part of a functional architecture (i.e., just providing the implementation components for a portion of the functional components in a functional architecture). The links from the implemented functional components to these unimplemented functional components should be disconnected. As some functional

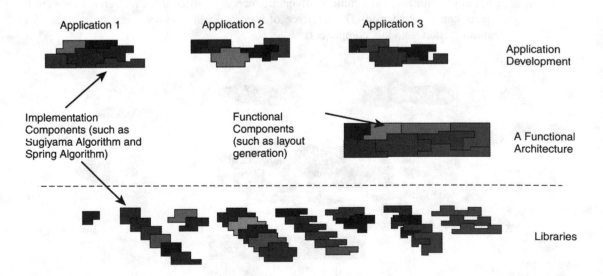

Figure 7.3 An object-oriented framework for information visualization. (*See color plate on page 319.*)

components depend on the existence of other functional components, there will be constraints on what links can be disconnected. The dependency of the functional components is maintained by the framework.

The scope of an application can be greater than what is covered in a single functional architecture. In this situation, the developer of a system is responsible for the interfacing of outside components with those specified in the functional architecture. We believe that a framework is specific for a particular application domain and there are different frameworks for different application domains. There could be overlaps of the functionality of these frameworks. This paper will not discuss this issue.

A functional component should specify two kinds of information: links and interfaces. A *link* defines how the implementation component of one functional component is connected with the implementations of other functional components. An *interface* defines the methods that can access the implementation component of another functional component.

A generic functional architecture can also be used as a basis for the management of a repository of implemented components. The functional architecture can be considered as pigeonholes, with each functional component representing a hole. During the development of an application, all the implementation components "drop" into the corresponding holes. When developing other applications, the designers can pick the developed implementation components from the repository to reuse them.

Kochhar and Hall (1996) developed a functional architecture for visualizing CAD/CAM presentations. We developed a generic functional architecture for 3-D geological modelling systems and implemented a framework—the GeoEditor Framework (Lin et al. 1995). The generic functional architecture discussed in this paper is an extension of that architecture.

A functional architecture can be used as the reference model for system integration. Figure 7.4 illustrates the development of an integrated environment based on a functional architecture. When developing an integrated environment, a designer identifies the components within the system to be integrated that are equivalent to functional components in the generic functional architecture. Then the designer matches the interfaces of this system with the corresponding functional components in the generic functional architecture and specifies the transformation of the data formats of this system to the data formats of the integrated environment. With the support of dynamic linking mechanisms, such as those in Smalltalk and Java environments, dynamic reconfiguration of the implementation components can be supported. The interface of a component also specifies the protocol for communication with this component.

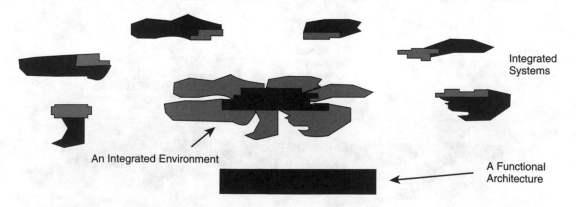

Figure 7.4   System integration based on a generic architecture.

A Generic Functional Architecture

As we discussed previously, the scope of a generic functional architecture should be neither too large nor too small. We chose as our scope a 3-D modelling system with full functionality for multiuser 3-D environments. This section presents a generic functional architecture for a 3-D modelling system with full functionality. This generic functional architecture is an extension of the generic functional architecture that we developed for simple 3-D modelling systems. In the next two sections we will discuss the extension of this generic functional architecture for system integration and the development of multiuser 3-D environments.

Basic Structure

The generic architecture can be decomposed into the following parts:

- the *functional layers,* where the functions are grouped semantically (e.g., rendering functions that display objects)
- the *entity layers,* which are derived from the functional layers (i.e., an entity layer is a collection of objects that are the "image" of the functions in the functional layer)
- the *transformation rules,* which are the high-level rules that determine how objects from one entity layer can be transformed into objects of another entity layer
- the *tools* are aggregations of functions in a particular functional layer together with the functions that transform external objects into entities of that layer and vice versa (e.g., graphical tools derived from the rendering functional layer)

Entity Pipeline

The functions supported by a 3-D modelling system with full functionality can be semantically decomposed into the four layers as illustrated in Figure 7.5:

- *Storing.* This layer stores the input and output models of the 3-D modelling system.
- *Data Management.* This layer maintains a consistent unified model. The functions include associating various types and versions of the data objects for the same 3-D space, converting the data between different data formats, and maintaining the consistency of the dataset in the entire spatial region.
- *Modelling.* This layer provides modelling and analysis operations. This layer also provides consistency checking based on both topology and the specific application context that is not supported by data management layer.
- *Rendering.* This layer provides functions to display the data in the modelling layer.

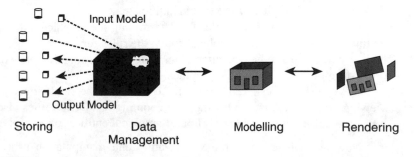

Figure 7.5   The layers for a 3-D modelling system with full functionality.

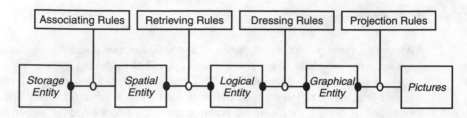

Figure 7.6   The entity pipeline and transformation rules.

A simple 3-D modelling system provides only the functions for the modelling and rendering layers.

To transfer data between the *storing layer* and the *rendering layer,* three entity layers are involved. These entities (illustrated in Figure 7.6) are spatial, logical, and graphical, which support the functions in the data management, modelling, and rendering functional layers, respectively. The entities in these three entity layers are derived from their associated functional layers. The storage entities are the data stored in either databases or files. The pictures are what are displayed on the screen. Only spatial, logical, and graphical entities are objects as they provide both data and operations that operate upon the data. In this paper, we concentrate only on these three entity layers.

The functions in the *data management* functional layer are supported by two kinds of entities: unit and spatial. A *unit* is a cell in the 3-D space. The cells can be regular and irregular. A *unit entity* contains information of the type and version of the data object associated with the unit and the storage devices for storing the data objects associated with the specific 3-D spatial region. The unit entities provide the indexes to associate the datasets that do not have direct connections. While there could be several versions for the same type of data object in the same cell, only one version can be active at a particular time. We call the model that consists of the active dataset for each cell in a specific 3-D region an active model of that 3-D region. *Spatial entities* are the data objects in the currently active model. The entity layers can be ordered: spatial, logical and graphical to form the *entity pipeline.*

Transformation rules act as the reference for the transformation between the entities in the different functional layers. As shown in Figure 7.4, transformation rules can be divided into four categories:

• *Associating rules* include the rules for associating the 3-D spatial space with the stored data and for converting between various data formats and the universal data format of spatial entities.

• *Retrieving rules* are the rules for retrieving the data for logical entities from the unified data model handled by the spatial entities.

• *Dressing rules* are the rules for determining the graphical attributes, such as color, texture, drawing style for a graphical entity, the rules for determining which aspects of the logical entities are to be converted to graphical entities and displayed, and the rules for converting between logical and graphical entities.

• *Projection rules* are the rules for controlling the rendering mechanisms, such as viewing angle, shading, shadowing, and zooming. Projection rules also include the rules for shape deformation and the portion of graphical entities to be displayed.

Now, we will describe the visualization and manipulation processes, respectively, based on this entity pipeline.

For visualization:

- A user first chooses a region. The unit entities contained in this region enable the system to list the possible input models. After the user chooses the an input model, the input model data will be extracted from either a file or a database system and the spatial entities generated using associating rules for data conversion.
- The user chooses a specific type of modelling (e.g., contact surfaces for structural geological modelling). Logical entities are generated from the spatial entities according to the appropriate retrieving rules.
- Using the dressing rules, the graphical entities are generated and displayed.
- The user can use projection rules to control the display.

For manipulation:

- The user can invoke a modelling operation to generate a piece of a model (e.g., to generate a surface from a set of nodes). After the modelling operation, some logical entities will be generated. The corresponding graphical entities will be generated according to the dressing rules. The changes in the spatial entities will be made according to the retrieving rules. If the changes are accepted in the spatial layer, then the system finally accepts the changes. Otherwise, the system will invoke an undo operation for these changes.
- The user can modify the graphical entities by direct manipulations of the pictures. According to the dressing rules, the changes made in the logical entities will be generated from the changes to the graphical entities. If the changes are accepted according to the consistency checking in the logical layer, the corresponding changes will be made in the logical entities. Otherwise, the changes will be omitted. If the changes are accepted in the logical layer, the changes in the corresponding spatial entities will be generated according to the retrieving rules. If the changes are accepted in the spatial layer, the system finally accepted the changes. Otherwise, the system will invoke an undo operation for these changes.

Tools

Tool kits, or more simply, *tools,* are the aggregation of functions in a particular functional layer together with the following components:

- A *glue* acts as a reference for the possible transformations between the entities in different functional layers. The glues do not give the result values themselves. The result values are determined by the "policies."
- A *policy* provides the mechanism to determine the result values for given entity values (e.g., the mechanism to determine the color value for a given visual entity).
- An *adjustor* is used to update, organize, and modify policies and entities.

In this architecture, the transformation rules are implemented by these three components. The mechanism for handling the color dressing rules is illustrated in Figure 7.7. The functions of the components involved are listed below:

- The *color dressing glue* takes the fields in the logical entity to be viewed and gives the color values as the output.
- A *color dressing map* provides a look-up table that gives the color value for a given value of the fields of the logical entity.
- A *color dressing policy* provides three functions: one is to search the color dressing map to give the color value according to the value of the logical entity; the second is to provide a mechanism on top of the color dressing map to organize access to the color

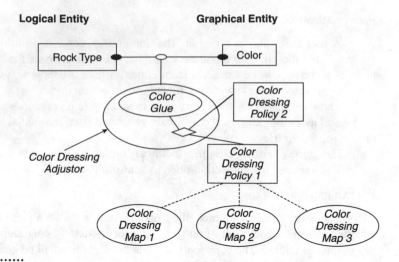

Logical Entity   **Graphical Entity**

Figure 7.7   Glue, policy, and adjustor for handling color dressing rules.

dressing map; and the third is to support users changing the color dressing map (including the use of different color dressing maps, changing the values in a color dressing map, and creating new color dressing maps).

• A *color dressing adjustor* has two functions: one is to choose the fields in the logical entity to be visualized, and the other is to change the color dressing policy to be called by the color glue.

To assign the color values for graphical entities, a color dressing glue is called. The color dressing glue itself does not provide the mechanism to give the color value, but calls the color dressing policy. The color dressing policy searches the color dressing map to decide the value for the given field value of the logical entity. Using a color dressing adjustor, one can change the color dressing policy, and with the color dressing policy, one can modify the color dressing rules.

Here, we use an example to further explain the relationship between glue, policy, and adjustor. There are two kinds of color dressing policy (Lin et al. 1995): *linear color dressing policy* and *hierarchical color dressing policy*. Figure 7.8*a* shows the linear color dressing policy, which functions as a look-up table and is the most commonly used mechanism for assigning color values. The hierarchical color dressing policy illustrated in Figure 7.8*b* allows users to flexibly change the abstraction level of the display according to the context of the application. Suppose that the logical entities with the attribute values *A, B, C,* and *D* can be abstracted as a group labelled 2. When users are interested in higher level abstractions of the logical entities, the hierarchical color dressing provides a mechanism for assigning the color *l* to the corresponding graphical entities of all logical entities in group 2. When users are more interested in the fine detail, the logical entities with values *A, B, C,* and *D* can be shown with the colors *a, b, c,* and *d,* respectively. Through this mechanism, a user can observe the logical entities at both levels of abstraction.

The tools in this architecture can be divided into five categories:

• A *picture tool* provides the functions for handling the images generated by graphical entities. A picture tool also provides policies and adjustors for handling projection rules. A picture tool can change only the "display settings" of an image, not the geometry of the graphical entities to be displayed.

Logical Attribute   Graphical

LAttri.	Color
A	a
B	b
C	c
D	d
E	e
F	f
G	g
H	h
I	i
J	j
K	k

(a) Linear color dressing policy

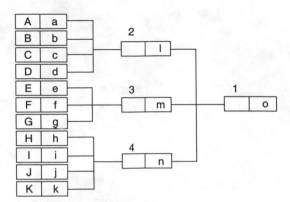

(b) Hierarchical color dressing policy

Figure 7.8   Color dressing policy.

- A *graphical tool* provides the functions for handling the graphical entities. A graphical tool also provides the glues, policies, and adjustors for handling dressing rules. A graphical tool can change the geometry of a graphical entity but not the application-related attributes of the corresponding logical entity.

- A *logical tool* provides the functions for handling logical entities, including the functions for modelling and analysis. The logical tools provide the glues, policies, and adjustors for handling retrieving rules. The logical tool can change all aspects of the logical entity including its geometry, application-specific attributes, and topological relationship with other logical entities.

- A *spatial tool* provides the functions for handling unit and spatial entities, and, in particular, there are functions for the spatial integration of entities from different logical tools; for example, functions for detecting different types of geometric and topological conflicts. The spatial tool provides the glues, policies, and adjustors for handling associating rules.

- A *storing tool* provides the functions to store the storing entities. A storing tool is normally a database or a file management system.

The separation of the functional components encapsulates the functionality of each particular functional component and greatly improves the reusability, flexibility, and extensibility of the system. This architecture can also support multiple tools in the same functional layer. These tools might differ from each other on their parameter settings or even their implementations. A possible tool configuration of a 3-D modelling system with full functionality is illustrated in Figure 7.9. The data exchanges between the tools are exactly the data exchanges between the corresponding entity layers. In most cases, a 3-D modelling system needs to support multiple tools in the same functional layer. In a 3-D modelling system with full functionality there is only one spatial tool to ensure geometric and topological consistency for the unique 3-D region being modelled, but there could be several tools in the other functional layers.

For a 3-D modelling system with multiple tools in one functional layer, a mechanism to coordinate the tools is required. Suppose that a logical tool *Ltool-1* links to a set of tools in the graphical layer, *Gtool-1, Gtool-2, ..., and* Gtool-n. These graphical tools can have different settings of the dressing rules or can be implemented differently (e.g., one might be implemented in Open Inventor, and the other one might be implemented in Open GL). Thus, the graphical entities and the dressing rules in these graphical tools could be

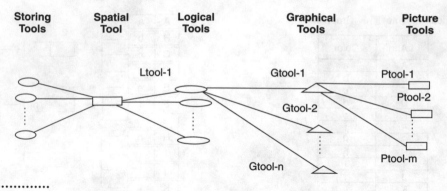

Figure 7.9 The configuration for a 3-D modelling system with full functionality.

different. For visualization, *Ltool-1* needs to broadcast the changes (if there are any changes in its logical entities) to the graphical tools linked with it, and the graphical tools will make the corresponding modifications. For manipulation, if a user makes some changes using a graphical tool (say, *Gtool-1*), a message describing the changes will be sent to the logical tool. The corresponding changes to the data will be made according to the data conversion rules between the graphical entity and the logical entity in the logical tool. Once the logical tool accepts the changes, it will broadcast the changes to the other graphical tools. However, the changes are accepted only by the system as a whole once the spatial tool accepts them.

The data formats of the tools in different functional layers are different. Even the data formats of the tools in the same functional layer but in different implementations can be different as well. In the example shown in Figure 7.9, suppose that the data formats in logical tool *Ltool-1* and the graphical tools *Gtool-1,* Gtool-2, ..., and *Gtool-n* are different. Through the dressing rules of *Gtool-1, Gtool-2,* ..., and *Gtool-n,* the data in *Ltool-1* is transformed into the data format of each tool in the graphical functional layer for a visualization process. In a manipulation process, *Ltool-1* receives the message from one graphical tool, and the data received from that graphical tool has been already converted into the data format of *Ltool-1* from the data format of that graphical tool.

System Integration

The last section described a generic functional architecture for a 3-D modelling system with full functionality. This section will discuss the extension of this generic functional architecture to enable the integration of 3-D modelling systems.

Integrating 3-D Modelling Systems

When designing an integrated 3-D environment, these questions should be asked:

- What kinds of 3-D systems are required?
- Can a particular 3-D system provide one of the functions required in an integrated 3-D system?
- If there is more than one 3-D system that can provide the same function, which system is the best?

The generic functional architecture for a 3-D modelling system with full functionality can be used as a reference model to answer the above questions. First, a designer needs to define the purpose of an integrated 3-D environment and then to use the functional architecture to specify in detail the functions required in the integrated environment. He or she can then review all available 3-D systems and identify those that can provide the required functions for the environment.

The role(s) of a modelling system can be identified by analyzing the data transferring between the system and the integrated modelling environment. Essentially, a modelling system will equate to a tool as shown in Figure 7.9. For example, a visualization system can act as a graphical tool, and a modelling system can act as a logical tool. A WWW browser that can display 3-D graphical entities is a picture tool in our generic functional architecture. To maintain the internal consistency of the integrated system it may be necessary to integrate different instantiations of the modelling system in respect of each tool type. However, to simplify, we say that a modelling system is equivalent only to a single tool.

Sometimes it is necessary to support multiple tools for the same functional layer in an integrated environment. Such an integrated environment is similar to a 3-D modelling system with multiple tools for the same functional layer, but the integrated systems could be located in different geographical locations.

Split Point

When a 3-D modelling system with full functionality as shown in Figure 7.9 supports multiple tools for the same functional layer, there exists a point that splits the tools in the adjacent layers. Such a point is called a *split point*. In the example illustrated in Figure 7.10, the logical tool *Ltool-1* has a split point where the tools in the graphical functional layer are joined. The information transferred between the linked tools is through the split point. There could be multiple split points in a 3-D modelling system.

To support multiple tools in the same functional layer, the mechanisms required at the split point are

- Coordination
- Data transformation

To maintain the data consistency of the data within different tools, a coordination mechanism is required. The data to be transferred between the tools can be divided into two categories: entities and transformation rules. When a tool in a lower functional layer

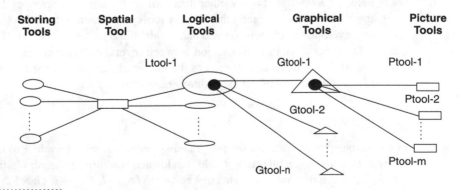

Figure 7.10 Split points in a 3-D modelling system.

makes a modification, the corresponding changes in the higher functional layer are gener-ated and appropriate checking and modification will be undertaken by the associated tool in the higher functional layer. Once the change is accepted in the tool of the higher func-tional layer, the process will be repeated to the next higher level functional layer. Finally, once the modification is accepted in the highest functional layer, it is broadcast to all lay-ers. In a 3-D modelling system with full functionality, the tool of the highest level func-tional layer is the spatial tool. In an integrated 3-D environment, only entity data needs to be transferred between the integrated systems.

In an integrated 3-D environment, it is possible that the integrated systems are dis-tributed through a network. Therefore, communication and distribution mechanisms are required at split points. In an integrated 3-D environment, it is necessary to transfer enti-ties between the integrated systems, and a coordination mechanism is required to integrate the datasets from the integrated systems to form a consistent dataset in the tool that con-tains that split point. The split point is always in the higher functional layer of the tools to be integrated. Through the split point, the integrated systems can be treated as these tools. The data formats of the tool that contains the split point and the data formats of the tools to be integrated are normally different. Therefore, data transformation between the data in different functional layers is required.

Data Transformation

In an integrated 3-D environment, data needs to be transferred between the integrated 3-D systems. As mentioned in the beginning of this chapter, the 3-D systems to be inte-grated are usually developed without considering the need for future integration, and so each system will have its own entity data format. Therefore, data transformation is required to make the data produced from one integrated 3-D system readable to the rest of the environment.

There has been much discussion on data transformations between different systems. The most effective approach (Power et al. 1995) is to have a universal data format for the exchange of data between different systems. Thus, to develop the data exchange function to another system, we need only to develop the data transformation function to change the data format in this particular system to the universal data format.

In our generic functional architecture, there are three entity layers, and three univer-sal data formats are required for these three layers. In our GeoEditor Framework (Lin et al. 1995) development, we chose VRML (VRML Architect Group 1996) as the universal data format for the graphical layer due to VRML being widely supported and data trans-lators between VRML and many other data formats having been developed. Thus, we can use many VRML browsers and other display tools as the picture tools in our environment. Also, we developed an object-oriented data model library (Power et al. 1995) that has been extended to support data transformation between several different commercial 3-D model-ling systems. We use this data format as the universal data format for both the spatial and logical functional layers of the GeoEditor Framework.

Distribution

In a distributed environment, the integrated systems that are equivalent to the tools in the generic functional architecture could be located in different geographical locations. Suppose that the integrated systems that equate to *Ltool-1, Gtool-1, Gtool-2, ...,* and *Gtool-n* are distributed via the network shown in Figure 7.11.

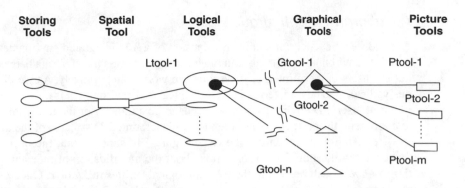

Figure 7.11   A distributed integrated environment.

A distributed mechanism is required to maintain the links between these integrated systems. Using a client-server architecture, the tool in a higher functional layer (such as *Ltool-1*) plays the role as a server, and the tools in the lower functional layer (such as *Gtool-1, Gtool-2, ...,* and *Gtool-n*) play the roles of a client. The split points are located in the tools of higher functional layers. The data formats in *Ltool-1, Gtool-1, Gtool-2, ...,* and *Gtool-n* could be different. For example, in a visualization process, *Ltool-1* sends data in its data format to *Gtool-1, Gtool-2, ...,* and *Gtool-n.* These graphical tools will convert this data into their own data formats. While in a manipulation process, a graphical tool (say *Gtool-1*) changes this data, and the changed data is converted into the data format of *Ltool-1* before being sent back to *Ltool-1.*

Environment Dynamic Reconfiguration

To make an integrated 3-D environment as "open" as possible, we need to be able to add or remove any of the integrated systems dynamically. It is difficult to support such a requirement in an integrated environment if the environment is developed in an *ad hoc* manner as it would be very difficult to define the interfaces for data exchange.

As we discussed previously, the major problems for the integration of 3-D systems are data transformation and communication. Current distributed computing environments (Orfali et al. 1996) can support the dynamic linking of two processes. Based on our generic functional architecture, the entities in each entity layer have a fixed universal data format. Once the module for the conversion of the data between the system to be integrated and the universal data format is developed, the integrated system can communicate with the rest of the environment. Since the universal data format is stable, a new system can be added to the environment at any time. Thus, the user of a 3-D integrated environment can easily switch on or off the use of a particular system within the environment.

Collaboration

In the previous section we described how a number of existing 3-D modelling systems can be integrated using the generic functional architecture for 3-D modelling systems with full functionality. Although these integrated environments are usually multiuser, additional features are required to enable effective collaborative working. This section will describe extensions to the generic functional architecture that are needed for multiuser 3-D environments.

Collaborative Configurations

A multiuser 3-D environment is an extension of a 3-D integrated environment. In a multiuser 3-D environment, not only can the integrated systems transfer data between each other, but the users of these integrated systems can work collaboratively. As we discussed previously, the data to be transferred in the 3-D integrated environment is only the entity data, but in a multiuser 3-D environment there needs to be a transfer of the transformation rule data to coordinate the perspective between the collaborating 3-D systems of the different users.

Referring to the example shown in Figure 7.8, suppose that the user using *Gtool-1* makes a modification. The information about the modification of relevant graphical entities in *Gtool-1* will be sent to the relevant logical entities in *Ltool-1*. Once *Ltool-1* accepts the change, *Ltool-1* will broadcast the change to *Gtool-2*, ...,and *Gtool-n*. The data transferred is the entity data. In a multiuser 3-D environment, one can also send the transformation rules between the integrated systems. For example, suppose that the user using *Gtool-1* finds an interesting geological feature at a particular geographical location, he or she will want to inform the other users who are doing geological modelling on the same rock mass. This can be done by broadcasting the viewing point and dressing rules.

The shared context and the way that users collaborate in a multiuser 3-D environment forms the *collaborative scenario* of a multiuser 3-D environment. The following are examples of the many possible collaborative scenarios:

- All the integrated systems have the same input model and produce the same output model
- All the integrated systems have the same input model but produce different output model
- All the integrated systems have the same input model and produce different output model, and one user monitors the other users' work

The collaborative scenario of a multiuser 3-D environment is implemented by the collaborative configuration of each individual integrated system. Changes of the collaborative scenario will cause changes to the collaborative configuration in each integrated system, and vice versa.

A collaborative configuration needs a number of objects to be defined, such as:

- The entity data broadcast message to be received
- The entity data modification to be sent
- The transformation rule broadcast message to be received
- The transformation rule modification to be sent

The generic functional architecture presented earlier supports multiple tools for the same functional layer, and the protocols for the transfer of both entity and transformation rule data between the tools are well specified based on the generic functional architecture. These protocols can now be used for the transferring of both entity and transformation rule data between the integrated systems in a multiuser 3-D environment. Since the protocols can cover all possible requests for data transferring, changes of the collaborative configurations at run-time can be supported.

Coordination

The aim of a multiuser 3-D environment is to develop a consistent and integrated output model. Therefore, it is necessary to coordinate the results produced by each individual integrated system. This coordination mechanism is needed at every split point.

A 3-D modelling system with multiple tools for the same functional layer requires a similar coordination mechanism. However, the multiuser 3-D environment is distributed and collaborative, and the split points of the multiuser 3-D environment will need to support both distributed communication and data coordination.

In the example shown in Figure 7.11, suppose that all the tools represented individual systems and the system that is in higher functional layer acts as a server and the one in lower functional layer acts as client in the multiuser 3-D environment. Because of latency, clients often need to keep a local copy of the output model. At a certain stage, the coordination mechanism at the split point needs to combine the central model maintained by the server and the local models developed by the integrated client systems. This is more complex than a multitool 3-D modelling system.

Conclusions

In the beginning of this paper, we described four problems that we believe need to be solved in order to effectively and efficiently construct multiuser 3-D environments:

- The easy integration of legacy systems that were developed originally without considering the need for integration
- The specification of the communication protocols of the shared context between the integrated systems in a multiuser 3-D environment
- The dynamic reconfiguration of implementation components in a multiuser environment
- The changing of collaborative configurations at run-time

Furthermore, we aimed to provide a structured approach to analyzing the role of new technologies, developing a balanced design, and managing a repository of the developed components.

These problems are not all solved in detail yet, but this paper describes a generic functional architecture for the 3-D modelling system with full functionality. This architecture explicitly decomposes the functions of such a system into some well-encapsulated functional components. The interfaces and the data transformation protocols between the functional components are well defined. This paper demonstrates that such a generic functional architecture can be used for the development of integrated 3-D systems and extended to enable the development of multiuser 3-D environments and, finally, collaborative multiuser 3-D environments.

Our generic functional architecture for the 3-D modelling system with full functionality is an extension of the generic functional architecture for simple 3-D geological modelling systems (Lin et al. 1995), which has guided the development of CSIRO's GeoEditor Framework. Further enhancements are being investigated to support virtual environments; in particular, immersive 3-D environments such as the EVL CAVE (Cruz-Neira et al. 1993) and "hands in" immersive environments such as the ISS Virtual Workbench (Poston and Serra 1994), as well as haptic devices such as SensAble Technologies' PHANToM (Massie and Salisbury 1994). The next major step we are planning will be the integration of a number of commercial mine scale modelling packages as the prelude to the development of prototype collaborative work environments and simulators for the Australian mining industry.

The development of multiuser 3-D environments can be very expensive if most of the environment has to be developed from scratch. Moreover, such a environment would be expensive to modify and maintain. A more efficient strategy is to build a framework based on a functional architecture. This allows both the reuse of multiuser environments and the rapid incorporation of new functionality.

Acknowledgments

The research presented in this paper was done in collaboration with CSIRO's Divisions of Radiophysics, and Exploration and Mining. We would like to thank Mike Sharrott and Chris Gunn for their contributions during the 3-D Collaborative Processing Project. Also, we had valuable discussions with Timothy Poston, Glynn Rogers, and Rhys Francis.

References

S. Benford and L. Fajlen, "A Spatial Model of Interaction for Large Virtual Environments." *Proceedings of ECSCW'93,* 1993, pp. 109–124.

R. Bentley et al., "Architectural Support for Cooperative Multiuser Interfaces," *Computer,* May 1994, pp. 37–45.

J. Calvin et al., "The SIMNET Virtual World Architecture." *Proceedings of IEEE VRAIS'93,* 1993, pp. 450–455.

C. Carlsson and O. Hagsand, "DIVE—a Multi-user Virtual Reality System," *Proc. IEEE VRAIS'93,* 1993, pp. 394–400.

C. Codella et al., "A Toolkit for Development Multi-user, Distributed Virtual Environments," *Proc. IEEE VRAIS'93, 1993, pp. 401–407.*

J. Cremer, J. Kearney, and H. Ko, "Simulation and Scenario Support for Virtual Environments," *Computers & Graphics,* Vol. 20, No. 2, 1996, pp. 199–206.

C. Cruz-Neira, D. Sandin, and T. DeFanti, "Surround-Screen Projection-Based Virtual Reality: The Design and Implementation of the CAVE," *Proc. SIGGRAPH'93,* 1993, pp. 135–142.

P. Dourish, "Developing A Reflective Model of Collaborative Systems," *ACM Transactions on Computer-Human Interaction,* Vol. 2, No. 1, 1995, pp. 40–63.

C.A. Ellis, S.J. Gibbs, and G.L. Rein, "Groupware: Some Issues and Experiences," *Communications of the ACM,* Vol. 34, No. 1, 1991, pp. 39–58.

C. Ellis and J. Wainer, "A Conceptual Model of Groupware," *Proc. CSCW'94,* 1994, pp. 79–88.

S. Greenberg, S. Hayne, and R. Rada, *Groupware for Real-Time Drawing: A Designer's Guide,* McGraw-Hill, 1995.

P. Hawryszkiewycz, "A Generalised Semantic Model for CSCW Systems," *Proc. DEXA'94,* LNCS 856, 1994, pp. 93–102.

R.D. Hill et al., "The Rendezvous Architecture and Language for Constructing Multiuser Applications," *ACM Transactions on Computer-Human Interaction,* Vol. 1, No. 2, 1994, pp. 81–125.

S. Kochhar and J. Hall, "A Unified, Object-Oriented Graphics System and Software Architecture for Visualising CAD/CAM Presentations," *Computer Graphics Forum,* Vol. 15, No. 4, 1995, pp. 229–248.

T. Lewis et al., *Object-Oriented Application Frameworks,* Manning Publications, 1995.

S. Li and W. Flowers, "Groupware Experiences in Three-Dimensional Computer-Aided Design, *Proc. CSCW'92,* 1992, pp. 179–186.

T. Lin et al., "From Databases to Visualisation—Providing a User Friendly Visual Environment for Creating 3D Solid Geology Models," *Proc. APCOM XXV,* 1995, pp. 11–20.

J.L. Mallet, "GOCAD: A Computer-Aided Design Program for Geological Applications in Three-Dimensional Modeling with Geoscientific Information Systems," *Mathematical and Physical Sciences,* Vol. 354 of Series C, 1992, pp. 123–141.

M. Mandviwalla and L. Olfman, "What Do Groups Need? A Proposed Set of Generic Groupware Requirements," *ACM Transactions on Computer-Human Interaction,* Vol. 1, No. 3, 1994, pp. 245–268.

T. Massie and J.K. Salisbury, "The PHANToM Haptic Interface: A Device for Probing Virtual Objects," *Proc. ASME Winter Annual Meeting,* 1994, pp. 295–302.

O. Orfali, D. Harkey, and J. Edwards, *The Essential Distributed Objects: Survival Guide,* John Wiley & Sons, 1996.

T. Poston and L. Serra, "The Virtual Workbench: Dextrous VR," *Proc. ACM VRST'94,* 1994, pp. 111–122.

W.L. Power, P. Lamb, and F.G. Horowitz, "From Databases to Visualization—Data Transfer Standards and Data Structures for 3-D Geological Modeling," *Proc. APCOM XXV,* 1995, pp. 65–70.

J.R. Rasure and C.S. Williams, "An Integrated Data Flow Visual Language and Software Development Environment," *J. Visual Languages and Computing,* Vol. 2, 1991, pp. 217–246.

W. Reinhard, J. Schweitzer, and G. Völksen, "CSCW Tools: Concepts and Architectures," *Computer,* May 1994, pp. 29–36.

C. Shaw et al., "Decoupled Simulation in Virtual Reality with the MR Toolkit," *ACM Transactions on Information Systems,* Vol. 11, No. 3, 1993, pp. 287–317.

G. Singh et al., "BrickNet: Sharing Object Behaviors on the Net," *Proc. IEEE VRAIS'95,* 1995, pp. 19–25.

D. Sriam et al., "An Object-Oriented Framework for Collaborative Engineering Design," *Proc. Computer-Aided Cooperative Product Development Workshop,* LNCS 492, 1989, pp. 51–92.

C. Upson et al., "The Application Visualization System: A Computational Environment for Scientific Visualisation," *IEEE Computer Graphics & Applications,* Vol. 9, No. 4, 1989, pp. 60–69.

VRML Architecture Group, *The Virtual Reality Modeling Language Specification: Version,* URL, http://vrml.sgi.com/moving-worlds/index.html (1996).

L.S. Wilson et al., "Computer-Aided Diagnosis Using Anatomical Models," *Medical Imaging Technology,* Vol. 14, No. 6, 1996, pp. 652–663.

• About the Authors •

Tao Lin is a Senior Research Scientist in CSIRO Mathematical and Information Sciences, Australia. His research interests include object-oriented frameworks for complicated interactive visualization systems, information visualization, human-computer interaction, and visual software development environments.

Lin received a B.Eng. degree from Beijing University in 1983, an M.Eng. degree from the University of Science and Technology, Beijing in 1986, and a Ph.D. degree from the University of Newcastle, Australia in 1993. He has published over 20 refereed international journal and conference papers. He is a member of the ACM.

Kevin Smith is a Principal Research Scientist in CSIRO Mathematical and Information Sciences, Australia. His research interests include the design and implementation of interactive modelling and visualization systems, pattern processing, and the application of massively parallel computing.

Smith received a B.Sc. (Hons.) degree from the University of Tasmania in 1974 and an M.Phil. from Murdoch University in 1977. After 2 years working for the British Gas Corporation on gas network modelling, he spent 10 years researching and developing scientific applications of massively parallel computing at Queen Mary and Westfield College (London) before joining CSIRO in 1990.

Chapter 8

Strategies for Mutability in Virtual Environments

Ben Anderson and Andrew McGrath

BT Laboratories

Abstract

Based on recent experiences of user trials, this paper suggests that in order to make virtual environments compelling, they must be mutable—able to and having a tendency to change. We outline five strategies that can provide mutability in virtual environments and provide examples of each from the various worlds we have developed. We review the utility of each of these strategies in the context of the design, implementation, and maintenance of virtual environments for significant user populations. We conclude that four of them—user-driven change, real-time representation of resources and objects, ecological responsiveness, and chaotic algorithms—offer some promising avenues for future development.

Introduction

There has been a considerable increase in the effort devoted to the development of virtual environments for human-human interaction and communication following the recent standardization of VRML 2.0.[1] This paper provides a reflective review of the development efforts at BT Laboratories that have resulted in a number of extensively discussed prototypes, culminating in one large-scale user trial of a "shared space" virtual environment. These experiences suggest that to be compelling, virtual worlds must be mutable—that is, they need to be able to, or tend to, change. People use, return, and inhabit virtual worlds for a number of reasons and one of the most crucial appears to be whether or not *things are likely to have changed.* The provision of mutable worlds in which the strata (rather than just the inhabitants) change over time is a serious design problem. Five approaches to the provision of such mutability are discussed: designer change, user-driven change, real-time representation, ecological responsiveness, and chaotic change. An example of these approaches from existing work is *The Mirror,*[2] BT Laboratory's experimental shared space

---

[1] VRML 2.0: http://vrml.sgi.com/moving-worlds/spec/
[2] The Mirror: http://www.bbc.co.uk/the_mirror
Done in conjunction with BBC Education, Sony and Illuminations.

running alongside BBC2's *The Net,* as well as in other virtual environments. The Mirror comprises shared spaces accessible to the public via the Internet as part of an experiment in creating an on-line community. Six spaces each illustrate different experimental themes. We conclude by discussing the issues and possibilities that these approaches afford.

Shortchanged in Virtual Environments

If we observe real life in comparison to virtual environments it is common to point out the difference in terms of frame rate, resolution, and field of view; however, one difference rarely noticed explicitly is that the real world is naturally mutable. Put simply, it changes. In fact it changes a great deal and it changes naturally, without relying on any human input. Virtual environments are fundamentally different in that they are not naturally mutable—yet. Most current virtual environments (for example, Blacksun[3] and Worlds Chat[4]) are static in the sense that very few of the objects within them change over time—there are plenty of attractive and often exquisitely detailed scenes. If users subsequently return, these scenes are still as attractive, still as exquisitely detailed, *but in exactly the same way as they were before.* There are two notable exceptions to this general rule:

- Users and their avatars
- Scripted objects

In the former case, the source of mutability is the movement of avatars through the virtual world, together with whatever communication (through text chat or internet audio) or user-driven changes in avatar representation that the system can support. Here then, the agents for change are the users themselves. The source of the unexpected, the "interesting," is, therefore, other people in the virtual environment.

Our experiences suggest that it is not sufficient for user's avatars to be dynamic in their representations or that it just be possible to meet and communicate with other users in this shared space. The problem is that this forces the onus of action onto the users themselves—with few users there will be little change and thus no reason to return. Current systems suffer because their usage has yet to reach a sufficient critical mass to ensure that there will *always be someone to talk to.* In this paper, we advocate strategies that may compel users to return to virtual environments *even if* there is no one to talk to.

Further, and perhaps more importantly, interaction and communication are never context-free—it is very rare that users will choose to communicate without some sort of "shared topic" of conversation. This suggests that designers of virtual environments must go beyond merely the provision of interaction and communication capabilities. It is critical that they provide engaging resources that will attract people to the same place (in order to concentrate what "mass" there is) and provide them with stimuli for shared discussion. It is critical, therefore, that we provide *contexts for conversations,* and we propose mutable objects as one way of achieving these goals.

In the case of scripted objects, the mutability is encoded into the VRML content of the world and may be triggered by user's actions or movements, or by timing events. Excellent examples include Protozoa Inc.'s animated creatures[5] and Silicon Graphics' *Floop.*[6]

[3]Blacksun: http://www2.blacksun.com
[4]Worlds Chat http://www.worlds.net/index.html
[5]Protozoa Inc.: http://www.protozoa.com/
[6]Floop: http://vrml.sgi.com/floops

However, our experiences of user trials suggest that "canned" animation and scripted objects, while initially diverting rapidly, pall in the way that repetitious events in real life also begin to annoy.

Virtual worlds can, and must, go beyond relying on the users to be the sole agents of mutability because this would imply that a sparsely populated virtual environment is necessarily an uninteresting place to be. Further, we suggest that it is insufficient to rely on static objects and canned animations with predictable repetitive behavior to maintain a user's interest. Virtual worlds will not be sufficiently compelling until objects within the environment are the agents of change themselves and that change is not repetitive.

However, the introduction of mutability must be viewed with caution as it raises three sets of interrelated issues.

From the user's point of view, change is not necessarily to be welcomed. Standard user interface design texts emphasize that users do not feel comfortable with interfaces that change in unexpected ways—they are comfortable with patterns of change, of rhythmic dynamics, of change that can be explained (Preece et al. 1994).

From the designer's point of view, mutability may be a two-edged sword. It may well encourage people to return to the virtual environment, but it will also assuredly demand increased design time to provide sufficient change over a period of time.

Mutability also has implications at the architectural level—which changes to the environment should remain local, and which should be automatically shared? In the real world, the default is that any changes in the environment are shared by everyone, but this need not be the case in a virtual environment. If local changes are not shared as a matter of course, the lack of a shared common view of an environment may well lead to communication confusion where one user can perceive and refer to a change but another cannot.

If, as providers of virtual environments, we aim to generate revenue from usage (be it by call, on-line, or subscription charges) we must ensure that *if we build them, people will come **again and again**.* However, we must also ensure that the overhead of managing the change of content is kept to a minimum. In the remainder of the paper, we discuss a number of associated strategies for supporting mutability in virtual environments and provide examples of their use. In particular, we focus on the ways in which they can provide support for mutability that does not necessarily increase the creation or management costs to the provider of the world.

Strategies for Mutability in Virtual Environments

We propose five approaches to the provision of mutability in virtual environments: designer change, user-driven change, system-driven dynamics, ecological responsiveness, and chaotic mutability. We discuss these in turn, providing examples drawn from ongoing work.

Designer Change

The easiest way for the maintainer of a virtual environment to provide change is to design new objects to be placed within it or to redesign those that are already there. Handcrafted worlds are the simplest to build and are the most suited to our existing methods of production. There is a strong temptation to handcraft compelling virtual worlds, and indeed, the quality of the worlds can be very high.

While this approach is initially attractive and certainly the easiest to initiate, it very rapidly becomes unmanageable because designing change into virtual worlds becomes a massive overhead because the methods of production used are expensive. As a result, they are suited to the broadcast approach to design where a virtual environment is to be used by a very large user population. However, when a team is managing the content for a number of individual worlds that only support communities of say 100–200 people, these high production costs become insupportable.

By way of a comparison, consider the overhead of managing a set of HTML pages to ensure that they are all suitably updated in order to maintain information freshness and are sufficiently compelling. Maintainers of virtual worlds can expect to increase this overhead by several orders of magnitude if they wish to alter or update a significant number of objects within the world. For example, the effort required to design and implement a bug-free animation in VRML 2 and Java is relatively high. Consider then wanting to add a new animation each week, or even each day. Consider the effort required to craft a virtual office space and then to rebuild that space whenever users required changes or expressed preferences.

In our experience, the real crux of the problem is that designing and handcrafting the content of a virtual environment is fundamentally a software engineering process. As Brooks (1995) has noted, such processes are subject to the law of diminishing returns—namely, that throwing more design hours at the design task serves only to increase the complexity of the communication about the design and actually has an adverse effect on the rate at which work proceeds. Since increasing the design hours is the only feasible way of increasing the rate of mutability of the virtual environment under this strategy, we conclude that it is inherently self-limiting. The logic is simple: maintaining sufficient mutability in a handcrafted virtual environment over a given period of time would require significant effort. However, increasing the effort by allocating greater design hours increases the time taken to complete the work. It is clear, therefore, that this strategy cannot scale if we wish to provide significantly dynamic environments.

Do It Yourself (DIY) Mutability

In the previous strategy, the virtual environment providers were responsible for its mutability. In contrast, the DIY strategy emphasizes the benefits of enabling users to be responsible for providing mutability by contributing their own objects and scripts to the virtual environment.

This strategy learns from the Internet community's experience of the growth of the World-Wide Web (WWW). The ease with which just about anyone with an Internet connection could create their own pages, and so become an information provider, has been one of the secrets of its success. The effort and overhead of maintaining fresh information and managing its change is therefore distributed throughout the user population. The WWW would be an extremely uninteresting place and would have grown slowly, if at all, if only a few teams with the relevant expertise and technical capabilities had been responsible for managing the information provided.

In the case of virtual worlds, this strategy, which is enshrined in many of the visions for VRML 2.0 and 3.0, proposes that users should be able to contribute to the environments they inhabit. Thus, worlds need to be configurable, and users need to be able to add their own objects so that they become content providers as well as content consumers. As has been shown in studies of configurable user interfaces, providing users with mechanisms to tailor their electronic environments is an extremely powerful way of encouraging a sense of ownership and of participation (MacLean et al. 1990). As importantly, it also enables users to adapt the infrastructure support, in this case the virtual environment, to their own specific uses. For example, studies of the development of CAD system tailoring skills have suggested that a process of role differentiation occurs such that those with the initial tailoring skills create and distribute "interesting" or "useful" macros (Gantt and Nardi 1992). Other users run these macros and then seek to tailor their functionality themselves. The original authors are then drawn into aiding this process, and a community of use develops that serves both to transfer knowledge and skills about how to write such macros and to encourage the creation of novel functionality. This DIY functionality is displayed in AlphaWorld,[7] where inhabitants can construct home spaces out of a list of generic shapes.

In The Mirror's Creation World, we have arranged for the participants to have access to a VRML modelling application and have invited them to send in their own artwork. In another part of The Mirror, "Memory World," we have invited participants to annotate the memory objects in the world with new memories of their own in the form of time capsules that leave text, sound, and images of personal thoughts for future visitors. This provides us with the opportunity to explore the possibilities of user authored systems, and early results suggest the same processes of en-culturation and self-help are evolving. The time capsules have been particularly successful with at least two or three being added each day.

However, user contribution to public virtual environments has also raised issues that are becoming familiar to on-line service providers:

- Exactly who is responsible for the content—the author or the service provider?
- When and why should contributions be monitored and/or censored?
- How can integration into the existing virtual environment best be managed?

While these problems are largely in the legal or regulatory domain beyond the scope of this paper, our experiences have also raised a number of user-oriented and technical issues:

- Users require persistence—contributions must always be available after submission although an author may choose to set an expiry date.
- Users want to be able to make changes to their contributions after submission, which raises issues of access and control—who can edit whose contributions, and when can they do so?
- User's contributions could, with relative ease, cause detrimental side effects to other users of the environment through dangerous code execution. This might also manifest itself in people building overly complex objects, adding them to the virtual world and therefore slowing down that world for other users.

Perhaps the most critical issue is ensuring that the skill required to surmount the contribution barrier is sufficiently low. This DIY strategy will succeed only when the skill required in crafting contributions to virtual environments is on a par with contributing to a usenet newsgroup or writing an HTML page. We note that current efforts aimed at developing VRML design and crafting tools are critically important in this regard.

We have sidestepped some of these issues in The Mirror's Play World, where users can play a number of games. One such game records a user's score and enters it on a ranked list in the same way as many arcade video games. Other users see these scores, then play against these rankings and so are added to the list. Interestingly, even this very simple form of contribution has proved extremely powerful. Users have returned a number of times to see if their scores have been overtaken and in doing so have been drawn into interaction with other players. This simple social focus, which provides users with a personal vested interest in the virtual environment, has therefore proved extremely valuable.

---

[7]Alphaworld: http://www.worlds.net/alphaworld/

In summary then, content can be generated by the users, allowing a sense of ownership, and has the advantage of parallel contribution to unpredictably mutable worlds. However, this can lead to low-quality work as a result of limited skills in participants, and it can also lead to offensive or inappropriate contributions. Finally, the strategy can be successful only if a sufficient user population is actually able to contribute.

Real-Time Representation

One of the oft-stated attractions of virtual environments in the corporate context is the ability to provide novel representations of an enterprise's knowledge and information resources and to make both these representations and the information itself available through a corporate intranet. There is considerable effort being devoted to designing new ways to represent information using three-dimensional visualization on the basis that it can support perceptual searching through visual cues, such as depth, as well as enabling experimentation with novel representational metaphors that can provide users with unexpected insights into patterns and links between information (Waterworth 1996). However, these efforts often appear to overlook the fact that information resources in an intranet context are ever changing simply because they are the objects with which we work on an hour-by-hour basis.

We propose that we should not view this constant change as a threat to the validity of a representation; in fact, virtual worlds should make use of this phenomenon and, when representing information objects, *must provide real-time representation of those objects.* Thus, when some characteristic of an information object changes, so must the representation of that object since to the user, they are one and the same thing. In an intranet context, this means that virtual worlds must be, to some extent, self-managing—the process that generates the VRML representation (for example) should be able to update a user's view of an object or set of objects.

Current approaches to the rejuvenation of representations tend to rely on the generation of complete new scenes through CGI scripts, which return a new VRML file on request, or through repeated downloading of a file that is continually recreated by a background process.

In contrast, BT Laboratories are developing an architecture that maintains a persistent representation of a set of information objects and which relies on triggers to indicate which objects should be updated and in what ways. This persistent repository then propagates these changes to browsers via a Java communicator applet, which inserts or deletes objects from the users' VRML viewer using the evolving External Authoring Interface or EAI.[8] This has allowed us to build persistent representations of a number of different information resources that have been immediately useful simply because they (a) are persistent and (b) are an accurate, real-time representation of the objects concerned (Crossley et al. 1997).

An example system has been implemented that demonstrates real-time growth of search trees based on text-based searches of Internet sites. BT Laboratories' Jasper agent provides a background search mechanism based on a range of resources such as users' bookmarks and history lists to generate a set of related on-line resources (Davies et al. 1995). Using the metaphor of information plants, we represent the set of related resources as stems and leaves of a plant that grows over a period of time. Groups of information plants therefore become an information garden, where they are "planted" and where they "grow" and subsequently "die" (see Figure 8.1 and also Davies et al. 1996). This metaphor

---

[8]External Authoring Interface: http://vrml.sgi.com/moving-worlds/spec/ExternalInterface.html

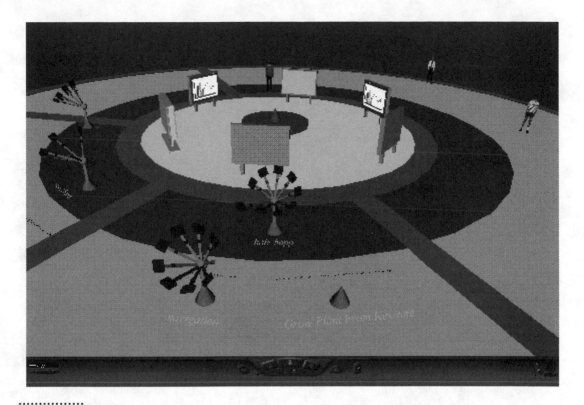

Figure 8.1 Growing Information Trees. (*See color plate on page 319.*)

we use has within it the idea of growth and change and decay (as well as taking cuttings!) and therefore are trying to ensure that this representation provides users with an intuitive idea of the relationships between different on-line resources. Furthermore, these gardens of information representation change over time, implying that users visit and revisit them, tending them as they might tend a real garden to ensure maximum information yield.

In this strategy we emphasize the use of *everyday* mutability—change that occurs as a matter of course—to drive mutability in a virtual environment. We are developing this strategy further by exploring its use in the real-time representation of document and other information stores. Early experiences with these systems suggest that they are extremely powerful in attracting users and use over a period of time simply because the visual structure is constantly evolving in meaningful ways.

Mutability in Responsive Ecologies

Our fourth strategy for providing mutability is inspired by the observation that the world is full of naturally occurring but apparently random events set against a background of gradual change. Real life is an *ecology* of events that are recycled in patterns that, while similar, are rarely identical—seasons are cyclical, but they never manifest themselves in exactly the same ways; night cycles in to day, but tonight may not be the same as last night.

Interestingly these changing resources and events help us to converse naturally, by providing much of the glue for holding conversations that we identified previously as a critical requirement. Furthermore, and perhaps critically, these events allow us as human beings to *situate* our memories—we remember what we were doing and what we said by association with unusual or memorable features. In most virtual environments to date, days and weeks would quickly blend into one, allowing us no way of distinguishing one time from another in our memory. By situating the user's experiences of the virtual environment in a milieu of apparently natural events, we provide them with cues for remembering. By having these events driven by scripts in virtual worlds, we can then allow users to search back for events (such as who they met) based on those cues. We can now, for example, search the world to recall what we said "when we were standing on the grass in the rain" or "the time it got dark by the lake."

We have been exploring these ideas by building virtual environments that draw on physical and biological analogues. Figures 8.2 and 8.3 show views of two such worlds.

The first, Identity World (again, part of The Mirror), shows a social environment that has been imbued with contextualizing phenomena. Identity World cycles between day and night with a periodicity of tens of minutes, and during these cycles, the shops that can be seen change from daytime to nighttime themes. Identity World is therefore an extremely different place to be when the environment gets dark, and we have noticed that users change the topic and tenor of their conversations accordingly.

The second, ChaosSphere, is an experimental virtual environment which has a number of biophysical properties:

• The world changes from dark to light (night to day) over a varying time period.
• There are two kinds of clouds that pass back and forth on varying paths, ensuring they vary their physical relationships to each other and do not appear to repeat too obviously.

Figure 8.2   Identity World (from *The Mirror*). (*See color plate on page 319.*)

- One cloud, which is soft in appearance, produces rain; the other more angular cloud produces hail.
- As these clouds pass overhead, the area beneath them is shaded, and a raining/hailing sound is heard.
- The color of the ground surface changes from dull brown to deep green and back again over a period of days.
- The world is inhabited by objects that are responsive to the proximity of the user—some will move away, others will move closer. Some of the objects will also respond to each other so that it is possible to watch a number of them following each other around apparently at random.

Our experiences with this world and other similar environments that are contributions to The Mirror suggest that users find such natural events not only extremely easy to comprehend but also extremely compelling. It has become common at some meetings for us to leave this world on in the visual periphery to form a backdrop to our discussions.

The key point in this strategy is that the events in the virtual environment, while scripted, are carefully designed to be responsive to different conditions within the world. Thus, the behavior of the objects can be dependent on the number of users in their locality. Once we make this step to making the responsiveness dependent on uncontrollable factors, the scripted behavior also becomes unpredictable. As a result, the addition of new responsive objects to the virtual environment can lead to unexpected patterns of change in

Figure 8.3 An ecologically responsive environment. (*See color plate on page 320.*)

the behavior of all the entities that are there. This is therefore an extremely powerful model for the provision of mutability because the interaction between a large number of relatively simple autonomous objects leads to extremely complex patterns of behavior. Making these objects responsive in simple ways to external stimuli, such as the presence of other objects, then produces extremely complex responsive behavior. Such environments appear to be endlessly fascinating to users because they change over time but in patterns that are both natural and explicable.

Once we have such an ecology set up in a world, we can start to use that ecology to help manage the world for us in an interesting way. One of the problems highlighted earlier was of worlds becoming too computationally expensive to run at acceptable frame rates on a standard computer. In an analogous way we tend to collect data but are bad at getting rid of old or useless data; in fact, redundant data can start to prevent us from finding useful data or at least slow the process down. If we have an ecology, we can use an agent to effectively recycle unwanted data. We can have an ecology where unused data situated in a temporary data area is reduced gradually until it disappears. We can image this as a graphic file where the lossy compression is gradually increased until it is only a thumbnail, or as a document where the information contained within it is gradually summarized automatically until it is gone. We can imagine, therefore, some aspects of the virtual worlds mutability being part of the everyday ecological life of a system trying to maintain an equilibrium of data quantity versus speed of access.

Chaos-driven Mutability

Earlier we suggested that encoded, canned animations do not provide sufficient dynamism to maintain user interest, and in the previous section we showed how making simple animation responsive to other events can make such scripted behavior much more engaging. However, algorithms do not always need to produce the same behavior—many of the algorithms that embody the mathematical principles of chaos never produce the same result, yet they exhibit a rhythmic behavior that lies midway between repetitive and random. Such behavior recurs in slightly different states over time so that an observer might perceive that *something of this kind* has happened before but *not in exactly the same way.*

Our final strategy then is to leverage algorithms that can provide random, rhythmic, or periodic change with no further intervention *at all.* By using periodic and rhythmic behavior, we ensure that users are not (often) presented with unexplainable change—if they wait long enough they will see variations on the theme, whatever it may be.

As an example, we have implemented a virtual environment, the AuB, which provides a combination of ambient background music and event-driven auditory icons (Gaver 1986). This environment, implemented using KoanMusic's plug-in,[9] provides an ambient auditory background based on a simple template. The plug-in replays this template and automatically varies a number of the parameters of the music so the same template never replays in exactly the same way. It does this using an implementation that relies on many of the principles of nonlinear mathematics, but it ensures that the output is reasonably musical rather than truly chaotic. Furthermore, the environment we have implemented is responsive to events and to the presence of other objects, which in turn are used to alter characteristics of the auditory background by controlling aspects of the plug-in's playback, such as which instruments can be heard, as well as their pitch, rhythm, and volume.

....
[9]KoanMusic: http://www.sseyo.com/

While we are just beginning to explore the possibilities of such chaotic algorithms, we have found that making their mappings responsive to other objects in the virtual environment can have effects that users find incredibly absorbing. For example, by linking the pitch, volume, and beat in our sound generator to the number of proximal objects, be they avatars or entirely autonomous, we create virtual environments that respond sonically as well as visually. Highly populated areas of the environment sound different, and the patterns of sound vary in ways that appear to be nonarbitrary—that is, they become more memorable. Users tend to cluster around these objects and experiment with the effects that their presence may have. In turn, this often leads to communication and conversations about the objects they are studying—we have therefore provided resources and contexts for conversation.

Conclusions

This paper has drawn together our experiences of implementing a range of virtual environments whose use ranges from corporate information intranets to public communication, leisure, and recreation spaces. A common theme that runs through these environments is the need for mutability—for an ability and a tendency to change over time. Early trials with our own virtual environments have suggested that without such mutability, virtual environments will at best be dull, sterile places that do not foster communication, while in the knowledge intranet context, they risk providing stale information resources. In either case, the value of the virtual environment is significantly decreased because users will tend not to return, communicate, or participate.

We have outlined five strategies that we see as key to the provision of mutable virtual environments and have outlined some examples of each. We have also discussed their attendant advantages and disadvantages. Our experiences in designing and implementing virtual environments for significant user populations have convinced us that only those strategies that provide maximum mutability for minimum design time are likely to be successful on an industrial scale.

As a result we are actively pursuing the use of real-time representation of on-line resources to provide virtual environments that are fresh and evolving. We are also experimenting with ecologies of simple objects that, by interaction with each other, produce richly varying patterns of behavior without a significant coding overhead. Finally, we are examining the potential uses of nonlinear dynamics to produce objects with simple animation scripts, but which can exhibit highly variable behavior.

Acknowledgments

Content for *The Mirror* was created by Andrew McGrath, Alison Coe, Marco Fauth, Amanda Oldroyd, Peter Platt, Matt Polaine, Tim Regan, and Matt Shipley.

The Information Garden was developed by Andrew McGrath, John Davies, Martin Crossley, Chin Wei Cheah, and Lewis Collins.

ChaosSphere was developed by John Dent, Nathan Laud, Bill Cameron, Andrew McGrath, and Peter Maydell.

This paper could not have been written without the contributions and efforts of *all* of the Shared Spaces team supported by the Virtual Business Campaign at BT Laboratories.

References

F.P. Brooks, *The Mythical Man-Month,* Addison-Wesley, 1995.

M. Crossley et al., "3D Internet," *BT Technol J.,* forthcoming, 1997.

N.J. Davies, R. Weeks, and M.C. Revett, "Information Agents for the World Wide Web, *BT Technol J.,* Vol. 14, No. 4, 1996.

N.J. Davies, R. Weeks, and M.C. Revett, "Jasper: An Information Agent for WWW," *Proc. 4th Intl. Conf. WorldWideWeb,* Boston, Dec. 1995; also available at http://www.w3.org/pub/Conferences/WWW4/Papers/180/

M. Gantt and B.A. Nardi, "Gardeners and Gurus: Patterns of Cooperation among CAD Users," *Proc. ACM CHI'92 Conference on Human Factors in Computing Systems,* 1992, pp. 107–117.

W.W. Gaver, "Auditory Icons: Using Sound in Computer Interfaces," *Human Computer Interaction,* Vol. 2, pp. 167–177.

J. Gleick, *Chaos: Making a New Science,* Viking Press, New York, 1987.

S.A. Levin, ed., *Studies in Mathematical Biology,* Mathematical Association of America, Washington, DC, 1978.

A. MacLean et al., "User-Tailorable Systems: Pressing the Issues with Buttons," *Proc. ACM CHI'90 Conference on Human Factors in Computing Systems,* 1990, pp. 175–182.

J.Preece et al., *Human Computer Interaction,* Addison-Wesley, 1994.

I. Stewart, *Does God Play Dice?,* Penguin, 1989.

J. Waterworth, "Personal Spaces: 3D Spatial Worlds for Information Exploration," *Proc. 3D & Multimedia on the Internet, WWW, and Networks,* 1996.

• About the Authors •

Ben Anderson is a research scientist in the Centre for Human Communications at BT Laboratories. He is currently researching flexible systems for human communication and collaboration, the implications of addictive user interfaces, and pursuing fundamental critiques of Human Computer Interaction using principles from the social sciences. He has a BSc (Hons) in Biology and Computer Science from the University of Southampton and a PhD in Computer Science from Loughborough University. E-mail: ben.anderson@ bt-sys.bt.co.uk

Andrew McGrath leads a team of designers and animators at BT Research Laboratories working on Shared Spaces projects. His most recent work has been on The Mirror—a collaboration between BT, Sony, and the BBC, exploring inhabited TV. Since leaving college in 1991, he has focused on 3-D user interfaces, including management of the Portal, an early VRML site. Andrew has a BA in 3-D Design from Glasgow School of Art and an MA in Product Design from Manchester Polytechnic. E-mail: andy.mcgrath@bt-sys.bt.co.uk

## Chapter 9

Bringing the MBone to Web Users

Adrian F. Clark

VASE Laboratory
Electronic Systems Engineering
University of Essex

• •

The Internet's "multicast backbone," or MBone, is intimately associated with continuous media services such as audio- and videoconferencing. However, the MBone operates independently of the World-Wide Web and, since it is totally distributed in nature, the best way to accommodate it within the Web's client-server approach is perhaps not obvious. This paper explores one such interface, namely, how the functionality of the MBone's "session directory" tool may be integrated with Web browsers, firstly to a browser written in the Tcl/Tk programming language and secondly as a plug-in for Netscape Navigator. Some difficulties with Netscape's plug-in API are noted and conclusions are drawn as to the relative ease of adding unusual functionality to interpreted and compiled browsers.

Introduction

The success of the World-Wide Web has already passed into computer folklore. So, we ask ourselves, what are the next "killer applications"? It is likely that in the coming few years one of them will be in the area of so-called continuous media: audio and video streams. Although there are quite a few "Internet phone" applications in use, none of them has proven particularly successful as yet. This is fortunate, since the data rates required would rapidly swamp the network with a relatively small number of users.

However, research into efficient mechanisms for disseminating these continuous media has been proceeding for a decade or more and has yielded a scheme that scales quite well to large numbers of users. This mechanism is known as the Internet's *Multicast backbone,* or simply, the *MBone*. This paper describes some experiments that have been carried out to investigate interfacing the MBone utilities to the Web. The first of these is an interface to a locally written Web testbed known as nib (for *Networked, Integrated Browser*), written in the Tcl/Tk programming language (Ousterhout 1994; Welch 1995), while the second is a plug-in for the ubiquitous Netscape Navigator. The ease of providing MBone functionality in both cases is assessed.

The paper proceeds by outlining the main features of the MBone, since this is likely to be unfamiliar to most readers and an appreciation of it is essential for what follows. The design and structure of nib are then outlined, followed by a discussion of how the MBone interface is achieved. Converting this interface to a Netscape plug-in is then discussed. Finally, some conclusions are drawn as to the relative ease of extending Web browsers.

The Multicast Backbone of the Internet

The Internet is, of course, a packet-switched network. When one sends an identical message to multiple recipients (an e-mail message, for example), separate packets containing identical data are routed to each recipient. Continuous media streams (audio or video) generate data rates of up to 128 kbit/s, even with the best coding techniques. Hence, the number of possible recipients of real-time audio or video would be severely restricted on wide-area networks. However, if one could send only one packet stream from an audio or video source, with the Internet's routing hardware and software taking care of replicating the packet streams and sending them to different recipients as and when necessary, the overall data rate would be much lower. (A good analogy is to think of the data stream as water flowing through a network of pipes, routers lying everywhere the pipe splits.) The term that describes this method of disseminating a packet stream is known as *multicasting,* and its instantiation on the Internet is termed the *multicast backbone,* or *MBone* (Kumar 1996; Macedonia and Brutzmann 1994).

A range of IP addresses have been set aside for the MBone, namely, "class D" addresses: those that have the first three bits all set. This means that the IP addresses 224.0.0.0 to 239.255.255.255 are multicast addresses. These multicast addresses are very similar to radio channels: anyone may send out information on any address, and the data will be received by anyone who happens to be listening.

Although the MBone may be used for any type of data traffic—for example, the author is engaged on a project that is implementing shared VRML 2 worlds on the MBone—it is closely associated with research into audio- and videoconferencing tools. (To be precise, only multicast addresses in the range 224.2.0.0} to 224.2.255.255 are dedicated to multimedia conferencing.)

There are some important subtleties concerning continuous data streams on the MBone. The first concerns the way in which packets are sent. Conventional traffic on the Internet is generally sent using the *transmission control protocol* (TCP): it ensures that if all packets do not, for some reason, arrive at the destination, they are retransmitted. However, if one is engaged in an interactive conference, packets that fail to arrive on the first try will cause the audio or video to break up; any retransmission due to TCP will not be of any use and will actually introduce additional load on the network, probably resulting in a further loss of quality. In other words, for interactive continuous media, timely delivery is more important than guaranteed delivery. For this reason, the *real-time proto-col* (RTP) (Schulzrinne 1997) is used in preference to TCP: RTP includes timing and sequencing information and is a connectionless service that drops packets that are not delivered at the first attempt.

The second subtlety concerns the way in which packet streams are disseminated. If every packet stream propagated over the entire MBone, it would soon become swamped with traffic. Hence, there are some ways of controlling the amount of the MBone over which a packet stream may travel. The first way is by giving each packet a *time to live* (TTL)—in practice, this is a parameter that the originator of the packet stream may set. Each time a packet passes through a router, its TTL value is reduced by one, and when the TTL reaches zero, the packet is simply discarded. Moreover, some thresholds are imposed on the TTL: for example, packets with a TTL less than 16 will not propagate outside an organization.

The next subtlety also concerns the propagation of packets. Packet streams are not, by default, sent to all possible recipients on the MBone; instead, when a multicast session is initiated, the MBone tool sends a message to its nearest router asking to be included in the relevant multicast group (identified by its multicast IP address). If the router is itself receiving that packet stream, it is simply sent to the requestor in addition to any other recipients; otherwise, the router asks the next router "upstream" to provide the packet stream to it. This process of asking the upstream router continues until a router that does receive the stream is located and the packet stream "tap" turned on. (This begs the question as to how one router knows which routers are upstream of it. The answer is that, on Unix machines at least, this information is provided when configuring the routing software.)

The final subtlety concerns the relationship of the MBone to the Internet itself. Although the MBone runs as a limited-data-rate service on the existing network, multicast routing is supported by only a small proportion of existing routers. (MBone routers are typically obsolete Sun workstations running special software.) If there are two multicast "islands" separated by a nonmulticast-aware link, it is obviously not possible to simply route multicast packets between them. Instead, multicast packets are put into a special IP packet "envelope" (the ability to do this has existed since the early days of the Internet) and sent over the link as conventional, unicast packets. The router at the receiving end removes the multicast packet from the envelope and forwards it appropriately. This multicast-in-unicast facility is known as a *tunnel*.

The MBone would, of course, be of little use if there were no services running upon it. The two most widely used tools are currently vat (for *visual audio tool*) for audio transmissions and vic (for *visual image codec*) for video, both originating from the Lawrence Berkeley Laboratory. The tools are multistandard, but a convention is emerging where audio is generally GSM coded (the same technique as used in cellular telephones) and video is H.261 coded. As the tools are started, the user passes information regarding the multicast address to use and the coding scheme to use. There are also tools for other media, such as a text tool and a shared whiteboard.

Of course, this begs the question as to how a user determines what sessions are current or planned. There is a further tool, UCL's sdr, which provides a "session directory" along with information regarding multicast address, coding mechanism, and so on. Since sdr is a visual tool, the user joins a session simply by clicking on the appropriate title line; there is no need to enter the information manually. Figure 9.1 illustrates these tools and Figure 9.2 shows a workstation screen when these tools are in use.

The nib *Browser*

nib is a Web browser written by the author in Tcl/Tk (Ousterhout 1994; Welch 1995). Tcl is a high-level scripting language with a syntax reminiscent of those of the Unix shells. However, Tcl code can run unchanged on all flavors of Unix, including Linux; Microsoft Windows 3.1, 95, NT; and the Macintosh. Though originally conceived principally as an "extension language" intended to improve the flexibility of large C programs, Tcl itself has proven to be adequate for developing some quite substantial applications.

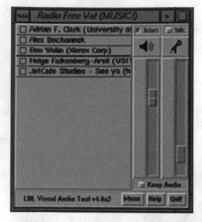

(a) vat, the audio tool

(b) vic, the video tool

(c) sdr, the session directory tool

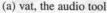

Figure 9.1   The MBone tools.

The Tk tool kit was devised as an extension to Tcl to enable graphical user interfaces (GUIs) to be built rapidly and changed easily; indeed, Tk has been adapted to several other programming languages, including Perl and Scheme. Tk was originally layered upon Xlib, the low-level interface to the X Window system (Scheifler and Gettys 1986), where it provided a Motif-like look and feel; presently, however, Tk has been ported to Microsoft Windows and to the Macintosh, where it provides a native look and feel.

Although Tcl is a purely interpreted language, and hence rather slow, the extensive set of built-in operators, which includes facilities such as regular expressions and socket access, makes for rapid development of substantial applications. Indeed, the author has

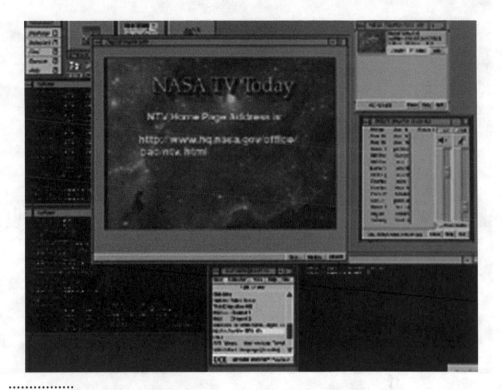

······················
Figure 9.2   The MBone tools in use on a workstation. (*See color plate on page 320.*)

used Tcl and Tk in a wide range of applications and, despite the syntactical quirkiness, has found them to be robust and well written—for example, attacking `wish`, the Tcl/Tk interpreter, with Purify (Pure Software Inc. 1997) yielded no memory leaks, unlike the X window system itself! Moreover, it is possible to extend the interpreter at run-time, either by defining new Tcl procedures or by dynamically linking compiled code. Also, "navel-staring" facilities are provided whereby a Tcl script may determine what facilities are available within the interpreter as it executes. For all these reasons, Tcl/Tk is actually very well suited for Web browser development. Furthermore, all the MBone tools shown in Figure 9.1 use Tcl/Tk for their GUIs.

`nib` is not the first browser written in Tcl and it is unlikely to be the last. The first, `TkWWW`, was written shortly after `Mosaic` and was the first browser to support HTML authoring. A second browser, `SurfIt`, was written in Australia at around the same time as `nib`; indeed, both browsers employ the same HTML-to-Tk conversion code, written by Stephen Uhler of the Tcl team at Sun Microsystems. `SurfIt` was interesting in that it supported downloadable Tcl applets, in much the same way as Sun's `HotJava` supports Java applets, the applets being executed in a safe Tcl interpreter (i.e., one in which commands such as `exec` and `open` have been removed). Most recently, Brent Welch, also of the Sun team, has produced `WebTk`, a high-quality tool that also supports HTML authoring. It is likely that the facilities described in this paper will eventually be extracted from `nib` and inserted into `WebTk`.

The appearance of the `nib` GUI is fairly conventional for a Web browser (Figure 9.3). The main feature of `nib` is that it is able to use the previously mentioned self-examination ("navel-staring") facilities of Tcl and adjust itself accordingly. There are three main ways

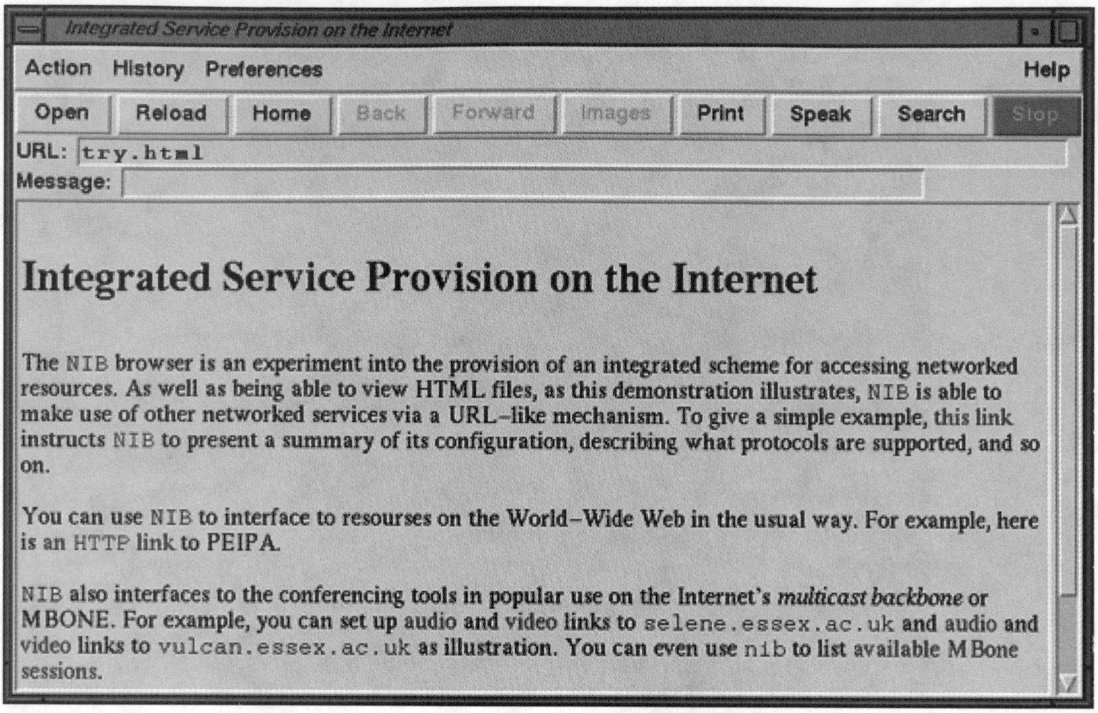

Figure 9.3 The nib Web browser.

in which this is used. First, Tcl is able to accommodate different networking facilities at run-time. This is because the core Tcl distribution has only recently included support for network programming (via the BSD "socket" abstraction); before then, there were two different (incompatible) networking interfaces, available in Neosoft's TclX extension and Berkeley's Tcl-DP package. As nib initializes, it examines the commands available and the version of Tcl. For recent versions of Tcl, the built-in networking support is used; otherwise, nib uses whichever of Tcl-DP or TclX is compiled into the interpreter; if neither is available, nib can view only local files.

The second area in which nib must perform self-examination is in image support. Tk normally supports only the PBMPLUS and GIF image formats, but other formats—notably JPEG—are widely used. The increasing adoption of the portable network graphics (PNG) format (Roelofs 1996) will soon give rise to a further type to be supported. Although JPEG and PNG are not guaranteed to be supported by Tk, extensions to support them are in widespread use. Hence, as nib initializes, it determines which image types may be used and adjusts its HTTP requests accordingly.

The third and final way in which nib uses self-examination is in its support for different transport protocols (HTTP, FTP, etc.). A convention is used for function naming that allows nib to determine at run-time whether a particular transport protocol is supported—and to allow new protocols to be supported simply by defining an appropriate (and appropriately named) procedure. This dynamic enhancement of the browser to support new protocols mirrors (and actually predates) those of Sun's HotJava browser, though the language employed is, of course, different.

When a user types a URL, it is parsed into four parts:

- The transport protocol
- The remote host on which the file is located
- The file name on the remote host
- The location within the file

The transport protocol extracted from the URL is used to construct the invocation of the Tcl function to perform the transfer. In addition to the usual protocols (HTTP and FTP), `nib` also supports `` `nib// ' ``, which accesses features internal to the browser (e.g., its help mechanism). This mechanism plays an important part in the MBone interface discussed in the next section.

`nib` is also able to adjust itself to its environment. For example, the GUI shown in Figure 9.3 has a button marked *Speak,* which makes the browser speak the Web page aloud. This is achieved using Nick Ing-Simmons' freely distributable `rsynth` package (Ing-Simmons 1994). As `nib` initializes, it attempts to find the program `say` (which is the text-to-speech converter from `rsynth`; if successful, the button is included in the GUI, and otherwise it is excluded.

Interfacing the MBone to `nib`

How best to bring the MBone to Web users? It is certainly possible to make the GUI produced by `vat` or `vic` appear within a Web browser, but is that the best way to interface them? The problem with so doing is that the Web user would have to keep the browser running in order to be able to access multicast sessions, and since Web browsers and the MBone applications are all substantial applications in their own rights, this would have a large memory requirement.

Bearing this in mind, the approach that has been taken is to produce a Web browser interface that is functionally equivalent to that of the `sdr` tool, namely, to list MBone sessions and allow the user to join them. (Actually, `sdr` is also capable of creating MBone sessions; that capability is not being reproduced here.) Each session has hypertext links that interface to the relevant Mbone tools (`vat`, `vic`, etc.).

Session announcements are naturally disseminated using the capabilities of the MBone. A session announcement is created by a user using `sdr`, and this is sent out on a particular multicast address roughly once a minute as long as `sdr` is running. When a user elsewhere on the MBone starts `sdr` to determine which sessions are available, the tool "listens" to this multicast address for announcements. This approach scales better to large networks than would any centralized database of information. However, it means that the program that listens for session announcements *must execute on the host that is running the Web browser,* which rules out using a central Web page or the conventional Web approach of CGI scripts.

Session announcements have a well-defined layout, termed the *session description protocol* (SDP) (Handley and Jacobson 1995). In essence then, all one has to do is to listen on the relevant multicast address and convert incoming SDP messages to HTML, and indeed this turns out to be the case. However, there is a subtlety here, one that will be returned to in the consideration of the Netscape plug-in in the following section. The software that listens for incoming SDP messages must run in a separate thread of execution from the GUI code, or the browser will simply hang. Tcl/Tk (and, come to that, Netscape) do not have built-in support for multithreaded applications. (Java, of course, does have multithreading support, so the code could, in principle, be written for Java-capable browsers; however, multicast sockets are not officially supported by Java as yet, and the security manager of the first generation of Java-capable browsers made accessing multicast addresses difficult.)

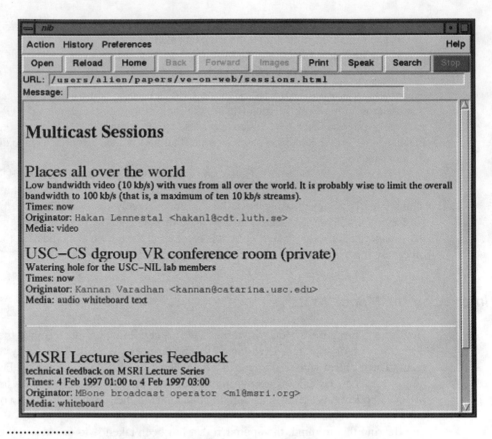

Figure 9.4 Display of MBone sessions in `nib`. (Note that this display is actually an edited listing of the sessions to show both live and future sessions.)

An additional problem, as far as `nib` is concerned, is that Tcl is basically string based and offers only limited support for binary input/output; since the first 8 bytes of a SDP message contain binary information, this causes difficulties with a Tcl-only solution. The solution that was eventually chosen was to write a short C program to listen for incoming SDP messages, to remove the first 8 bytes, and to forward the remainder to `nib` for parsing and display. The C program does little more than open a socket and listen, and is a little over a single page in length. Tcl/Tk have the capability of executing a command when input is present on a pipe, so that is used to parse packets and display them.

As `nib` initializes, a pipe to the listener program is instigated; nothing is displayed, however, until the user elects to view MBone sessions, which can be done via a pull-down menu item or via the URL.

```
nib://MBoneSessions
```

Each session announcement contains information such as title, start and end times, media employed, and a short description. `sdr` lists sessions in alphabetical order of title; however, the `nib` code lists them in chronological order of start time, with a horizontal bar separating current sessions from those yet to start (Figure 9.4).

The links created by the SDP-parsing code for "`audio,`" "`video,`" and so forth, make use of `nib`'s ability to extend the protocol list, having the basic form

```
(tool)://(multicast address):(port)/#(settings)
```

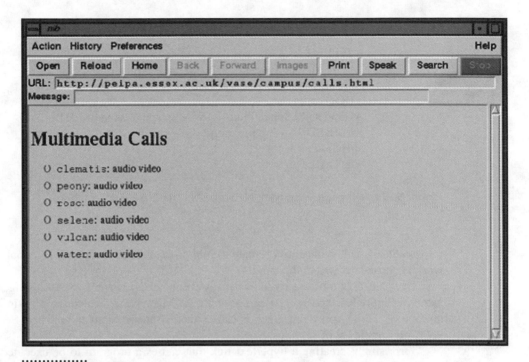

Figure 9.5 Web page used with `nib` for initiating audio- and videoconferencing sessions between machines.

to enable it to be parsed in the same way as a conventional URL. Here, (*w*) represents `vat`, `vic`, and so on.

The ability to represent MBone sessions with a URL-like notation proves useful in other ways too. For example, in the author's laboratory there is a static Web page (Figure 9.5) that can be used to initiate an audio- or videoconferencing link from the current machine to any other in the laboratory; this is often more convenient than making a telephone call!

Producing an MBone Plug-in for Netscape Navigator

The ability to join MBone sessions by means of links on a Web page proved so useful that the author decided to bring equivalent functionality to a mainstream browser, Netscape Navigator ("Netscape" hereafter). Netscape has the concept of a plug-in, which consists of code to handle a particular MIME type that is linked dynamically when that MIME type is encountered. These plug-ins are widely used for handling new image and sound formats, VRML viewing, and so on. In fact, there is even a Tcl plug-in, though, at the time this interface was constructed, it was incapable of providing the required interface. If Tcl cannot be used, the obvious language is Java, but, as mentioned previously, the security manager that prevents rogue applets from destroying information gets in the way here. It was eventually decided to construct the plug-in entirely in C.

A plug-in normally operates by being passed information to decode or unpack, either generating the display directly or passing HTML back to Netscape for display. The Mbone plug-in is a little different, it receives no data to unpack but instead must listen on the appropriate multicast address. Incoming session announcements are converted to HTML and passed to Netscape. However, the plug-in runs in the same thread of execution as the

```
<HTML>
<HEAD>
<TITLE>
    The MBone Plug-in for Netscape Navigator
</TITLE>
<BODY>
<H1>The MBone "Plug-in for Netscape Navigator</H1>
<EMBED TYPE="plugin/mbone" ACTION=list_sessions>
</BODY>
</HTML>
```

....

Figure 9.6   Web page to activate Netscape Navigator's MBone plug-in.

browser itself, and so must itself create a child process to listen; indeed, essentially all the work is carried out within the child.

The plug-in is most easily activated by means of Netscape's <EMBED> tag, as indicated in Figure 9.6. "plugin/mbone" is the MIME type for which the plug-in is registered; the "list_sessions" action causes the MBone sessions to be listed. See Figure 9.7 for a sample list of sessions.

Surprisingly, creating a hypertext link that causes a program to be executed locally does not seem to be a regular requirement for HTML browsers. The approach that the author has taken here is to generate a short file in a temporary directory, similar to that in Figure 9.6, for each link. The same plug-in is activated, but the ACTION field contains the command required to activate the MBone tool. In other words, there is a separate file for each vat, vic, and so forth, invocation.

At the time of this writing, there are two irritations regarding the plug-in. The first is that the plug-in is not activated until the previously mentioned <EMBED> tag is encountered; a way of activating the plug-in automatically upon starting Netscape would be helpful here. Of course, users can always put the tag into their home pages. The fact that the plug-in has to listen for announcements means that the list of sessions is slow to appear; hence, the plug-in creates a cache file of session announcements when it exits, rereading it as soon as it is activated and using that as the basis of the list.

The second problem is actually more serious. Since the announcements are passed to Netscape as HTML, they form a sequential list. When new announcements are heard, they cannot be stitched into the list chronologically as they can with nib; instead, they must appear at the end. What the plug-in does is to store all announcements in memory so that they can be put into the correct order when the page is reloaded, but this means that the user must look in two places or keep reloading the page—not an ideal solution.

Discussion and Conclusions

The MBone is a very active area of computer network research and is likely to evolve into an invaluable service in the coming few years, at least for those with adequately fast Internet connections. However, there remains the question as how best to interface both existing and forthcoming MBone tools to the user. This paper has explored the practicalities of bringing some of the session management capability to a Web browser, since this provides a familiar and easy-to-use interface to the user.

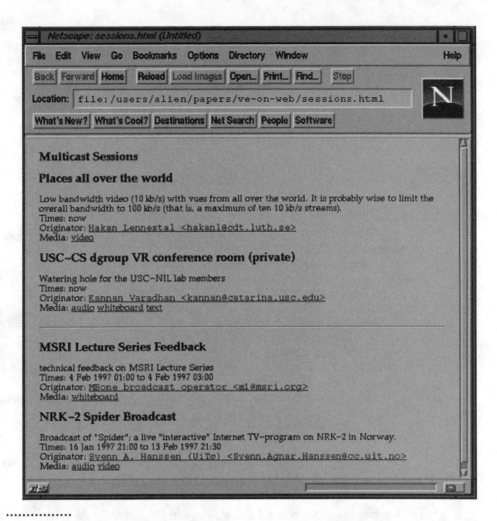

Figure 9.7   Listing of MBone sessions with Netscape Navigator plug-in. (Note that this display is actually an edited listing of the sessions to show both live and future sessions.)

The author's "`nib` browser, being written in Tcl/Tk, proved very simple to modify to accommodate the required MBone functionality. The ability to add a URL-like mechanism for interfacing to the MBone tools was shown to be useful for other purposes too. On the other hand, the lack of multithreading in Tcl and the interpreter's poor support for binary input/output meant that parts of the interface had to be written in C—which rules out the possibility of writing a downloadable Tcl "applet" for listing MBone sessions.

It was thought initially that the plug-in for Netscape Navigator would be able to provide complete functionality, albeit not as a downloadable applet. However, the fact that the plug-in must execute asynchronously and would naturally make incremental modifications to the displayed information makes an effective interface difficult. The only way complete functionality of the plug-in could be achieved would be to take total control over the whole of the display window and mimic the appearance of the normal display using low-level X calls, but this is far from an ideal solution. If Netscape's plug-in API were more sophisticated, to allow limited editing of the displayed HTML, interfacing to asynchronous or continuous data streams would be much easier.

Perhaps the message that comes from this work is that Web browsers, which have not been with us for all that many years, are fairly easily configured to do things that their designers have thought of, but are not very flexible in supporting things that they have not. In terms of the session directory plug-in itself, a truly satisfactory solution has not yet been found. However, as the Java language and environment continue to develop, multicast support within Web browsers such as Netscape will improve—so it is only a matter of time before an effective session directory applet is available.

References

M. Handley and V. Jacobson, *SDP: Session Description Protocol,* tech. report, Internet Engineering Task Force, 1995.

N. Ing-Simmons, `rsynth,` available from `ftp://svr-ftp.eng.cam.ac.uk/pub/comp.speech/synthesis/` (1994).

V. Kumar, *MBone: Interactive Multimedia on the Internet,* New Riders, 1996.

M.R. Macedonia and D.R. Brutzmann, "MBone Provides Audio and Video across the Internet, *IEEE Computer,* Vol. 27, No. 4, April 1994, pp. 30–36.

K. Ousterhout, *An Introduction to Tcl and Tk,* Addison-Wesley, 1994.

Pure Software Inc., Purify. `http://www.pureatria.com/` (1997).

G. Roelofs, *Portable Network Graphics Home Page,* tech. report, available from `http://www.wco.com/~png/` (1996).

R.W. Scheifler and J. Gettys, "The X window system," *ACM Trans. Graphics,* Vol. 5, No. 2, April 1986, pp. 79–109.

H. Schulzrinne, *Real-Time Protocol,* tech. report, `http://www.cs.columbia.edu/~hgs/rtp/` (1997).

B.B. Welch, *Practical Programming in Tcl and Tk,* Prentice-Hall, 1995.

• About the Author •

Dr. Adrian F. Clark obtained a BSc degree in Physics from the University of Newcastle upon Tyne in 1979 and a PhD in image processing from the University of London in 1983. He has been with the Department of Electronic Systems Engineering at the University of Essex since 1988, where he is now a Senior Lecturer. He has worked on image restoration, analysis, compression and synthesis, and on algorithms and software techniques for image processing, using both serial and parallel hardware. Presently, he has become involved in networked virtual reality through Essex's *Virtual Applications, Systems and Environments* Laboratory, which he heads. Dr. Clark acted as Convenor of a British panel of experts involved in the development of an International Standard for Image Processing. He is a former secretary of the British Machine Vision Association and has chaired Technical Committee 5 (Benchmarking and Software) of the International Association for Pattern Recognition.

## Chapter 10

Handling of Dynamic 2-D/3-D Graphics in Narrow-Band Mobile Services

C. Belz, H. Jung, L. Santos, R. Strack

Computer Graphics Center (ZGDV)

P. Latva-Rasku

Nokia Research Center (NRC)

Abstract

The factors limiting the efficient delivery and presentation of multimedia material in the cellular environment are, among others, the low bandwidth of the transmission channel(s) and the modest capabilities of truly mobile terminals. The ACTS[1] project MObile Media and ENTertainment Services (MOMENTS) is leveraging the usage of state-of-the-art techniques for the handling of vector graphics and animation contents, obtained through very low data transmission channels, in order to provide very attractive multimedia services in that environment. This paper focuses on the achievements obtained within MOMENTS in regard to the handling of dynamic 2-D/3-D graphics for the projected wireless multimedia services.

Key Terms

mobile computing   animation   plug-in
mobile services   information visualization   MOMENTS

....
[1]Advanced Communications Technologies and Services.

Introduction

Global information management systems—such as the World Wide Web (WWW)—show that the on-line access to vast amounts of distributed multimedia information is possible not only for the expert, but also for the "naive" end user. At the same time, relatively cheap and widely available wireless data communication services are available to the mass market.

Within the coverage of the cellular network, the vision of information access for "everyone, anytime, anywhere" has become reality. However, due to the low bandwidth of wireless narrowband wide-area networks (such as the 9.6 Kbit/s of GSM, DCS-1800) and the limited resources of mobile hardware, in comparison to stationary systems, the handling of distributed multimedia applications and services faces severe problems.

Users of mobile systems will expect to get access to all of the multimedia data making up modern information applications including time-dependent data such as speech, sound, animation, or video—at least within the limits defined by the mobile system's input and output capabilities. Also, they have become accustomed to comfortable, easy-to-use interactive multimedia systems based on the concepts of direct manipulation. A step backwards in interactivity will cause a serious acceptance problem. Therefore, in order to build "everyone, anytime, anywhere" information applications that will be *of use,* the problem of how to make these services interactive across a slow data link has to be solved. Suitable concepts have to be identified or developed that guarantee the effective usability of mobile multimedia information and services.

The overall objective of MOMENTS is to demonstrate the technical feasibility and business viability of a wireless media highway for the distribution of advanced multimedia products. The aim of the project is to contribute significantly to the understanding of the users' perception of the values of wireless multimedia services, identify how commercial exploitation of the services using third generation systems can be accelerated, create new enabling technologies, and make a valuable contribution to standardization.

This paper focuses on the handling of dynamic 2-D/3-D graphics and animation within the projected wireless multimedia services.

Presentation of Graphical Information

Within MOMENTS, the work on the media presentation has started with the evaluation of the available standards of media formats and handling tools for binary images, vector graphics, and animation. In respect to the media formats, the analysis was concentrated on the appropriateness of the standard for representing the content and the compactness of the encoding.

• Binary images: The advantage of binary images is that they are small in size, and thus, the downloading time from the server to the mobile client is minimal. Binary graphics can be used to display financial information, maps, cartoons, and so forth.

• Vector graphics and animation: 2-D vector graphics are suitable for the presentation of specific material (for instance, map material) that requires lossless magnification: 3-D data formats are suitable to describe realistic and animated scenes. Existing standards and emerging formats were selected for the 2-D and 3-D vector graphics and animation (Autodesk Inc. 1992; Bell et al. 1996; Henderson and Mumford 1993).

For the media handling tools, the mandatory analysis criteria were their suitability to the restricted operation environment of mobile computing (slow and latent data availability) and means to offer a comfortable usage of the service. New tools would be developed if necessary.

Concepts and Realization Issues

Two investigation fronts were formed in respect to media handling towards the achievement of the project goals and respective specifications. One task group was concerned with concepts on data representation and manipulation, the other, with the displaying processes. Also, due to the commercial nature of the project, the establishment of innovative key features was of great importance to reinforce the value and distinctiveness of the final product.

In regard to data manipulation and code optimization, several tools were conceived and implemented. Some of them have the specific purpose of reducing the size of the data file. Other tools were created on behalf of specifications from the displaying processes, such as progressive display, or future system enhancements.

Concerning displaying tools, new tools were developed as well. These tools possess outstanding features regarding adaptation to the mobile environment and the content display itself.

For the utilization of those tools and correct relationship with other processes within MOMENTS (see MOMENTS architecture, Figure 10.4) several agreements and cooperative work have been performed between content provider, service provider, and teams dedicated to the enduser.

Binary Images

A converter was developed to modify a selected binary image format in such a manner that it makes possible, afterwards, to display the images progressively by the user displaying tool. This converter will be installed at the content provider site.

For the user terminal site, a viewer has been developed to display that binary image format inside Netscape Navigator browser (Figure 10.1). The main features of this viewer include the following:

- Display embedded binary images
- Multiple images per page
- Progressive displaying of images

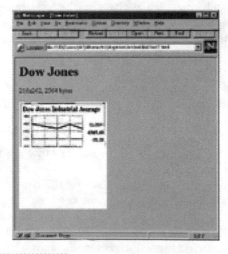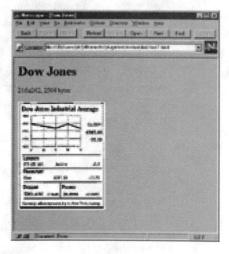

Figure 10.1   The progressive handling of binary images.

In respect to the displaying process, parts of the images are displayed on the terminal as soon as they can be decoded (progressive image buildup). The idea of undertaking this process is to relieve the users' tolerance threshold while waiting for the content presentation.

Vector Graphics

Appropriate software mechanisms were developed to adapt the 2-D and 3-D scene description languages to the restrictions posed by the mobile operation environments and some other project specifications. The adaptations encompass, among others,

- Selection and support of the most appropriate (content specific) state-of-the-art vector graphics format and its encoding, respectively
- Elimination of irrelevant features within the vector graphics formats
- Reference of templates within the vector graphics formats
- On-the-fly generation of vector graphics formats via service specific generators or converters
- Standard conformance alteration of the encoding in order to guarantee copyright issues for the service content

Some of the developed tools will be of use on the content provider site, and others, on the service provider site.

A very important data manipulation request has been formulated and implemented at this point: the handling tool at the user terminal should be flexible enough to be able to receive further enhancements to support other vector graphics formats: an integrated approach supporting different standardized formats for the different projected services.

Animation Vector Graphics

The term *animation* can be used in various ways, mainly in regard to the kind of data that are transmitted to the user's mobile data terminal (MDT) and the handling of data on the MDT.

On the one hand, it can refer to the playback of animated sequences on the MDT. Changes within an original graphical scene are rendered and recorded as video images on the server side. The resulting video sequences are transmitted to and displayed on the MDT.

On the other hand, animation can be regarded as changing the appearance of a graphical scene where the graphical description of the scene and the behavior of the objects (i.e., the nonrendered images) are transmitted to the terminal. The rendering of the data is then performed on the client side.

Within MOMENTS, the latter definition of the term animation is used. Transmitting graphical description of the scene and the behavior of the objects is in many cases suitable for mobile services, and, furthermore, user interactivity is applied to animation. This interactivity comprises user-oriented navigation through scenes, as well as the possibility to change the appearance of the scene interactively. The geometric objects of a scene are defined by a standardized graphical modelling language. The behavior of the scene's objects is realized by a script that references the objects of the scene. This script can be either directly hooked into the scene description itself or can be stored separately. The advantage of this concept is the potential to define several object behaviors or animations in the context of one scene description. Furthermore, the animation script may encompass additional static components that update or extend the current scene description (Belz et al. 1996). Figure 10.2 illustrates this principle.

On the MDT the following steps are performed to present the animation to the user:

1. The scene description (those parts needed) is transmitted and successively parsed.
2. The scene is rendered (at least partially) as a starting point of the animation.

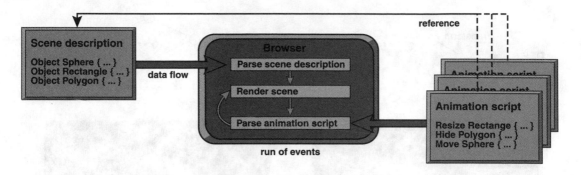

Figure 10.2   Principle of animation presentation on the MDT (within a browser); several animation scripts can refer to one scene description.

3. The animation script is executed step-by-step, whereby the changes of the scene are continuously rerendered. Those objects of the scene that are needed and that were not transferred so far to the MDT are transmitted and parsed.

The overall scene is managed on the MDT via a scene graph that can be updated and extended according to events triggered by the user or the system (e.g., predefined updates at distinct time slots that constitute parts of the animation). The following example illustrates this mechanism.

Let us suppose an initial scene description contains elements like a globe (Figure 10.3). For every service the user is able to select, an external script is available. These scripts contain information about the elements that represent a 3-D presentation of the service selected and their behaviors (e.g., rotation commands). If the user selects a service (e.g., financial information), an event is triggered that initiates the respective external script. The script is responsible for including the elements—describing the 3-D presentation of the service selected (e.g., financial information)—in the scene graph (Figure 10.3). The scene graph additionally contains a subtree of nodes that starts, for example, with *DEF Financial Information*—representing the 3-D presentation of the financial information service selected by the user (DEF is the keyword for naming nodes). The newly included elements will then be animated (e.g., letting the 3-D presentation of the financial information service orbit the globe (Figure 10.3).

Establishment of the M3D Plug-in

Within MOMENTS, the greatest development achievement that concerns this paper was the design and implementation of a complete system for a combined handling and display of different vector graphics formats and animation scripts, as well as attractive user interaction media at the MDT.

Based on the concepts outlined in the previous sections, the system was realized in the form of a Netscape plug-in (M3D) to be called by pertinent contents of MOMENTS service pages on WWW. Through M3D, the user is able to access and view binary images, 2-D or 3-D graphical presentations, and run 3-D animations (see Figures 10.5 and 10.6). That means it can be used both as a "plain viewer" or as a "scripting" environment that offers active buttons and objects.

The gain for the content authors and users is enormous due to the possibility of having a combination of those media formats on the same page.

In regard to the sort of contents to be presented, the M3D plug-in is open for almost any 2-D and 3-D vector graphics standards. The current set of supported standards provides

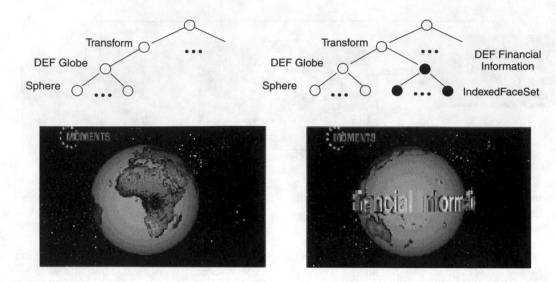

Figure 10.3 Example of an animated presentation of the service currently selected by the user. (*See color plate on page 320.*)

all of the necessary functionality to efficiently represent dynamic 2-D and 3-D data. That set includes distinct encodings and versions of CGM and VRML.

Besides the open development approach, the M3D plug-in possesses features such as

- Multiple/integrated display and handling of vector graphics and animation per plug-in
- Multiple vector graphics/animation instances per HTML page
- Support of basic viewing principles (such as zooming and scrolling)
- Support of scripting functionality

The M3D plug-in is invoked when a reference to a MOMENTS vector graphics or animation is encountered in an HTML document. If it is specified in any service, buttons are made available by the plug-in for further invocations.

Projected Services within MOMENTS and System Architecture

The projected services to be offered by MOMENTS cover the following areas:

- Financial information: Economic indicators, interest rates, and so forth.
- Location-dependent information: City, travel, and traffic information
- Education services
- Customer service and support
- Entertainment: Leisure information, appetizers,and so forth.
- Advertisement

In all of those areas, a content shaped by means of a congruence of media formats (multimedia) is of great significance; mainly, the visual message is enhanced through attractive and comprehensive 2-D/3-D graphics and stirring objects. Moreover, from the nature of the services, they must provide to the users a great interaction possibility for the refinement of the content selection and as a participant on the service events.

In order to achieve the provision of a service suited for commercial use and render it satisfactorily to the customers, a chain of service operations has to be established. Those

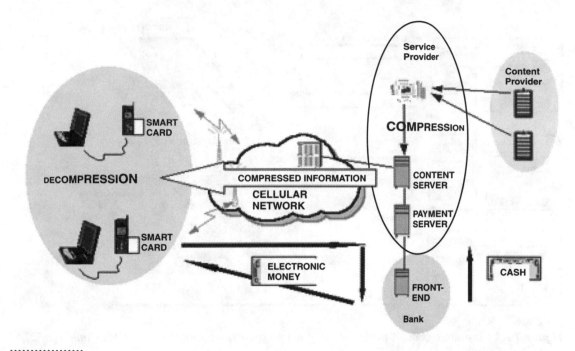

Figure 10.4   The overall MOMENTS architecture. (*See color plate on page 321.*)

operations go from multimedia content provisioning up to the supporting of the users' proceedings, and they include network providers, payment, and so forth. Figure 10.4 shows the overall architecture of the service operation that constitutes MOMENTS services. Within MOMENTS, that chain is guaranteed by the group of involved parties.[2]

User trials will be carried out in three different European countries—Germany (E-Plus), United Kingdom (Orange), and Italy (Omnitel) —with at least 100 participants and a duration of 16 months. These user trials will allow a realistic assessment of wireless multimedia services and verify the identified business opportunities.

System enhancements will be developed and demonstrated in the Technology Demonstrator to access the benefits of wider band user channels and to determine how they can be incorporated into UMTS. New techniques that will be evaluated and applied include presentation technologies, optimized for the inherently narrow cellular user channels, and wide-band user channels allowing transmission rates of the third generation mobile systems (n × 9.6 Kbit/s, n ≥2).

Example of Services Based on the Established Concepts

City Information

The first of the projected services for MOMENTS under implementation refers to city information. It has been conceived as a service to offer to the calling user all of the typical information concerning the physical composition (maps, buildings, tourist places) of a city.

[2]The partners are Nokia Telecommunications Oy (FI), Prime Contractor, with ZGDV e.V. (D) as subcontractor, Bertelsmann AG (D), Citicorp Kartenservice GmbH (D), DataNord Multimedia SrL (I), E-Plus Mobilfunk GmbH (D), Gemplus (F), Omnitel Pronto Italia S.p.A (I), Orange PCS Ltd. (UK), and Reuters AG (D).

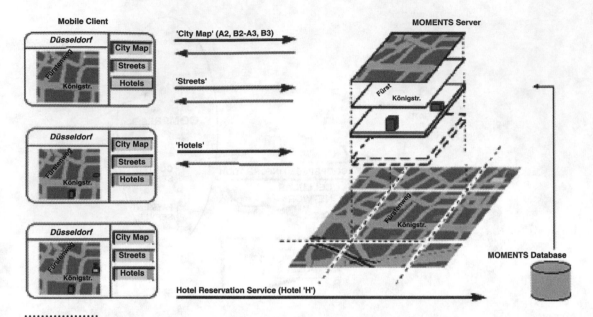

Figure 10.5 Dynamic content presentation (city information service). (*See color plate on page 321.*)

This service has been primarily chosen because it encompasses several service features and respective implementation aspects, suiting the operation constraints of a mobile system environment. Those solutions will be of most use on further executions.

First, the information content of a service has been stratified in layers of similar items. The consequent procedure is that the content will be presented gradually to the user as he/she continuously requests other information. Therefore, apart from a basic presentation scene pertinent to the service, information is available at the service page by supplementary requests only. Also, when suitable, the content will be accompanied with hyperlinks to other relevant services. On request, the respective complement will be prepared and supplied by the MOMENTS server and combined with the already available material. This service feature reduces utilization costs for the user by alleviating transmission operations. Thus, it brings a valuable asset for the system as an application for a mobile system environment.

Another feature of the system concerns the possibility of combining visual information of different formats, for example, one for the map material and another for icons of hotels, restaurants, cinemas, and so forth. Also, animation scripts can be hooked, in principle, to any graphical object, therefore allowing the service authors to give some "quicks of life" to content presentation.

Having the city information as service scenario, Figure 10.5 illustrates the dynamic content presentation methodology, and Figure 10.6 presents the MOMENTS user interface and illustrates the usage of the M3D plug-in for a combined media content.

Entertainment

This service area is and has to be very fertile in new offers and innovations, mainly due to the high demanding characteristics of the users. The first service to be delineated was the presentation of cartoons (binary images).

For its implementation, again it had to suit the operation constraints of a mobile system environment. The time to receive and present binary image content would be intoler-

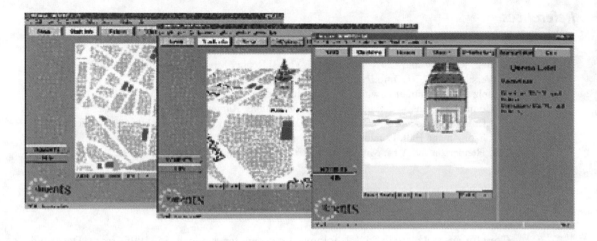

Figure 10.6   Snapshots of the M3D plug-in within the MOMENTS user interface. (*See color plate on page 321.*)

able by a mobile user. The solution applied was to modify the encoding in such a manner that a progressive process of interpretation and display of the image is possible afterwards.

Another aspect of entertainment is the use of animation as appetizers in order to increase the attractiveness of presented information for the user. Animations can be applied to leisure information services, which inform the user about music, movies, TV programs, and so on. Furthermore, animation as an appetizer can entertain the user during waiting periods, for example, while downloading data from the server to the MDT.

Summary and Conclusions

This paper has presented technological approaches to handle binary images, vector graphics, and animation contents of multimedia services to be used in a mobile environment. Specialized system features imposed by the limitations of a wireless network and the mobile data terminals, according to the projected services, have been developed and implemented. Additionally, typical application scenarios illustrate the value of those services and the technological solutions applied.

The separation of one overall scene description into several components and the special treatment given to the encodings (according to the data formats and forms of presentation) depict the concepts on the modelling level of the information content. Also, features of the display tool, such as the multiple handling of media format, ratify the innovative and optimized characteristics of the system to be used in a mobile environment.

Acknowledgments

This work has been performed in the framework of the project ACTS AC002 MOMENTS, which is partly funded by the European Community. The authors would like to acknowledge the contributions of their colleagues from: Nokia Telecommunications Oy, Bertelsmann AG, Citicorp Kartenservice GmbH, DataNord Multimedia SrL, E-Plus Mobilfunk GmbH, Gemplus, Omnitel Pronto Italia S.p.A, Orange PCS Ltd., Reuters AG, and Computer Graphics Center (ZGDV e.V.).

The authors would also like to thank Hewlett-Packard, the information technology provider of the project.

References

Autodesk Inc., *Drawing Interchange and File Format,* Release 12, 1992.

G. Bell, R. Carey, and Chris Marrin, "The Virtual Reality Modeling Language Specification, Version 2.0," URL: http://vag.vrml.org/VRML2.0/FINAL/ (August 1996).

C. Belz et al., "Animation within Mobile Multimedia On-line Services," Mobile Communications—*Technology, Tools, Applications, Authentication, and Security,* J.L. Encarnação and J.M. Rabaey, eds., IFIP World Conference on Mobile Communications, Chapman and Hall, London, June 1996, pp. 98–108.

L.R. Henderson and A.M. Mumford, *The CGM Handbook,* Academic Press, 1993.

• About the Authors •

Constance Belz received a diploma in computer science in 1995 from Darmstadt Technical University (THD). Since November 1995, she is working as a research assistant at the Computer Graphics Center (ZGDV) in Darmstadt. Her work as a staff member of the R&D Dept. of Mobile Information Visualisation concentrates on mobile multimedia applications in general and the modeling and handling of animation for a mobile environment in particular.

Harald Jung received a diploma in computer science in 1996 from Darmstadt Technical University (THD). Since May 1996, he is working as a research assistant at the Computer Graphics Center (ZGDV) in Darmstadt. His work as a staff member of the R&D Dept. of Mobile Information Visualisation concentrates on mobile multimedia applications in general and the modeling and handling of animation for a mobile environment in particular.

Luiz Santos holds a first degree in aeronautical-mechanical engineering from the Aeronautic Institute of Technology, a postgraduate diploma in business administration from the School of Business Administration, and a master degree in CAD/CAM from Cranfield Institute of Technology. He has worked as system analyst in private companies, as a research assistant in the Advanced Technical Institute (University of Lisbon), and, since October 1993, he works as a researcher at the Computer Graphics Center (ZGDV), Darmstadt.

Rüdiger Strack received a diploma in computer science in 1990 from Darmstadt Technical University (THD). He gained his PhD (Dr.-Ing.) at THD in 1995. From 1990 to March 1995, he worked as a research assistant at the Fraunhofer Institute for Computer Graphics (Fh-IGD) at Darmstadt. His work concentrated on the areas imaging and distributed multimedia environments. Since 1993, he was acting as a project manager at Fh-IGD. Since April 1995, he is head of the R&D Dept. of Mobile Information Visualisation at the Computer Graphics Center (ZGDV) in Darmstadt. At ZGDV he is responsible for research and development in the areas of mobile computing and distributed multimedia information systems. This includes the coordination of several industrial and scientific projects.

Chapter 11

Realistic Avatars and Autonomous Virtual Humans in VLNET Networked Virtual Environments

Tolga K. Capin and Daniel Thalmann

Computer Graphics Laboratory
Swiss Federal Institute of Technology

Igor Sunday Pandzic and Nadia Magnenat Thalmann

MIRALab
University of Geneva

Abstract

Networked Collaborative Virtual Environments (NCVE) have been a hot topic of research for some time now. However, most of the existing NCVE systems restrict the communication between the participants to text messages or audio communication. The natural means of human communication are richer than this. Facial expressions, lip movements, body postures, and gestures all play an important role in our everyday communication. Part of our research effort in the field of Networked Collaborative Virtual Environments thrives to incorporate such natural means of communication in a virtual environment. This effort is mostly based on the use of realistically modeled and animated virtual humans. This paper discusses several ways to use virtual human bodies for facial and gestural communication within a virtual environment.

Introduction

The pace in computing, graphics, and networking technologies, together with the demand from real-life applications, made it a requirement to develop more realistic virtual environments (Ves). Realism not only includes believable appearance and simulation of the virtual world, but also implies the natural representation of participants. This representation fulfills several functions:

- The visual embodiment of the user
- The means of interaction with the world
- The means of feeling various attributes of the world using the senses

The realism in participant representation involves two elements: believable appearance and realistic movements. This becomes even more important in multiuser networked virtual environments (NVE) as participants' representation is used for communication. A NVE can be defined as a single environment that is shared by multiple participants connected from a different host. The local program of the participants typically stores the whole or a subset of the scene description, and they use their own avatars to move around the scene and render from their own viewpoint. This avatar representation in NVEs has crucial functions in addition to those of single-user virtual environments:

- Perception (to see if anyone is around)
- Localization (to see where the other person is)
- Identification (to recognize the person)
- Visualization of others' interest focus (to see where the person's attention is directed)
- Visualization of others' actions (to see what the other person is doing and what she/he means through gestures)
- Social representation of self through decoration of the avatar (to know what the other participants' task or status is)

Using virtual human figures for avatar representation fulfills these functionalities with realism, as it provides the direct relationship between how we control our avatar in the virtual world and how our avatar moves related to this control. Even with limited sensor information, a virtual human frame that reflects the activities of the user can be constructed in the virtual world, and this increases the sense of presence in this virtual world.

NVEs with virtual humans are emerging from two threads of research with a bottom-up tendency. First, over the past several years, many NVE systems have been created using various types of network topologies and computer architectures. The practice is to bring together different previously developed monolithic applications within one standard interface and consists of building multiple logical or actual processes that handle a separate element of the VE. Second, at the same time, virtual human research has developed to the level to provide realistic-looking virtual humans that can be animated with believable behaviors in multiple levels of control. Inserting virtual humans in the NVE is a complex task. The main issues include

- Selecting a scalable architecture to combine these two complex systems
- Modeling the virtual human with believable appearance for interactive manipulation
- Animating the virtual human with minimal number of sensors to have maximal behavioral realism
- Investigating different methods to decrease the networking requirements for exchanging complex virtual human information

In this paper, we survey problems and solutions for these points, taking the VLNET (Virtual Life Network) system as a reference model. The VLNET system has been developed at MIRALab at University of Geneva, and Computer Graphics Laboratory at Swiss Federal Institute of Technology, Lausanne. In VLNET, we try to integrate artificial life techniques with virtual reality techniques in order to create truly virtual environments shared by real people and autonomous living virtual humans with their own behavior, which can perceive the environment and interact with participants (Capin et al. 1997a; Noser et al. 1996).

Realistic Avatars and Autonomous Virtual Humans: Design Issues

Realistic avatars, representing participants, and autonomous virtual humans share similar design issues and techniques, while they have a number of differences. In this paper, we overview these differences and similarities.

In networked virtual environments, it is crucial to differentiate between user embodiments and other objects in the scene at first glance. The realistic human bodies are easily differentiable from the other 3-D models and objects in the scene, hence using them as participant representation allows someone to identify them easily. Using this embodiment regularly, the participant has a bounded, authentic, and coherent representation in the virtual world. In addition, by changing decoration of the body through clothes and accessories, the representation also has an emergent identity.

Various applications have different requirements from avatar representation. For example, for a computer-supported collaborative work application, the most realistic representation and animation of the body and the face might be the goal. On the other hand, a 3-D chat application might have different modeling and animation requirements. For example, the current simple multiuser chat applications hide the real identity of the user, and this likely increases the interaction as it eliminates the problem of shyness. The animation in these chat applications can be as simple as playing predefined gestures of the body and the face, and can be as complicated as including interactions with the environment using the embodiment (e.g., grasping objects).

Including autonomous virtual humans that interact with participants increases the real-time interaction with the environment. Therefore, it is likely to increase the sense of presence of the real participants in the environment. The autonomous virtual humans are connected to the VLNET system in the same way as human participants and also improve the usage of the environment by providing services such as replacing missing partners and helping in navigation. As these virtual humans are not guided by the users, they should have sufficient behaviors to act autonomously to accomplish their tasks. This requires building behaviors for motion, as well as appropriate mechanisms for interaction.

Our autonomous virtual humans are able to have a behavior, which means they must have a manner of conducting themselves. Behavior is not only reacting to the environment but should also include the flow of information by which the environment acts on the living creature, as well as the way the creature codes and uses this information. Behavior of autonomous virtual humans is based on their perception of the environment.

These varieties in requirements for virtual human representation necessitate different approaches to modeling and animation depending on the application. We have developed various applications, as described later, for investigation of these techniques. However, this variety in techniques should not eliminate the fact that these different types of virtual humans should be present together within a single environment in a standard way. In VLNET, we try to create truly virtual worlds with autonomous living actors with their own behavior, where real people should be able to enter and meet their inhabitants. The ultimate objective is to build intelligent autonomous virtual humans able to take a decision based on their perception of the environment and their interaction with the real participants.

Control of Virtual Humans

Whether avatar or autonomous, there is a need for the virtual human to interact with the environment: it should be animated using motion control techniques, and it should be provided enough information about the environment for perception.

Previously, we introduced three types of virtual humans according to the techniques to control them: direct-controlled, user-guided, and autonomous (Capin et al. 1997a).

• Direct-controlled virtual humans: The joint and face representation of the virtual human is modified directly (e.g., using sensors attached to the body) by providing the geometry directly.

• User-guided virtual humans: The external driver *guides* the virtual human by defining tasks to perform, and it uses its motor skills to perform this action by coordinated joint movements (e.g., walk, sit).

• Autonomous virtual humans: The virtual human is assumed to have an internal state that is built by its goals and sensor information from the environment, and the participant modifies this state by defining high-level motivations and state changes (e.g., turning on vision behavior).

All these motion-control techniques can be used for avatars representing real participants, as well as autonomous virtual humans. For example, a participant can control her representation by directly updating the body with magnetic trackers or by guiding her embodiment through commanding the actor to sit or walk, or she can control by changing the high-level state of her embodiment. Similarly, the same control techniques can be used for autonomous actors, too. Then what makes the avatars and autonomous actors different? The main difference is not in how to control the embodiment, but in what kind of information that the networked virtual environment provides to the participant or the autonomous actor program. For example, consider navigation. The real participant sees the rendered images on her display or HMD and decides which direction to walk. In this case, the NVE provides to the participant only the rendered image. Ideally, the same image should be the only information to the autonomous actor so that it can navigate by avoiding collisions with the environment. However, this would require excessive computing power for vision techniques for processing this image, and it could be done without losing the realism of the motion, as all the information for the scene is already stored. For this purpose, Renault et al. (1990) proposed a synthetic vision technique, where the environment is rendered through the eyes of the embodiment using the z-buffer hardware. Instead of storing the color of each pixel in the z-buffer, a pointer to the object is stored with the depth value. This simplified information should be given to the autonomous actor instead of the normal image. Therefore, the information provided to the real and autonomous actors is not the same. Typically, perceptions to the autonomous actor should be coded in this simplified way in order to avoid additional computation. We are continuing work on how to provide general information about the scene to external applications controlling autonomous actors, particularly in the visual, auditory, and tactile senses and behavioral coding of the virtual environment. The next section describes the VLNET system, which our work is founded on.

The VLNET System

Typically, the virtual environment simulation systems are complex software systems, therefore, a modular design imposes itself. It is appropriate to design the run-time system as a collection of cooperating processes, each of which is responsible for its particular task. This also allows easier portability and better performance of the overall system through the decoupling of different tasks and their execution over a network of workstations or different processors on a multiprocessor machine. In VLNET, we use this multiprocess approach. Figure 11.1 shows the architecture of the VLNET client. We separate the two types of processes: the core VLNET processes, and external driver processes.

VLNET Core Processes

Within the VLNET core, the main process executes the main simulation and provides services for the basic elements of VEs to the external programs, called drivers. The display, cull, and database processes are standard IRIS Performer processes and allow asynchronous loading and display of the scene with the main simulation. The main process consists of four logical units, called engines. The role of the engine is to separate one main function in the VE to an independent module and to provide an orderly and controlled allocation of VE elements. Moreover, the engine manages this resource among various programs that are competing for the same object. The communication process is responsible for receiving and sending messages through the network and uses incoming and outgoing message queues to implement asynchronous communication.

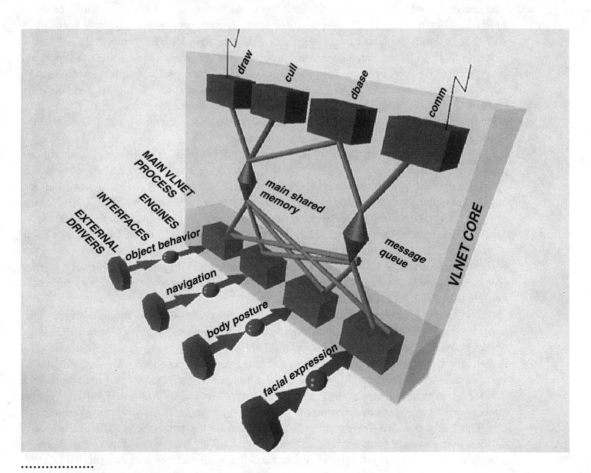

Figure 11.1   Client architecture of the VLNET system and its interface with external processes. (*See color plate on page 322.*)

The object behavior engine is responsible for the requests for changing or querying the object definition and behaviors in the scene, and collision detection among them. The navigation engine connects the user input to the navigation, picking, and manipulation of objects. The input is in the form of relative and absolute matrices of the global position of the body and the requests for picking or releasing an object.

Similarly, the face and body representation engines are specialized for the virtual human figure. The face engine is responsible for bridging between VLNET and external face drivers. The engine obtains the camera video images or face model parameters, discussed later, from the external face driver, and places in VLNET internal shared memory and outgoing message queue.

The body representation engine has an external interface for the body posture, including joint angles or global positioning parameters and high-level parameters to animate the body. The role of this engine is to provide possibilities to define multiple levels of control for the human body and to merge the output of different external body drivers to a single final posture.

External Drivers

The drivers provide the simple and flexible means to access and control all the complex functionalities of VLNET: *Simple,* because each driver is programmed using a very small API that basically consists of exchanging crucial data with VLNET through shared memory; and *flexible,* because using various combinations of drivers it is possible to support all sorts of input devices, ranging from the mouse to the camera, with complex gesture recognition software, to control all the movements of the body and face using those devices, to control objects in the environment and stream video textures to them, and to build any amount of artificial intelligence in order to produce autonomous or semiautonomous virtual humans in the networked virtual environment.

The drivers are directly tied to the engines in the VLNET main process. Each engine provides a shared memory interface to which a driver can connect. Most drivers are optional, and the system will provide minimal functionality (plain navigation and manipulation of objects) without any drivers. The drivers are spawned by the VLNET main process on the beginning of the session, based on the command line where all combinations of drivers can be specified. The drivers can be spawned on the local host or on a remote host, in which case the transparent networking interface processes are inserted on both hosts. In a simple case, as with most drivers shown in Figure 11.1, a driver controls only one engine. However, it is possible to control more than one engine with a single driver, ensuring synchronization and cooperation.

The **facial expression driver** is used to control expressions of the user's face. The expressions are defined using the minimal perceptible actions (MPAs) (Kalra 1993). The MPAs provide a complete set of basic facial actions, and by using them, it is possible to define any facial expression. Examples of existing facial expression drivers include a driver that uses the video signal from the camera to track facial features and map them into the MPAs, describing expressions, and a driver that lets the user choose from a menu of expressions or emotions to show on his face. The facial expression driver is optional.

The **body posture driver** controls the motion of the user's body. The postures are defined using a set of joint angles corresponding to 75 degrees of freedom of the skeleton model used in VLNET. An obvious example of using this driver is direct motion control using magnetic trackers (Molet et al. 1996). A more complex driver is used to control body motion in a general case when trackers are not used. This driver connects also to the navigation interface and uses the navigation trajectory to generate the walking motion and arm motion. It also imposes constraints on the navigation driver; for example, not allowing the hand to move further than arm length or take an unnatural posture. This is the standard body posture driver that is spawned by the system unless another driver is explicitly requested.

The **navigation driver** is used for navigation, hand movement, head movement, basic object manipulation, and basic system control. The basic manipulation includes picking objects up, carrying them and letting them go, as well as grouping and ungrouping of objects. The system control provides access to some system functions that are usually accessed by keystrokes; for example, changing drawing modes, toggling texturing, and displaying statistics. Typical examples are a spaceball driver, tracker+glove driver, extended mouse driver (with GUI console). There is also an experimental facial navigation driver that lets the user navigate using his head movements and facial expressions tracked by a camera (Pandzic et al. 1994). If no navigation driver is used, internal mouse navigation is activated within the navigation engine.

The **object behavior driver** is used to control the behavior of objects. Currently, it is limited to controlling motion and scaling. Examples include the control of a ball in a tennis game and the control of graphical representation of stock values in a virtual stock exchange.

The **video driver** is used to stream video texture (but possibly also static textures) onto any object in the environment. Alpha channel can be used for blending and achieving effects of mixing real and virtual objects/persons. This type of driver is also used to stream facial video on the user's face for facial communication (Capin et al. 1997a).

VLNET Server

A VLNET server site consists of an HTTP server and a VLNET connection server. They can serve several worlds, which can be either VLNET files or VRML 1.0 files. For each world, a world server is spawned as necessary (i.e., when a client requests a connection to that particular world). The life of a world server ends when all clients are disconnected.

Figure 11.2 schematically depicts a VLNET server site with several connected clients. A VLNET session is initiated by a client connecting to a particular world designated by a URL. The client first fetches the world database from the HTTP server using the URL. After that it extracts the host name from the URL and connects to the VLNET connection server on the same host. The connection server spawns the world server for the requested world if one is not already running and sends to the client the port address of the world server. Once the connection is established, all communication between the clients in a particular world passes through the world server.

In order to reduce the total network load, the world server performs the filtering of messages by checking the users' viewing frusta in the virtual world and distributing messages only on as-needed basis. Clients keep the possibility of contouring this mechanism by requesting a higher delivery insurance level for a particular message (e.g., for heartbeat messages of a dead-reckoning algorithm) (Capin et al. 1997b).

Communication in VLNET

Natural human communication is based on speech, facial expressions, and gestures. Ideally, all these means of communication should also be supported within a Networked Collaborative Virtual Environment. This means that the user's speech, facial expressions, and hand/body gestures should be captured, transmitted through the network, and faithfully

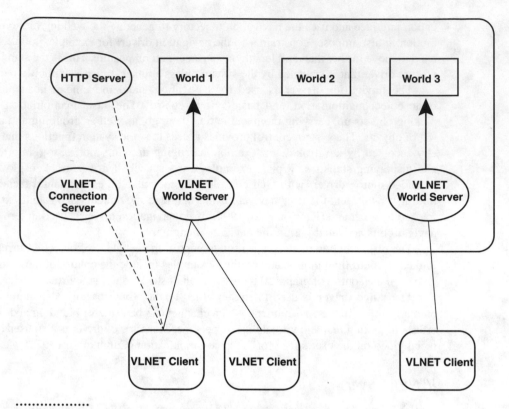

Figure 11.2 Connection of several clients to a VLNET server site.

reproduced for the other participants on their sites. The capturing should be done in a nonintrusive way to increase interaction.

Obviously, the way to a complete system as described previously is long and paved with problems. Capturing facial expressions or gestures nonintrusively and with enough precision is a complicated task. The synthesis of realistically looking human bodies and faces and their animation in real-time is also very demanding. Communication protocols must ensure that the multimodal data is transmitted to all the participants, and in the final synthesis the multimodal outputs have to be synchronized.

We are trying to solve some of these problems within the Virtual Life Network system and to provide solutions, leading to the complete communications as described.

So far, our work is not particularly concentrated on the audio (speech) communication. We use public-domain audio conferencing tools (VAT) to integrate this capability in the VLNET system. Therefore, audio communication is not discussed in this paper.

The next two sections will present several solutions for the facial communication, as well as some solutions for the gestural communication of the body.

Facial Communication

Facial expressions play an important role in human communication. They can express the speaker's emotions and subtly change the meaning of what was said. At the same time, lip movement is an important aid to the understanding of speech, especially if the audio conditions are not perfect or in the case of a hearing-impaired listener.

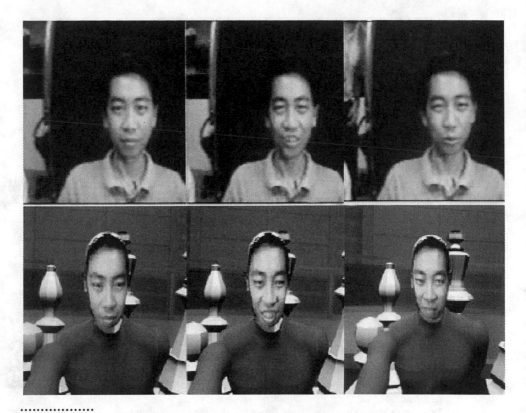

························
Figure 11.3   Video texturing of the face. (*See color plate on page 322.*)

We discuss four methods of integrating facial expressions in a Networked Collaborative Virtual Environment: video texturing of the face, model-based coding of facial expressions, lip movement synthesis from speech, and predefined expressions or animations.

Video Texturing of the Face

In this approach, the video sequence of the user's face is continuously texture mapped on the face of the virtual human. The user must be in front of the camera, in such a position that the camera captures the head and shoulders. A simple and fast image analysis algorithm is used to find the bounding box of the user's face within the image. The algorithm requires that head and shoulder view is provided and that the background is static (though not necessarily uniform). Thus, the algorithm primarily consists of comparing each image with the original image of the background. Since the background is static, any change in the image is caused by the presence of the user, so it is fairly easy to detect his/her position. This allows the user a reasonably free movement in front of the camera without the facial image being lost. The video capture and analysis is performed by a special facial expression driver.

Figure 11.3 illustrates the video texturing of the face, showing the original images of the user and the corresponding images of the virtual human representation.

Model-Based Coding of Facial Expressions

Instead of transmitting whole facial images as in the previous approach, in this approach the images are analyzed, and a set of parameters describing the facial expression is extracted (Pandzic et al. 1994). As in the previous approach, the user has to be in front of the camera

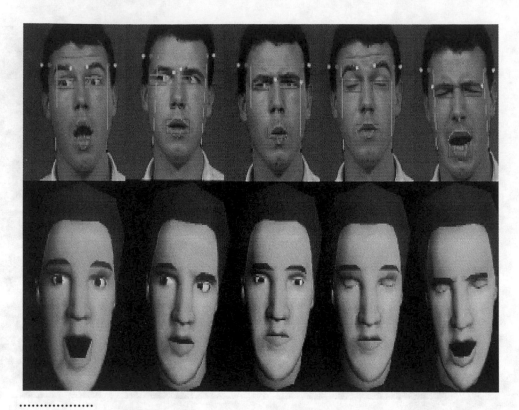

Figure 11.4 Model-based coding of the face: original and synthetic face. (*See color plate on page 322.*)

that digitizes the video images of head and shoulders type. Accurate recognition and analysis of facial expressions from video sequence requires detailed measurements of facial features. Currently, it is computationally expensive to perform these measurements precisely. As our primary concern has been to extract the features in real time, we have focused our attention on recognition and analysis of only a few facial features. The set of extracted parameters includes vertical head rotation (nod), horizontal head rotation (turn), head inclination (roll), aperture of the eyes, horizontal position of the iris, eyebrow elevation, distance between the eyebrows (eyebrow squeeze), jaw rotation, mouth aperture, and mouth stretch/squeeze.

The analysis is performed by a special facial expression driver. The extracted parameters are easily translated into minimal perceptible actions, which are passed to the facial representation engine, then to the communication process, where they are packed into a standard VLNET message packet and transmitted.

On the receiving end, the facial representation engine receives messages containing facial expressions described by MPAs and performs the facial animation accordingly. Figure 11.4 illustrates this method with a sequence of original images of the user (with overlaid recognition indicators) and the corresponding images of the synthesized face.

This method can be used in combination with texture mapping. The model needs an initial image of the face, together with a set of parameters describing the position of the facial features within the texture image in order to fit the texture to the face. Once this is done, the texture is fixed with respect to the face and does not change, but it is deformed together with the face, in contrast with the previous approach where the face was static and the texture was changing. Some texture-mapped faces with expressions are shown in Figure 11.5.

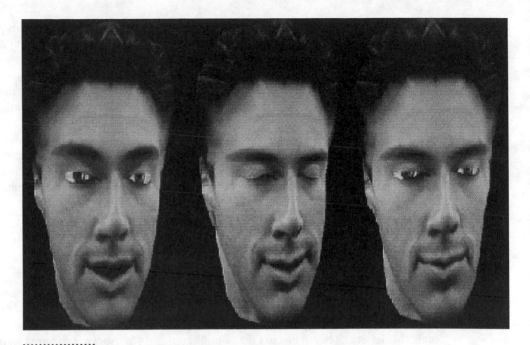

Figure 11.5   Predefined facial expressions: surprise, sleep, boredom.

Lip Movement Synthesis from Speech

It might not always be practical for the user to be in front of the camera (e.g., if he doesn't have one or if he wants to use a HMD). Nevertheless, the facial communication does not have to be abandoned. It is possible to extract visual parameters of the lip movement by analyzing the audio signal of the speech. An application doing such recognition and generating MPAs for the control of the face can be hooked to VLNET as the facial expression driver, and the facial representation engine will be able to synthesize the face with the appropriate lip movement. An extremely primitive version of such a system would just open and close the mouth when there is any speech, allowing the participants to know who is speaking. A more sophisticated system would be able to actually synthesize a realistic lip movement, which is an important aid for speech understanding.

Predefined Expressions or Animations

In this approach, the user can simply choose between a set of predefined facial expressions or movements (animations). The choice can be done from the keyboard through a set of "smileys" similar to the ones used in e-mail messages. The facial expression driver in this case stores a set of defined expressions and animations and just feeds them to the facial representation engine as the user selects them. Figure 11.5 shows some examples of predefined facial expressions.

Gestural Communication

Gestures play an important role in human communication. Using the body, many messages can be communicated. The body movements can be roughly divided into three groups:

• *Instantaneous gestures:* Most of the time, often even unconsciously, we accompany our speech with gestures. They stress the speech and give emphasis on particular words.

They also very often have a meaning in themselves. The whole body posture also conveys information about the person's state and possibly emotions. For example, from the posture it can be determined if the person is tired, tense, or relaxed.

• *Gesture commands:* These are gestures that the user makes to specify some action. For example, the sign "come here" can be specified by raising the arm. These movements can change from one person or culture to another; therefore, there is no well-defined set of rules for the meanings.

• *Rule-based sign language:* These are gestures, for example, used by deaf people, that essentially follow well-defined rules to specify words or sounds. The signs typically work as a metaphor for defining other objects or language. The gestures can also be used by the software to define special tasks (e.g., showing forward direction to initiate a walk).

All these types of gestures can be controlled by two different methods: direct tracking and predefined postures or gestures. The type of control might be suitable for a specific type of gestures; however, a combination of them can be used for different tasks.

Direct Tracking

A complete representation of the participant actor's body should have the same movements as the real participant body for more immersive interaction. This can be best achieved by using a large number of sensors to track every degree of freedom in the real body. Molet et al. (1996) discuss that a minimum of 14 sensors are required to manage a biomechanically correct posture. However, this is generally not possible due to limitations in the number and technology of the sensing devices, as it is either too expensive to have this many sensors or it is too difficult for the participants to move with so many attached sensors. Therefore, the limited tracked information should be connected with behavioral human animation knowledge and different motion generators in order to "interpolate" the joints of the body that are not tracked (Capin et al. 1997a). The main approaches to this problem include inverse kinematics using constraints, closed form solutions, and motor functions.

The raw data coming from the trackers has to be filtered and processed to obtain a usable structure. The software developed at the Swiss Federal Institute of Technology (Molet et al. 1996) permits the conversion of the raw tracker data into joint angle data for all the 75 joints in the standard HUMANOID skeleton used within VLNET (Boulic et al. 1995; Capin et al. 1997a), with an additional 25 joints for each hand. As shown in Figure 11.1, this software is viewed as body posture driver by the VLNET system, and VLNET communicates with it through the body posture interface. VLNET body representation engine obtains this joint table from the body posture interface and uses such data to produce deformed bodies ready for rendering. The posture data in the form of joint angles fits into a VLNET message packet by reducing each angle to 8 bits, with a maximal error of 1.4 degrees in 360 degrees. This error rate is sufficient enough to provide body postures that are visually similar to the real body. By coupling the flock of birds driver with VLNET, we can obtain full gestural communication in a very direct, though intrusive, way.

Predefined Postures or Body Gestures

In a similar fashion as for the facial expressions, the body postures or gestures can also be predefined and chosen by a metaphor. For example, the smileys normally used within e-mails can be used to set a subset of the joints in the current body using the keyboard. The body posture driver just stores the predefined postures and gestures (i.e., animated postures) and feeds them to the body posture engine as the user selects them. Figure 11.6 shows some examples of predefined postures.

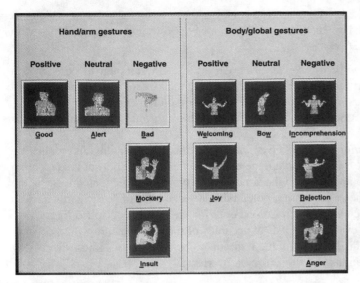
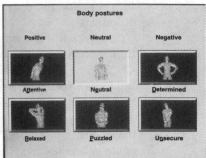

Figure 11.6 Basic set of gestures and postures. (*See color plate on page 323.*)

The main difference between direct control and predefined postures/gestures is that the direct control provides more correspondence to the real posture of the participant. Therefore, it is expected to provide more immersive feeling. However, the predefined postures can increase the communication among participants in networked environments in the absence of enough number of trackers. These two types of control can be combined to animate the participant's body for different types of gestures. There is a need to investigate and define a set of tools that provides sufficient proprioceptive information for instantaneous gestures, while providing easy and natural control for rule-based signs and sign gesture commands.

Networking

The articulated structure of the human body, together with the face, introduces a new complexity in the usage of the network resources because the size of a message needed to convey the body posture is greater than the one needed for simple, nonarticulated objects. This might create a significant overhead in communication, especially as the number of participants in the simulation increases. In order to reduce this overhead it is possible to communicate the body postures in more compact forms, accepting some loss of accuracy in the posture definition. This is not the only trade-off to be considered when choosing the optimal approach. Conversions between different forms of posture definition require potentially expensive computations, which might induce more overhead in computation than was reduced in communication. The choice will also depend on the quality and quantity of raw data available from the input devices, the posture accuracy required by the application, and the projected number of participants in the simulation.

For the networking analysis, we separate the discussion into body and faces, as they use different control methods and as they use different channels for communication. In any case, we can decompose the communication into three phases: coding, transmission, and decoding of the data. The transmission lag for a message will be the sum of the lag of all these phases. In addition, each message type contains an accuracy loss of data, which is a trade-off to decrease the lag. In this section, we analyze different message types with respect to the following aspects:

• *Coding computation at the sender site:* We evaluate the amount of computation needed in order to convert the input data into the message to be sent at the sending site.

• *Bitrate requirements:* We evaluate the bandwidth requirements for different parameters to describe the motion. We assume a minimum limit for real-time computation as 10 frames/second.

• *Decoding computation at the receiver site:* We evaluate the amount of computation needed to interpret the message and obtain the body posture(s) for display at the receiving site. The weight of this computation on the simulation is typically more than the one at the sender site because the messages from a potentially large number of participants have to be processed contrary to coding, which is done only for the locally controlled figure.

• *Accuracy loss:* We evaluate the loss of accuracy of the body posture with respect to the original input data. This is typically the trade-off to be considered against decreasing coding/decoding computations and transmission overhead.

We compare these issues in Figure 11.7 for various types of human body motion control. We consider four types of message packets that can be used to convey the body posture information:

• *Global positioning parameters:* The global positioning of 17 body parts can be sent as 3 rotation and 3 translation values. This data can be used directly to display the body, bypassing the conversion to joint angles.

• *Joint angles:* These values are the degrees of freedom comprising the body, each represented by a floating point value. We evaluate three possibilities for the message type: the actual floating point representation of the angle, and 2-byte integer and 1-byte angle information, discretized between 0 and 360. This data has to be transformed into global positioning for display.

• *End-effector matrices:* A 4×4 floating point matrix is used to determine the position of the end effectors; in this example, the head and the right hand. An inverse kinematics with two end effectors (head and hand) has to be applied to the received message to obtain the final posture.

• *State information:* Only the high-level state information is conveyed, which makes the messages small. Moreover, the messages are sent only when state changes. The computation complexity involved to produce the posture(s) from the state information can range from quite simple (in the case of predefined static postures such as sitting or standing) through medium (in the case of predefined dynamic states such as walking or running) to very complex (in the case of more complex dynamic states such as searching an object). In this evaluation, we take the medium-level walking action as an example.

We analyze three typical situations with respect to different real-time control data: (1) complete body posture data is available, (2) only head and hand end-effector data is available, (3) walking motion guiding data is available from an external driver.

Figure 11.7 shows the bitrate requirements, coding and decoding computations per frame, and accuracy loss for different message types. Figure 11.7a is calculated with the assumption of a minimum real-time speed of 10 frames/second. We see that the bitrate requirement varies between 4 Kbytes/sec and 240 bytes/second for one human body. Figures 11.7b and 11.7c show the coding and decoding results on an SGI Indigo2 Impact workstation with 250 MHz processor. The results show that there are a wide range of

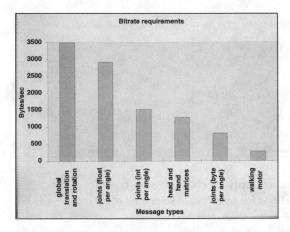
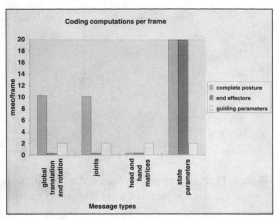
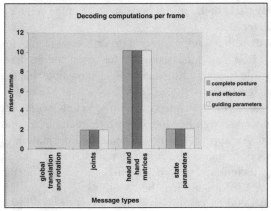
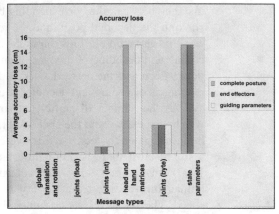

Figure 11.7 Networking body data. (*a*) Bitrate requirements for each message type. (*b*) Coding computations for each body posture. (*c*) Decoding computations for each body posture. (*d*) Accuracy loss with respect to original input data. (*See color plate on page 323.*)

possibilities to define and transmit human figure information with respect to computation and bandwidth requirements. The choice of the control and message type will depend on the particular application requirements. Where high accuracy is needed (e.g., medical training applications) the transfer of body part matrices or at least end-effector matrices will be required; in the large-scale simulation with numerous users it might be efficient to convey small messages containing the state information and use filtering and level of detail techniques to reduce the computational overhead. We chose the joint angles transfer as the optimal solution covering a wide range of cases since it offers fair or good results on all criteria, balancing the network and compression/decompression computational overhead (the traversal of the human hierarchy to convert joint values to transformation matrices of body parts, to be rendered on display) and accuracy loss. Using the state parameters for high-level motions decreases the network bandwidth, however, it requires a fast decoding process at the receiving site. The walking motor example showed the possibility of decreasing bandwidth requirements for sending motion data. Figure 11.7*d* shows the accuracy loss of posture data with varying message types and input methods. The results were computed by averaging the Euclidean

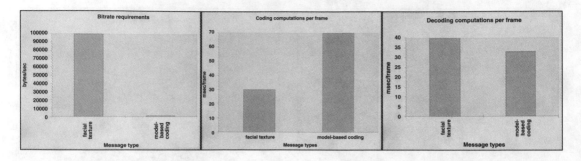

Figure 11.8   Networking face data. (*a*) Bitrate requirements for each message type. (*b*) Coding computations for face data. (*c*) Decoding computations for face data. (*See color plate on page 323.*)

distance between the corresponding body parts in the initial body posture at the sender's site and the decoded posture at the receiver's site.

Similarly, Figure 11.8 shows the bandwidth requirements and experimental results for coding and decoding of the face. Figure 11.8*a* shows that using model-based coding drastically decreases the network overhead, while Figures 11.8*b* and 11.8*c* demonstrate that the overheads of coding and decoding are insignificant.

As the number of participants increases, the transmission and decoding overheads will be excessive, and the speed might decrease significantly. Therefore, methods should be investigated to decrease this overhead. An approach is not to send the information to a site at all if there is no or little interaction, using filtering techniques (Pandzic et al. 1997). In addition, the *dead-reckoning* techniques may be applied to extrapolate the human information from the last received information of body and face, and the last speed. The initial results on human body dead reckoning have shown that up to 50 percent of the traffic can be decreased by applying simple predictive filtering on the joint angles (Capin et al. 1997b).

Conclusions and Future Work

The human figure representation in networked virtual environments is not an easy task. First, we presented an easy architecture showing how they can be included in a complex NVE system. Next, we showed the different levels of controlling the human body and compared them. Then, we presented different possibilities to send human information and discussed the load they put on the sender, network, and receiver, as well as accuracy loss. The human figure information can put a load on the computational and networking resources; and the best control, representation, and transmission form should be selected depending on the application and the resources.

Virtual human representation and communication in networked virtual environments is in an early stage. We will further our research on compression of virtual human models and networking techniques to decrease communication requirements. The initial results are promising, and we hope to achieve very low bitrate virtual human communication.

Moreover, we will work on human motion control for direct, guided, and autonomous virtual humans control. We are currently investigating techniques for providing simplified perceptual information about the virtual environment to the autonomous actors, and interaction between real participants and autonomous actors.

Acknowledgments

This research is financed by "Le Programme Prioritaire en Telecommunications de Fonds National Suisse de la Recherche Scientifique" and the TEN-IBC project VISINET.

Numerous colleagues at LIG and MIRALab have directly or indirectly helped this research by providing libraries, body and environment models, scenarios for applications; in particular, Elwin Lee, Eric Chauvineau, Hansrudi Noser, Marlene Poizat, Laurence Suhner, and Jean-Claude Moussaly. We would like to thank Mireille Clavien for designing body gestures and postures, and Anthony Guye Vuilleme for implementing this user interface.

References

R. Boulic et al., "The Humanoid Environment for Interactive Animation of Multiple Deformable Human Characters," *Proc. Eurographics '95,* 1995.

T.K. Capin et al., "Virtual Human Representation and Communication in VLNET Networked Virtual Environment," *IEEE Computer Graphics and Applications,* March 1997a.

T.K. Capin et al., "A Dead-Reckoning Algorithm for Virtual Human Figures," *Proc. IEEE VRAIS'97,* IEEE Computer Soc. Press, 1997b.

P. Kalra, *An Interactive Multimodal Facial Animation System,* PhD thesis, nr. 1183, EPFL, 1993.

T. Molet, R. Boulic, and D. Thalmann, "A Real-Time Anatomical Converter for Human Motion Capture," *Proc. Eurographics Workshop on Computer Animation and Simulation,* R. Boulic, ed., Springer, Wien, 1996, pp. 79–94.

H. Noser et al., "Playing Games through the Virtual Life Network," *Proc. Artificial Life'96,* Chiba, Japan, 1996, pp. 114–121.

I.S. Pandzic et al., "A Flexible Architecture for Virtual Humans in Networked Collaborative Virtual Environments," *Proc. Eurographics'97,* Budapest, Hun., 1997.

I.S. Pandzic et al., "Real-Time Facial Interaction," *Displays,* Vol. 15, No. 3, 1994.

O. Renault O., N. Magnenat-Thalmann, and D. Thalmann, "A Vision-based Approach to Behavioral Animation," *J. Visualization and Computer Animation,* Vol.1, No.1, 1990.

D. Thalmann, J. Shen, and E. Chauvineau, "Fast Realistic Human Body Deformations for Animation and VR Applications," *Proc. Computer Graphics Int'l '96,* Pohang, Korea, 1996.

Chapter 12

Interactive Cloth Simulation: Problems and Solutions

Pascal Volino and Nadia Magnenat Thalmann
MIRALab University of Geneva

Abstract

Simulating synthetic garments interactively requires more than the implementation of a simple dynamic system designed only for simple cloth objects. Many issues must be resolved, pertaining to accuracy and computational efficiency, robustness, stability, collision detection and response, and constraint handling. The purpose of this paper is to survey these problems and to describe the solutions embodied in our new garment simulation system.

Key Terms

cloth animation	collision detection	constraints
mechanical simulation	self-collisions	stability
particle sytems	collision response	interaction

Introduction

Far away from the "academic" situation of a rectangular cloth hanging like a flag or laying on a table and wrapping around its edges, generating animations involving dressed bodies raises heavy problems that should not be underestimated. First, the shape of the garment cloth object is quite complex, and its modelization as a deformable surface requires quite numerous elements, impeding computation time. Secondly, collision interaction with the environment (for instance, the body that wears the cloth, or other superposed cloth) is essential for determining the actual shape and animation of the garments.

Thinking about an interactive clothing system, additional problems are raised, such as how to optimize the computation speed without generating numerical instabilities, and to build a system that is robust enough to cope with the inaccuracy of the input devices and recover efficiently from unrealistic situations.

Concerning cloth simulation, most of the literature has put interest on exploring different mechanical models for cloth simulation, such as continuous systems and finite differences (Weil 1986; Terzopoulos et al. 1987; Terzopoulas and Fleischer 1988), finite elements (Collier et al. 1991; Kang and Yu 1995; Eischen et al. 1996), particle systems (Breen et al. 1992; Breen et al. 1994; Eberhardt et al. 1996), or miscellaneous techniques using different formalisms or geometrical methods (Haumann and Parent 1988; Promayon 1996). Improvements have been made in various directions for dealing efficiently with specific problems (Provot 1995; Hutchinson 1996). The problematic of building and animating garments worn by actual synthetic actors was mainly developed in Lafleur et al. 1991, Magnenat-Thalmann and Thalmann 1991, Carignan et al. 1992, and Volino et al. 1996, along with the study of related problems (Yang and Magnenat-Thalmann 1993; Werner et al. 1993; Volino and Magnenat-Thalmann 1994; Volino et al. 1995).

In this study, we intend to explore several issues raised by garment mechanical simulation:

- How to fix the compromise between accuracy and computation speed, and to choose an efficient mechanical modelization of the physical behaviors of the material and environment
- Choosing an efficient way to integrate the mechanical model
- Coping with stability problems: How to reduce instabilities caused by the modelization, and to obtain simulation time steps as big as possible without compromising stability
- Speeding up computation using coarse discretization meshes
- How to detect collisions efficiently, particularly self-collisions
- Dealing with the surface orientation problem when computing collision response
- How to integrate collision response efficiently when collisions are quite numerous, and integrate other kinds of constraints as well

A good study of these issues should lead to the implementation of an efficient garment simulation system, suited for interactive cloth manipulation, along with dressing realistically animated actors.

Building a Mechanical Simulation System: Some Considerations and Related Issues

The mechanical simulation system is the core of a cloth animation application. Most of the performance and realism of the simulation relies on it. In this part of the paper, we discuss the different kinds of mechanical simulation systems, their main potentialities and weaknesses, and how they can be integrated into cloth simulation systems. The next part will describe our choices in the actual implementation of our garment software.

Choosing an Adapted Simulation System

The simulation system computes the evolution of the object state using a modelization of its mechanical behavior. Basically, two families of systems may be considered: continuous systems and finite elements, and particle systems.

Continuous Systems and Finite Elements

These systems use continuous modelizations of the object properties as their state varies continuously. The resulting equations are then integrated using numerical methods. These models can efficiently describe materials that have complex mechanical behavior, and they do not necessarily involve working on discretized objects. These models have been used for cloth simulation as described in Weil 1986, Terzopoulous et al. 1987, Terzopoulous and Fleischer 1988, Yang and Magnenat-Thalmann 1991, Magnenat-Thalmann and Thalmann 1991, Carignan et al. 1992, and Yang and Magnenat-Thalmann 1993.

Independently from the model equations, the numerical resolution problem has to be considered. It mostly involves discretization of the equation terms and their derivatives (finite differences).

More recently in the story of mechanical simulation techniques, the finite element systems were developed mainly thanks to the recent developments of computing facilities able to cope with their heavy numerical calculations. Basically, such systems consider decomposing deformations on the system as a sum of "elementary deformations" applied on each of its elements, which can be triangular or square polygons, associated with a given interpolation (quadratic, cubic, and so on). Each one of these elementary deformations, or degrees of freedom, is associated to an "energy" computed from the mechanical parameters of the simulated material. Continuity between the elements leads to dependency equations between these degrees of freedom, which also take into account "boundary conditions" where some elements have their position, speed, or acceleration imposed. All these equations bring a huge linear system, which is the size of the total number of degrees of freedom. Solving it using iterative sparse matrix techniques yields the amplitude of each degree of freedom, minimizing the global energy of the system.

These techniques may be used in a wide variety of contexts and yield accurate results with complex mechanical behaviors. However, they require intensive computation, mainly for solving numerically the huge linear system, and thus often become incompatible for systems that should be responsive enough for real time or interaction.

For garment animation, the main drawbacks of these methods are the constraints generated by a clear formulation of the boundary conditions, which in this case mainly represent the numerous and complex contacts caused by collision between the cloth and the body. The resulting nonlinearities cannot be formulated efficiently into the finite element problem. Furthermore, optimizing the "shape" of the sparse linear system requires an adequate numbering of the elements, which is difficult to integrate into an interactive system where the object topology can be constantly altered by shape edition or adaptive refinements.

Some applications, mainly discussed in Collier et al. 1991, Eischen et al. 1996, and Kang and Yu 1995, use finite element methods for cloth simulation. Quite accurate and realistic for simulating precisely the behavior of fabric, they however restrict the context to small fabric rectangles with very simple interaction, and the performance ratings clearly show the difficulty of using these methods for interactive applications.

Particle Systems

The most simple and intuitive way of designing a mechanical simulation system is to consider the object as being discretized into a set of vertices that interact with each other through elastic forces. A time discretization process then updates numerically the position and speed of each vertex and yields the evolution of the system. By opposition to continuous systems, particle systems work on explicit discretizations of the simulated objects.

Based on this simple idea, a huge category of particle-system-based simulation techniques have been worked out, which mainly differ from one another by the way the forces between the particles are computed.

The simplest models, called spring-mass models, consider a triangular mesh where the vertices are masses and the edges are springs with constant rigidity and optional viscosity. These models yield very simple computations but are not very accurate for simulating deformable surfaces, as an array of springs cannot represent exactly the elastic behavior of a plain elastic surface. Another problem is to model the surface curvature forces accurately, which can be solved either through computing the surface angle between the polygons adjacent to an edge, or more simply through adding cross springs. Depending on the chosen model, a regular mesh may be necessary.

Then the motion has to be computed. Several methods are available, going from direct integration of the accelerations through the Euler formula to more evolved methods involving Lagrange equations or variational calculation. The choice of the technique depends on the complexity of the model and the way the equations can be formulated (force, energy, work, and so on), as well as on the ability to integrate constraints. These different techniques have been widely used for cloth simulation, as described in Breen et al. 1992, Breen et al. 1994, Provot 1995, Volino et al. 1995, Eberhardt 1996, Hutchinson et al. 1996, and Volino et al. 1996.

Particle System Models: The Simulation Loop

In a particle system that considers a surface discretized into a set of punctual masses, performing the simulation is usually done by computing at given time steps the evolution of the position and velocity of these masses according to their interactions between each other and with the environment.

In the most common approach, the simulation loop may be divided into two main steps:

- Computing vertex accelerations from the current positions and velocities of the object using mechanical parameters and laws
- Integrating the accelerations in order to compute the new vertex positions and velocities one time step later.

Let $X(t)$ and $X'(t)$ be the position and velocity vectors of all the vertices taken globally at the time t. The first step is then to compute the acceleration vector $x''(t) = t(x(t), x'(t))$ using the mechanical laws. The second step is to integrate the first order ordinary differential system $(x, x') = (x'', x'')$ along the variable t in order to obtain the evolution of X and X' with time.

Designing an Efficient Mechanical Model

The definition of the model contains a compromise between the accuracy of the simulated model and the computation efficiency. The system is considered to be in a given state at a given time, described by the vertex positions and velocities, plus some remnant parameters such as current plastic deformations. The mechanical model will then compute vertex accelerations at this given time from this data as well as from the mechanical parameters of the surface and some other external parameters such as gravitation or wind.

The surface behavior can be described using several mechanical properties, which are formal modelizations of the real behaviors of the considered material and its reactions against external interactions. The main mechanical parameters include

- The elasticity parameters (elongation elasticity, Poisson coefficient, bending elasticity)
- The viscosity
- The plasticity

The parameters describing external influences mainly include

- The gravitation acceleration
- The friction forces with the air (wind, turbulence, and the related viscosity forces)
- The contact forces with colliding external objects (reaction forces, friction forces)

A usual implementation of a mechanical model will compute a chosen set of these parameters using more or less sophisticated modelizations, depending on the desired realism and the computation speed.

Working on triangular meshes treated as a particle system, the deformation state of the surface is usually computed from the geometrical deformations of individual triangles, which is extracted from its edge length variations. Several formalisms then allow to compute the forces applied on the vertices, some based on energy variations, some others computing forces directly from a spring representation of the elements on the vertices.

The simplest method, suitable for interactive application, is to use the spring force approach. If a precise modelization of the elasticity is required, the method could use an exact evaluation of the deformation in each triangle element. Elasticity properties are then used to compute the constraints along the edges of the triangle, which are then converted into equivalent forces applied on the vertex masses. Such a method was, for example, described in Volino et al. 1995. On the other hand, a very approximate but fast technique is to compute forces directly from the elongation of the edges, considered as springs. It is, however, impossible to reflect exactly the elastic behavior of the material, especially when dealing with important deformations and using very irregular meshes.

The use of regular triangular meshes may simplify the computations efficiently. In such cases, using an unified spring network can model elongation as well as curvature elasticity (Hutchinson et al. 1996). However, regular meshes may be a heavy constraint against complex model shapes for realistic garments.

When using a realistic mechanical model, unrealistic deformations may result in unrealistic forces and accelerations. This can be a drawback for these models when dealing with situations where the objects cannot be accurately positioned, such as using inaccurate design and positioning virtual reality devices in interactive applications. Furthermore, the considered integration methods may perform the acceleration evaluations on trial points outside the realistic deformation range. A robust and fast simulation would thus require the sacrifice of some model realism in order to remove all the singular configurations where the forces are infinite and provide a response as continuous as possible.

Choosing a Good Numerical Integrator

As stated before, the simulation process is, in fact, the resolution of an ordinary first order differential system where the state of the system at a given time is a vector containing the position and the velocity of all the vertices. The derivative of this vector contains the velocity and the computed acceleration of all the vertices at the same given time. Solving this differential system can be performed using several numerical integration techniques (Press et al. 1992).

The simplest and most known technique, which in the context of mechanical simulation refers to the Euler method, computes the state of the system at the next time step by linearly integrating the derivative of the current time step. This method necessitates the computation of the accelerations for each time step one time, and its precision is linear (the computation error per time unit decreases linearly with the time step).

More advanced integration techniques yield quadratic precision, such as the midpoint method, which requires two times the computation of the acceleration for each time step. One step beyond is the fourth order Runge-Kutta integration, requiring four acceleration evaluations for each time step.

The higher the order is, the better the precision, and generally, the higher time steps may be considered for efficiently integrating the differential system. Practice has shown the Runge-Kutta method as being one of the best adapted techniques for dealing with our problem (Eberhardt et al. 1996). However, a high-order system with a time step set too high may also lead to big simulation errors. Thus, it is important to implement an efficient time-step control system. An adaptation of the Runge-Kutta method allows, with slightly additional computation, to evaluate an estimation of the error for each time-step iteration.

Stability and Simulation Time Step

For dynamical systems simulated by a time-step iterative process, a key issue is to maintain the stability of the system. The simulation inaccuracy increases as the time step increases, with different orders depending on the chosen integration method, and the risk of instability caused by diverging errors, also increases. The difficulty is to maintain the time step at a reasonable value, high enough for correct computation efficiency, but without any risk of instability.

Considering an object as a mesh of masses interconnected by springs, "vibration modes" may be observed in the system. Each mode is characterized by its frequency. Generally, modes involving global displacement of the structure are low frequency and reflect the modeled object elasticity. At the opposite, local vibration modes have high frequency and depend only on how the object was discretized. For the system to be simulated accurately, all the modes should be simulated accurately. In particular, local vibration modes should be damped enough to remove discretization vibration artifacts. Thus, the time step of the simulation should be comparable to the maximum frequency of all the vibration modes.

The mesh discretization amount influences the local vibration frequencies. For instance, if the element size is two times bigger, the masses are four times heavier and the springs two times longer and two times more rigid for modeling a surface of similar rigidity. Thus, the vibration frequencies are two times lower, enabling time steps two times higher. Furthermore, such a mesh requires four times less elements to process. We then clearly see the interest of using coarse meshes for fast and stable mechanical simulation, the global simulation time varying cubically with the mesh size.

One naive idea is to use spring viscosity to increase local vibration damping and prevent instability. However, the higher the damping, the higher the accelerations resulting from the damping and the higher the frequencies generated by these forces. Very high damping, by increasing the system's global stiffness, is more likely to cause instability than to prevent it.

The only efficient approach is to use an adaptive time step. This time step may be computed using a simulation error evaluation that some integration methods provide, as discussed earlier. If the error exceeds a given threshold, the step is recomputed with a smaller time step. If the error is low enough, the time step is slightly increased.

Rough Meshes and Model Accuracy

Mechanical simulation is often a compromise between accuracy and computation speed. For interactive applications, the model should be able to compute several frames per second at least.

In order to reduce the computation cost, rough discretizations are used for modeling cloth objects. Not only the amount of data to be processed is reduced, but the increased stability allows enlargement of the computation time step, as discussed earlier.

However, a rough discretization cannot reflect exactly the behavior of a fabric surface, mainly because the elements cannot bend enough to reflect the surface curvature. Not only wrinkles smaller than an element would not be modeled at all, but the mesh itself would prevent further curvature to appear. For instance, some surfaces, such as paper, can be considered almost inelastic for elongation or compression but can be bent very easily. However, an irregular triangular mesh with inelastic edges cannot be bent at all. This curvature rigidity has to be compensated by allowing the mesh edges to be elastic.

Thus, correct simulation of significantly flexible surfaces using rough triangular meshes implies lowering the rigidity of the mesh to values that can be much lower than the real elongation rigidity. Some evolved techniques may also be used, such as dynamically lowering the edge rigidity according to the current curvature.

The Mechanical Simulation Engine: Our Solution

The basic spring-mass particle system is one of the simplest and fastest models for simulating elastic surfaces. By considering the mesh edges as the only support of the interaction, minimum geometrical computation has to be performed to compute internal constraints. Unfortunately, this model is also quite inaccurate as it cannot exactly reflect real elasticity when the elements get deformed, and the Poisson coefficient is quite badly represented.

In an attempt to optimize computation speed without sacrificing away too much of the realism in simulating the elastic behavior of a surface. we have improved this simple particle system to efficiently model surface elasticity and the Poisson coefficient using a geometrical approach that compensates the major approximations.

An Efficient Elastic Model

Let's consider a triangle (P_a, P_b, P_c) in which deformations have elongated its edges from rest length (L_a, L_b, L_c) to the current length (l_a, l_b, l_c) (Figure 12.1). In a simple elastic spring-mass model, each edge would attract its vertices to reach its rest length and impose a displacement along its main direction, proportionally to the amount of elongation from the rest length.

Quite easy to compute, this model does not however reflect the actual forces when a "full material" triangle gets deformed. Each deformed edge will produce a force component along its direction, which is usually not the deformation direction, as in the example shown in Figure 12.2. The resulting effect is an extra orthogonal deformation similar to the one produced by the Poisson coefficient, but which produces unrealistic effects especially when the triangles are not equilateral (irregular meshes or high deformations).

The main idea of our new model is to recompute the individual elongation contribution of each edge of the triangle by taking into account the interdependence of the displacements that would be generated by each of them in their respective directions. Thus, the combined effect of the edge forces based on these corrected displacements will produce a more accurate constraint situation.

In the situation shown in Figure 12.3, if we suppose that the length of the edge J varies by an amount of d_j, its extremity points P_i and P_k will be displaced in its direction by an amount

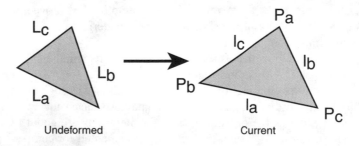

Undeformed   Current

Figure 12.1 Deforming a triangle element.

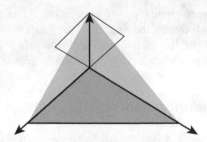

Figure 12.2 Vertical compression stretches the triangle horizontally.

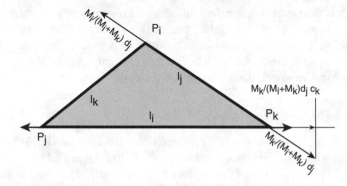

Figure 12.3 Computing the elongation effect of the edge *J* on the direction of the edge *I*.

proportional to d_j, weighted in according to the values M_i and M_k, the inverse mass of P_i and P_k. The elongation contribution on edge I is then the displacement of the point P_k multiplied by the cosine of the angle between the two edges, c_k. We linearize the problem by supposing that the edge angles do not vary significantly; (i, j, k) are all the permutations of (a, b, c).

As we would like the final length variation of the edge I to be the value l_i-L_i, we equal this to the sum of the elongation contributions of the three edges I, J and K individually would elongate at an amount of d_i, d_j and d_k. Doing this on the three edges simultaneously yields a linear system of three equations with three unknowns (d_a, d_b, d_c).

Solving the system leads to values of (d_a, d_b, d_c), that, if applied altogether, would contribute to deform the triangle to its equilibrium state.

When working with almost regular meshes, we can make the approximation that all the vertices of the mesh have approximately the same mass, which yields very simplified expressions of the force values. This assumption has proved experimentally not to alter much the behavior of the simulated elastic material in most usual situations dealing with cloth simulation.

The main positive aspect of this model is its simplicity, as it is yet able to compute realistic elastic forces in irregular triangle meshes. It involves very few vector operations by directly computing force contributions along the edge direction without the need of any local coordinate system. Experimental tests have shown that, included in a resolution system such as the one discussed later, iterations are about twice as slow as the most basic spring-mass system (which was implemented for comparison), but this model still provides a realism similar to what was obtained in our previous work (Volino et al. 1995), which involved much more complicated geometrical evaluations.

The Numerical Integration Method and Stability Issues

Among the different numerical integration methods available in the literature, we have chosen to implement a fourth order Runge-Kutta method, adapted in order to provide an evaluation of the simulation errors for each simulation step (Press et al. 1992). A study (Eberhardt et al. 1996) has made several comparisons of this method with several more evolved methods, showing clearly its performance superiority on simple examples.

The error evaluation is an important clue for efficiently determining an optimal time step that ensures the stability of the model. Our implementation has shown to be stable and robust whatever the geometrical initial situations, making this model suitable for interactive tools involving direct (and inaccurate) manipulations using VR devices.

Efficient Collision Detection

Collision detection is also a very critical point for designing a fast cloth-simulation system. It is essential to the realism of the garment simulations, as the final shape of the cloth is mostly determined by the way they interact with the underlying body. This part gives a description of the collision detection algorithms implemented in our garment software.

The Collision Detection Problem

Given a cloth object constituted of a polygonal mesh containing many hundreds of polygons, and given some scene objects (for instance, the body) modeled the same way, the problem of collision detection is to determine efficiently which of the polygons are in contact. It is obviously out of the question to test geometrically each possible polygon couple, that process being unrealistically proportional to the square of the total number of polygons.

Several algorithms families have been developed in the literature for dealing efficiently with that problem. Among them can be found algorithms using space subdivision techniques (voxelization or octree) to isolate spatially colliding elements. Another family of algorithms builds bounding box hierarchies out of the elements to speed up collision evaluation.

While the former methods are very efficient for numerous small objects moving independently in space, the latter methods are more adapted to our problem as the proximity and adjacency of the polygon elements remain constant within a surface, thus allowing the building of proximity hierarchies to be performed once as preprocessing.

Implementing Hierarchical Collision Detection

The construction of the hierarchy is performed during preprocessing. Each object polygon is a leaf of the hierarchy tree, and the other tree nodes are built recursively by grouping two or three adjacent elements. The root node representing all the object polygons is finally obtained.

Each time collision detection has to be performed with an object, bounding boxes are evaluated recursively for all the nodes of the tree. Bounding box intersection test is then performed with the bounding boxes of the nodes starting from the root node of the object. If there is no intersection, test with the child nodes is not performed. Geometrical collision evaluation with a polygon is performed only if the bounding box test with the corresponding leaf node is performed and is successful. Using this algorithm, average collision detection time is proportional to $M \log N$ where M is the number of collisions and N the total number of elements. The preprocessing time is proportional to $N \log N$.

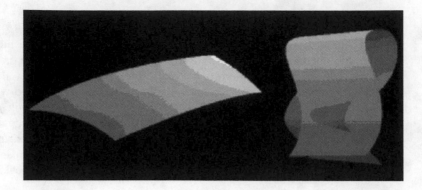

Figure 12.4   Self-collision only occurs in curved surfaces. (*See color plate on page 324.*)

This algorithm has shown to be very efficient for detecting collisions between the polygons of two surface meshes, as the detection time is almost only proportional to the number of colliding elements. Yet, the constructed bounding box hierarchy may be used for a variety of additional applications, such as additional collision detections or conditional display.

Detecting Self-Collisions Efficiently

Self-collision detection between the polygons of a single object remains a problematic task. This is, however, an important issue for handling folded cloth or buckle contacts, for example. Though the former algorithm still applies, adjacent elements would be considered as colliding according to their bounding boxes. All the adjacent elements of the polygonal mesh would then be detected as colliding, severely trashing the performance of the algorithm.

To remedy this problem, the former algorithm has been efficiently extended. The basic idea is to consider that there may be self-collisions within a surface region only if that region is curved enough to allow them (Figure 12.4). Therefore, self-collisions should not be looked for within flat surfaces.

We adapt these curvature tests in our algorithm by checking self-collisions between elements of one hierarchy node only if its surface is curved enough to have self-collisions. Furthermore, instead of using bounding boxes, we test for collisions between the elements of two adjacent nodes only if the surface curvature of the union of the two node surfaces is curved enough. For further details about this collision detection algorithm, refer to Volins and Magnenat-Thalmann (1994).

This technique provides a self-collision detection algorithm that keeps an efficient computation time proportional to $M \log N$. As in regular situations the total number of self-collisions remains small, the extra cost given by self-collision detection using this method does not alter the global performance of the system (Figure 12.5).

Maintaining Collision Consistency

Correctly simulating the collisions between multilayer garments or between the numerous wrinkles of a crumpled cloth requires techniques for robustly maintaining collision orientation consistency. This part describes the approach implemented in our system.

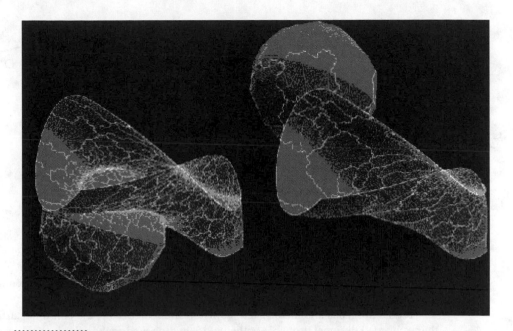

Figure 12.5   Hierarchical detection at work for collisions and self-collisions. The considered hierarchy domains are displayed for the current geometrical configuration. (*See color plate on page 324.*)

The Surface Orientation Problem

Collisions are considered when two elements get geometrically close enough to consider them in interaction. Such proximity obviously is interpreted as a contact between the considered objects, generating reaction effects that will prevent the objects from getting any closer.

However, with the important number of collisions to be considered, it is practically impossible to handle all the collisions perfectly. (It could be possible, but the complexity and the cost of the required geometrical evaluations would be far beyond what is affordable for fast simulation.) Thus, it is always possible that approximate collision response would let the objects cross and interpenetrate each other. A robust system should not bother these problems and should let the objects come back to normal as soon as possible. This would require efficient detection of the resulting "inverted" collisions, in order to handle them properly.

When dealing with solid objects, orientation is most of the time not a problem: The object surfaces have an inside and an outside. When a point is detected in the vicinity of such a surface, its side relative to the surface will determine the proper collision response behavior. However, when dealing with cloth objects, the "inside" of their surfaces does not exist, as any object may logically be in contact with both sides of the surface (Figure 12.6).

Solution: Globally Orienting Collision Contact Regions

The basic idea of our technique initially presented in Volino et al. (1995) and implemented into our new system is to keep track of all the collision areas of the scene. Thus, we group all detected collisions that are part of the same "contact region," and we compute the most probable orientation for this collision region by considering statistically the individual collision orientations. The common collision orientation is then given for each collision.

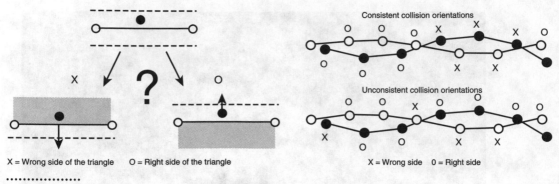

X = Wrong side of the triangle   O = Right side of the triangle   X = Wrong side   0 = Right side

Figure 12.6   The collision orientation ambiguity.

The most important difficulty is to track the collision regions efficiently. We use an incremental process, based on the labeling of the remnant collisions, using neighborhood walking along the colliding surfaces.

If an ambiguous and inconsistent collision situation occurs, the colliding surfaces will then "choose" the most probable collision orientation, and each individual collision within the considered area will then behave in a consistent way.

Combined with remnant algorithms, which keep track in memory the collision orientations between successive frames, this orientation consistency correction gives us a very robust system, able to recover most of the inconsistencies concerning colliding nonoriented surfaces.

Collision Response, High Damping and Other Constraints

Collision effects, as well as other external effects that may take part in an interactive cloth simulation system, are geometrical and cinematical constraints that should be handled efficiently, though they may be numerous and interacting between each other. This part discusses different kinds of constraints and how they are handled in our simulation system.

The Problem of Integrating Constraints in a Mechanical Simulation

There is an essential context difference between a simple application simulating a basic rectangular cloth animation and an actual garment software intending to simulate realistic dresses on an animated actor. In the latter situation, collision response is a very essential factor determining the realism of the animation, as the cloth shape is mainly defined by the way it is interacting with the underlying body. Almost all the cloth components are thus involved in collisions.

Traditional collision response uses a potential repulsion field to modelize the reaction forces, which are intense and highly discontinuous. In an approach introduced in Volino et al. (1995), we handled geometrical constraints, such as those generated by collisions, using kinematical correction on the constrained elements: Positions and velocities were corrected according to the mechanical conservation laws to fit the constraints precisely. This approach allowed to skip the potential walls used to enforce the constraints. More recently, Eberhardt et al. (1996) used a similar way to handle friction effects.

An Efficient Solution: Kinematical Correction

In our new system, we generalize the former approach by giving a procedure for dealing with several types of constraints, such as collisions, but also interaction tools, such as seaming elastics or some types of dissipative forces, such as friction or high damping.

As stated before, the first motivation is to limit the value of the forces that are enforcing these constraints. These forces may be intense, thus participating to important energy transfers, and their discontinuity may lead to important simulation errors concerning their effects. Thus, instead of enforcing constraints using forces, we directly perform kinematical corrections on the current state of the object (vertex positions and velocities), and we ensure stability by correcting accelerations as well, after their evaluation using the mechanical model.

Constraint corrections are split into three components:

1. An *immediate correction of the position and velocities* of the concerned vertices, taken into account before the dynamical simulation process, aimed at reflecting the immediate effects of the constraint. Though they are less accurately simulated, the damping effects integrated here are ensured to be perfectly dissipative, whatever their intensity, without altering the simulation.

2. An *acceleration correction* in the dynamic simulation that will attenuate or cancel the acceleration difference between the constrained vertices, in order to maintain the imposed kinematical constraints.

3. An *acceleration contribution* in the dynamic simulation, when long-range constraints participate actively in the dynamics with continuous forces and durable effects. Along with force correction, it may be used for modeling an imposed acceleration constraint. Several types of constraints can be implemented as follows.

Collisions

Collision response is mainly a geometrical constraint imposing a minimum distance between two surface elements in contact. Position and velocity immediate correction (1) is performed to put the elements in an acceptable position and prevent their speeds to push them further together. Force correction (2) then enforces minimum collision distance between iterations. Friction effects, which are usually intense, are simulated by velocity correction (1) and force correction (2) to simulate solid Coulombian friction.

Elastics

Elastics are interaction tools introduced in Volino et al. (1995) to permit seaming of cloth panels in a garment simulation system. An elastic attracts two vertices together and produces an attach point holding two elements together. In our previous work, they were simulated by adaptive forces that pulled the vertices together. However, their behavior was thus difficult to control and could lead to unpredictable results when interacting with collisions.

Simulating them as kinematical constraints is the main improvement of our work. We use a combination of force correction and contribution (2) (3) to handle them as speed constraints, producing smooth accelerations toward the goal, parametrized by only one user defined time constant. They produce predictable results whatever the kind of objects they are attached to, and handling them along with collisions in a unique system makes them interact smoothly.

High Damping

Damping is usually modeled by a force contribution (3) opposed to a speed difference. However, if damping is high (for example, when dealing with nonlinear models or solid friction), its effect will have time constants that may become much smaller than the simulation time step, perturbing the simulation efficiency and accuracy.

Purely dissipative effects may then be simulated directly by a velocity damping (1) that is guaranteed to be dissipative whatever its intensity.

Results: Our New Garment System

By implementing our efficient cloth mechanical model, integrating it using fast collision detection, and using some adapted constraint handling techniques, we obtain a powerful cloth animation system suited for animating garments on animated actors. This system is, for the time being, at the top of the evolution of several generations of garment software developed in our lab (Lafleur et al. 1991; Yang and Magnenat-Thalmann 1993; Volino et al. 1995; Volino et al. 1996). Handling constraints leads us to a unified way of dealing smoothly with collision response as well as with the seaming "elastics," as shown in the following example.

Compared to our previous approaches, the main benefit of our new system is the computation speed. The cloth assembly shown in Figure 12.7 was performed in only 90 seconds, and the garment animation shown in Figure 12.8 (5 seconds) was computed

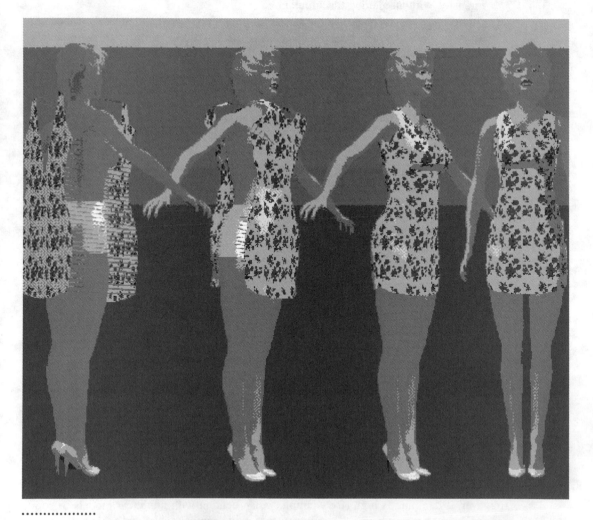

Figure 12.7   Dressing a virtual actor. (*See color plate on page 324.*)

in 18 minutes on 200MHz R4400 SGI Indigo II, taking into account mechanical simulation computations as well as full collision and self-collision detection. As soon as the cloth begins to fit the body, collision detection becomes the major computation weight, which reaches more than 70 percent in this example. Implementing incremental collision detection algorithms could reduce this.

In an attempt to use our clothing system with virtual reality devices, we adapted a flock-of-bird type tracking system to move cloth objects held by some of its vertices. Direct manipulation of the object is then possible in real time, with the six degrees of freedom. Several trackers may be used simultaneously to manipulate several objects or several parts of a same object, enabling for example, two-hand manipulation. An example of interactive seaming is shown in Figure 12.9.

On a 250MHz R4400 SGI Indigo II, the "feeling" of interactive manipulation remains very good with objects of less than 300 triangles, for which the display rate is about 10 frames per second, and a good quality real-time simulation is reached with objects of less than 100 triangles, where the display rate exceeds 40 frames per second. These computation times include full collision handling. Yet, the system accommodates very well to noisy tracking signals and remains stable even when the noise becomes too high for interaction use. The system is robust enough to cope with a few erroneous values, which would send the object far away during a few frames.

Figure 12.8 Cloth animation on a virtual actor. (*See color plate on page 325.*)

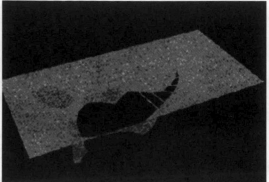

Figure 12.9   Interactive cloth manipulation: Interaction, cutting and seaming. (*See color plate on page 325.*)

We are now developing more natural manipulation tools using, for example, data gloves. A grasping system is being implemented enabling two-hand cloth manipulation, and by this way, many exciting new possibilities for an interactive clothing system. We would then like to take advantage of the potentialities of our new model to push further in the direction of interactive clothing applications. First, a powerful set of virtual tools would allow us to design, assemble, and manipulate garments in a very natural way, enabling us to visually "feel" the fabric material in the 3-D space. Then, using our VLNET system (Pandzic et al. 1997), we are preparing tools and techniques for a collaborative interactive system where distant partners together design and fit a common dress on a virtual body.

Acknowledgments

We are grateful to the Swiss National Research Foundation (FNRS) for funding this project, to Stéphane Carion who is developing the VR tools and interface, and to Marlène Poizat for her garment design work.

References

D.E. Breen, D.H. House, P.H. Getto, "A Physical-Based Particle Model of Woven Cloth," *Visual Computer,* Vol. 8, No. 5–6, 1992, pp. 264–277.

D.E. Breen, D.H. House, M.J. Wozny, "Predicting the Drape of Woven Cloth Using Interacting Particles," *Computer Graphics, SIGGRAPH Proc. 1994,* Orlando, Fla., July 1994, pp. 365–272.

T.K. Capin et al., "Virtual Human Representation and Communication in VLNET Networked Virtual Environments," *IEEE Computer Graphics and Applications,* Special Issue on Multimedia Highways, 1997.

M. Carignan et al., "Dressing Animated Synthetic Actors with Complex Deformable Clothes," *Computer Graphics, SIGGRAPH Proc. 1992,* 1992, Vol. 26, No. 2, pp. 99–104.

J.R. Collier et al., *"Drape Prediction by Means, of Finite-Element Analysis,"* J. Textile Inst., Vol. 82, No. 1, 1991, pp. 96–107.

B. Eberhardt, A. Weber, and W. Strasser, "A Fast, Flexible, Particle-System Model for Cloth Draping," *Computer Graphics in Textiles and Apparel, IEEE Computer Graphics and Applications,* Sept. 1996, pp. 52–59.

J.W. Eischen, S. Deng, T.G. Clapp, "Finite-Element Modeling and Control of Flexible Fabric Parts," *Computer Graphics in Textiles and Apparel, IEEE Computer Graphics and Applications,* Sept. 1996, pp. 71–80.

D.R. Haumann and R.E. Parent, "The Behavioral Test-Bed: Obtaining Complex Behavior with Simple Rules," *Visual Computer,* Springer-Verlag, Vol. 4, 1988, pp. 332–347.

D. Hutchinson, M. Preston, and T. Hewitt, "Adaptative Refinement for Mass-Spring Simulations," *Eurographics Workshop on Animation and Simulation,* Poitiers, France, Aug. 1996, pp. 31–45.

T.J. Kang and W.R. Yu, "Drape Simulation of Woven Fabric Using the Finite-Element Method," *J. Textile Inst.,* Vol. 86, No. 4, 1995, pp. 635–648.

N. Magnenat-Thalmann and D. Thalmann, "Complex Models for Visualizing Synthetic Actors," *IEEE Computer Graphics and Applications,* Vol. 11, No. 5, 1991, pp. 32–44.

B. Lafleur, N. Magnenat-Thalmann, and D. Thalmann, "Cloth Animation with Self-Collision Detection," *IFIP Conf. on Modeling in Computer Graphics Proc.,* Springer, 1991, pp. 179–197.

I.S. Pandzic et al., "A Flexible Architecture for Virtual Humans in Networked Collaborative Virtual Environments," *Eurographics Proc. 1997.*

W.H. Press et al., *Numerical Recipes in C,* 2d ed., Cambridge Univ. Press, 1992.

E. Promayon, P. Baconnier, and C. Puech, "Physically Based Deformations Constrained in Displacements and Volume," *Computer Graphics Forum, Eurographics Proc. 1996,* Poitiers, France, Vol. 15, No. 3, Aug. 1996, pp. 155–164.

X. Provot, "Deformation Constraints in a Mass-Spring Model to Describe Rigide Cloth Behaviour," *Graphics Interface Proc. 1995,* Quebec City, Canada, May 1995, pp. 147–154.

D. Terzopoulos, J.C. Platt, and H. Barr, "Elastically Deformable Models," *Computer Graphics, SIGGRAPH Proc. 1997,* Vol. 21, 1987, pp. 205–214.

D. Terzopoulos and K. Fleischer, "Modeling Inelastic Deformation: Viscoelasticity, Plasticity, Fracture," *Computer Graphics, SIGGRAPH Proc. 1988,* Vol. 22, 1988, pp. 269–278.

S.P. Timoshenko and J.N. Goodier, *Theory of Elasticity,* 3d ed., McGraw-Hill, 1982.

P. Volino and N. Magnenat-Thalmann, "Efficient Self-Collision Detection on Smoothly Discretised Surface Animation Using Geometrical Shape Regularity," *Computer Graphics Forum, Eurographics Proc. 1994,* Oslo, Norway, Vol. 13, No. 3, Sept. 1994, pp. 155–166.

P. Volino, M. Courchesne, and N. Magnenat-Thalmann, "Versatile and Efficient Techniques for Simulating Cloth and Other Deformable Objects," *Computer Graphics, SIGGRAPH Proc. 1995,* Los Angeles, Aug. 1995, pp. 137–144.

P. Volino et al., "An Evolving System for Simulating Clothes on Virtual Actors," *Computer Graphics in Textiles and Apparel, IEEE Computer Graphics and Applications,* Sept. 1996, pp. 42–51.

J. Weil, "The Synthesis of Cloth Objects," *Computer Graphics, SIGGRAPH Proc. 1986,* Vol. 24, 1986, pp. 243–252.

H.M. Werner, N. Magnenat-Thalmann, and D. Thalmann, "User Interface for Fashion Design," *IFIP Trans. Graphic Design and Visualisation,* 1993, pp. 197–204.

Y. Yang and N. Magnenat-Thalmann, "Techniques for Cloth Animation," *New Trends in Animation and Visualisation,* John Wiley & Sons, 1991, pp. 243–256.

Y. Yang and N.Magnenat-Thalmann, "An Improved Algorithm for Collision Detection in Cloth Animation with Human Body," *Computer Graphics and Applications, Pacific Graphics Proc. 1993,* Vol. 1, 1993, pp. 237–251.

## Chapter 13

The Blob Tree: Implicit Modelling and VRML

Brian Wyvill and Andrew Guy

Department of Computer Science
University of Calgary

● ●

Abstract

We describe our implicit modelling system, which includes an interactive editor for building models defined by blending, warping and boolean operations. The object-oriented editor uses the Open Inventor API. The editor can output polygonal models and ray-traced images. We also propose a definition of such models as an extension to VRML.

Key Terms

blending	warping	VRML
implicit surfaces	CSG	

Introduction

Methods for implicit surface modelling have been developing since the early 1980s (Blinn 1982). The major advantage of implicit surface modelling systems has been the use of automatic blending between skeletal elements; however, recent developments incorporate ideas from constructive solid modelling (CSG) and space warping (Wyvill et al. 1997). Although implicit modelling is not yet used in industry to any large extent, the usefulness of such an approach is rapidly becoming apparent.

In the 1990s 3-D computer models have been used increasingly on the World-Wide Web using VRML (virtual reality modelling language) (Hartman and Werecke 1996). One of the key features of VRML is that it can be a very compact description of a solid model. When transmitted across the Web between remote sites, the data should remain compact and the VRML browser at the remote site can use local computing power to realize the model from the VRML description.

Polygonal models, however, do not fit this scenario nor do models based on bi-cubic patches since the data required to describe the model are often very large and cannot be effectively compressed. Skeletally based implicit models, on the other hand, are ideal for this kind of transmission. The descriptions are very compact and can be transmitted at high speed. At the receiving end, local computer power is used to realize the model either by building a polygonal mesh or direct ray tracing. If a polygonal mesh is desired, the resolution of the voxel grid dictates the mesh size. This could be controlled by a user or possibly by the system, choosing a coarse grid for distant views and re-polygonizing at a finer resolution for near views. The mesh is merely a representation of the underlying implicit model.

In this paper we describe an implicit modelling system based on skeletal models and CSG and warping operations. The system includes an interactive editor based on Inventor (Wernecke 1996a), which can be easily translated into the VRML language. We also present a proposal for extending VRML to include implicit primitives, and we describe how a VRML browser may be extended to realize the model.

Previous Work

Jim Blinn introduced the idea of modeling with skeletal implicit surfaces as a side effect of a visualization of electron density fields (Blinn 1982). Such models have various desirable properties including the ability to blend with their close neighbors. These models have been given a variety of names: *Blobby Molecules* (Blinn), *Soft Objects* (Wyvill) (Wyvill et al. 1986), and *MetaBalls* (Nishimura) (Nishimura et al. 1985). Jules Bloomenthal pointed out that these models could be grouped under the more general heading of *implicit surfaces,* defined as the point set: $F(P) = 0$ (Bloomenthal 1988).

A system that includes blending, warping, and boolean operations was introduced by Pasko et al. (1995), who based the operations on functionally defined primitives. This functional approach, called the theory of *R—Functions,* also encompasses an analytical description of blending and boolean operations. Our system is similar except that we use skeleton-based primitives (Bloomenthal and Wyvill 1990) instead of functional primitives, structured using the *Blob Tree.* As yet we have not done a comparison of these two approaches; however, working with skeletons has enabled us to build an interactive modeler providing intuitive control to the operator.

Our warping uses the deformations introduced by Barr (1984), that is, the operations of *twist, taper,* and *bend.* Our polygonizer is a modification on the software described in Wyvill et al. (1986). We used a uniform space subdivison algorithm rather than an adaptive algorithm since it was readily available. Adaptive polygonizers have been described for both CSG systems (Woodwark and Bowyer 1986) and implicit surface systems (Velho 1996).

Implicit Surface Models

Skeletal implicit surface models are constructed from combinations of geometric skeletal elements. An implicit model A is generated by summing the influences of N_A skeletal elements, whose potential field will be denoted F_{A_i}, which together define a scalar field F_A. The global potential field $F_A(x,y,z)$ of an object, we call the implicit function, may be defined as

$$F_A(x, y, z) = \sum_{i=1}^{i=N_A} F_{A_i}(x, y, z)$$

The surface of the object may be derived from the implicit function $F_A(x,y,z)$ as the points of space whose value equals a threshold denoted by T_A.

$$\Sigma_A = \left\{ M(x, y, z) \in \mathbb{R}^3, F_A(x, y, z) = T_A \right\}$$

Each component of the implicit function $F_A(x,y,z)$ may be split into a distance function $d_{A_i}(x,y,z)$ and a field function $f_{A_i}(r)$, where r stands for the distance to the skeleton (Blanc and Schlick 1995). We will refer to the following notation:

$$F_{A_i}(x, y, z) = f_{A_i} \circ d_{A_i}(x, y, z)$$

Note that traditionally in implicit surface systems, skeletal elements do not have any structure, which means that all elements blend in the same way. In this work we introduce a new approach using a tree structure to store the relationships between the skeletal elements.

Visualizing the surfaces can be done either by direct ray tracing using an algorithm similar to that described in Kalra and Barr (1989) (a more recent technique is described in Hart 1996) or by first converting to polygons (Wyvill et al. 1986).

The Blob Tree

In our system models are defined by expressions that combine implicit primitives and the operators $\cup$ (union), $\cap$ (intersection), – (difference), + (blend), and w (warp). The Blob Tree is not only the data structure built from these expressions but also a way of visualizing the structure of the models. The operators just listed are binary with the exception of warp, which is a unary operator. In fact, it is more efficient to use n-ary rather than binary operators.

Nodes of the Blob Tree

Throughout this paper, T will refer to the Blob Tree. N will be a node in the tree, and its left and right leaves will be referred to as $L(N)$ and $R(N)$, respectively. Thus, the field created by a node in the tree will be denoted as $F(N)$. In the following sections we describe the nodes of the tree.

Boolean Operators

We recall that the union and intersection of primitives may be respectively defined as

$$\begin{cases} F_{A \cup B} = max(F_A, F_B) \\ F_{A \cap B} = min(F_A, F_B) \end{cases}$$

The difference operator may be expressed in terms of a negation and an intersection: $F_{A-B} = F_{A \cap (-B)}$. However, those functions show discontinuities, and several other C^n functions have been proposed (Pasko et al. 1995).

$$\begin{cases} F_{A \cup B} = \left(F_A + F_B + \sqrt{F_A^2 + F_B^2} \right)\left(F_A^2 + F_B^2 \right)^{\frac{n}{2}} \\ F_{A \cap B} = \left(F_A + F_B - \sqrt{F_A^2 + F_B^2} \right)\left(F_A^2 + F_B^2 \right)^{\frac{n}{2}} \end{cases}$$

Warp Operators

A useful tool in our system is the ability to distort the shape of a surface by warping the space in its neighborhood. A warp is a continuous function $w(x,y,z)$ that maps $\mathbb{R}^3$ into $\mathbb{R}^3$. Sederberg and Parry (1986) provide a good analogy for warping when describing free-form deformations. They suggest that the warped space can be likened to a clear, flexible plastic parallelepiped in which the objects to be warped are embedded. A warped element may be defined as

$$F_{A_i}(x, y, z) = f_{A_i} \circ d_{A_i} \circ w_{A_i}(x, y, z)$$

Throughout this paper, a warp function will be denoted as $w(x,y,z)$. A warped element may be fully characterized by the distance to its skeleton $d_{A_i}(x,y,z)$, its potential function $f_{A_i}(r)$, and eventually its warp function $w_{A_i}(x,y,z)$. Such elements will be denoted as $\{f_{A_i}(r), d_{A_i}(x,y,z), w_{A_i}(x,y,z)\}$.

In some cases we may want to compute the gradient of the potential field (e.g., for normal computation). We recall that the gradient of a C^1 bijective function $F(x,y,z)$ is

$$\nabla F(P) = f' \circ d \circ w(P) \times J_w^{-1}(P) \times \nabla d \circ w(P)$$

Thus, for each component A_i, we need to be able to compute both $w_{A_i}(P)$ and the Jacobian $J_{w_{A_i}}(P)$ for each warp function $w(x,y,z)$.

Affine Transformations

The affine transformations can be applied as warp functions. Although the effect of this is no different from applying the same transformations to the skeletons in normal space, the advantage is that a skeletal element can be defined in its canonical position and orientation and a warp used to transform it into the world space, as is common practice in many ray tracers. The advantage in our system is that warping and transformations may be treated in a consistent fashion.

Affine transformations can also be composed with other warps. For efficiency, consecutive affine transformations can be concatenated.

Barr Deformations

The Barr deformation operators, *twist, taper,* and *bend* have been implemented as warps. As indicated in Barr (1984), the deformations can be nested, producing models such as shown in Figure 13.1. In the figure, three blended cylinders have been twisted and tapered; each sample point $P(x,y,z)$ is first transformed into warp space using $w_{Taper} \circ w_{Twist}(P)$. The group of three cylinders has then been unioned with two blended spheres.

Barr applies the warp function to wire-frame models and thus uses the warp function w_{A_i} to change the coordinates of the vertices. In our case, we wish to warp space, thus we use the inverse warp function $(w_{A_i})^{-1}$.

The inverse twisting operation is a twist with a negative angle, and the inverse tapering operation is a taper with the inverse shrinking coefficient. The inverse of bend cannot be produced by modifying the bend parameters (see Barr 1984) for details of the inverse of the bend operation).

Generic Warps

In practice, any kind of warp may be used, but the Jacobian may be difficult or impossible to compute if the warping is not bijective. In that case, we rely on a discrete approximation of the gradient, which will be discussed in a later section.

The Leaf Nodes

The leaf nodes of the Blob Tree contain skeletal implicit primitives. The following have been defined:

Figure 13.1 Boolean combination of warped and blended cylinders and blended spheres. (*See color plate on page 325.*)

- Points: ellipsoids and super-ellipsoids
- Lines: cylinders capped with hemispherical ends
- Circles: torii
- Polygons: offset surfaces
- Polyhedra: offset volumes
- Plane: offset plane

Example of a Blob Tree

An example of a model built from these nodes is shown in Figure 13.2. Starting at the left-most leaf node, the three spheres are blended (+ node). A horizontal plane is intersected with the model, leaving only the lower half of it (∩ node). This model is then blended with a *bent* cylinder. A second cylinder has a *twist* applied to it, and it is then blended to the result. Finally, a global *twist* is applied to the entire tree to produce Figure 13.3. As can be seen from this example, nodes can be added to the Blob Tree in any order.

Traversing the Blob Tree

Polygonization and ray-tracing algorithms need to evaluate the implicit field function at a large number of points in space. The function $F(N, M)$ returns the field value for the node N at the point M, which depends on the type of the node:

Function $F(N, M)$:

1. Primitive : $F(M)$.
2. Warp : $F(L)(N), w(M)$.
3. Blend : $F(L)(N), M) + F(R)(N), M)$
4. Union : $\max(F(L(N), M), F(R(N), M))$

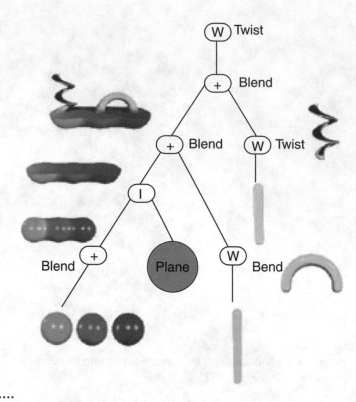

······················
Figure 13.2 The Blob Tree for Figure 13.3. (*See color plate on page 326.*)

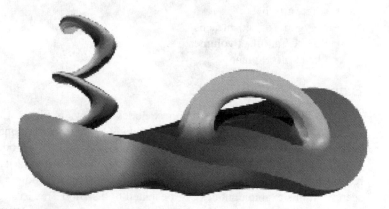

······················
Figure 13.3 Implicit surface model including boolean operations and warping. (*See color plate on page 326.*)

5. Intersection : $\min(F(L(N), M), F(R(N), M))$
6. Difference : $\min(F(L(N), M), - F(R(N), M))$

Computing the Normal and Surface Properties

Exact normals may be computed given the normal of each primitive. The normal for a blend node is the weighted average (i.e., weighted by the value of the implicit function) of the normals computed from each of the child nodes. For a CSG node, normals are computed from the appropriate child. If the node is a warp, a method for computing the normal must be provided with the warp function. This will either use the Jacobian and the normal from the child node or a numerical approximation. In the case of the Barr warps, details of the Jacobians are given in Barr (1984). A numerical approximation to the gradient is

$$\nabla F(x, y, z) = \frac{1}{2\delta} \begin{pmatrix} F(x + \delta, y, z) - F(x - \delta, y, z) \\ F(x, y + \delta, z) - F(x, y - \delta, z) \\ F(x, y, z + \delta) - F(x, y, z - \delta) \end{pmatrix}$$

Both techniques have pros and cons. Discrete evaluation requires the computation of the potential field at six locations around the given points, whereas the exact computation involves the computation of the Jacobians of the warps, which may be expensive (e.g., the Jacobian of the twist function requires a sine and cosine evaluation).

For attributes such as color, reflection, refraction coefficients and so forth, the resultant is computed as follows: for a blend node, the resulting attribute values are calculated as the weighted average of the attributes of the children. For a CSG node, the attributes are taken from the attribute of the appropriate child (e.g., for union take $max(L(N), R(N))$).

Currently, the warp nodes do not alter the attribute values although this possibly opens up an avenue for some special effects with warped texture spaces in the future.

An Implicit Modelling Editor

We have designed and implemented an implicit modelling editor, called *softed,* which is based on Open Inventor (Wernecke 1996a). The following sections introduce the ideas of Open Inventor and describe our extensions.

Overview of Open Inventor

Open Inventor is a set of C++ classes that provide a high-level interface to the OpenGL library. Open Inventor defines a scene graph that consists of nodes in a directed acyclic graph (i.e., a tree with shared subbranches).

Property nodes affect the properties of subsequent nodes that are visited in the graph traversal. Other node types are *shape,* which designates primitives, and *group,* which define groups of nodes.

In Figure 13.4, the nodes cube and sphere are of type *SoShape,* the nodes red and green are material nodes of type *SoMaterial,* and the transformation information is held in nodes *transCube* and *transSphere,* which are of type *SoTransform.*

The nodes labelled as *SoSeparator* are used to group and provide a stack for the current state, including material transformation information. During traversal when a SoSeparator node is encountered, the current state is pushed onto the stack and after completing traversal of the child nodes, the stack is popped.

SoftEd

SoftEd allows the user to interactively edit an inventor scene graph. The user interface consists of two windows: a scene viewer in which the 3-D model can be viewed and manipulated (Figure 13.5 and 13.6) and a graph viewer window that shows the scene graph and allows the user to modify the structure of the scene, including cut-n-paste and drag-n-drop of nodes and subtrees (Figure 13.7).

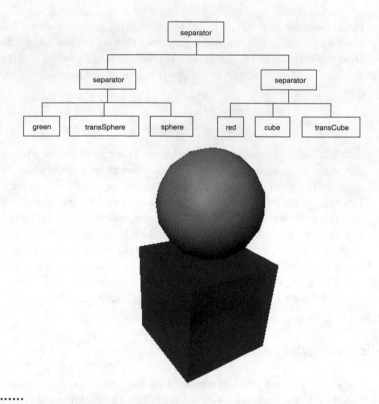

Figure 13.4 Inventor Scenegraph.

Figure 13.5 Stamingo designed with SoftEd. (*See color plate on page 326.*)

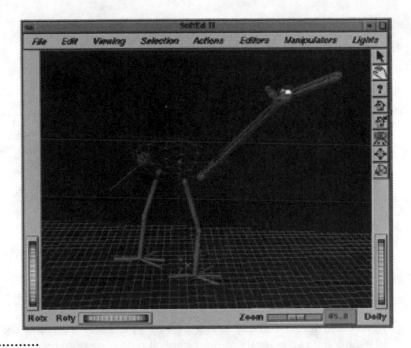

Figure 13.6   3-D view of the Stamingo. (*See color plate on page 327.*)

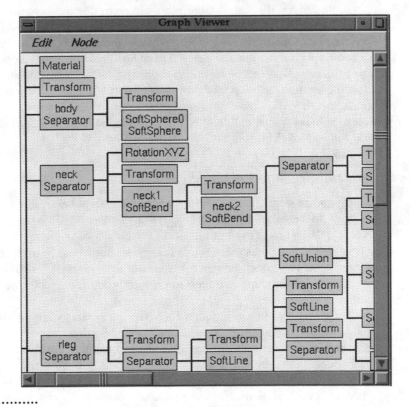

Figure 13.7   Part of the scene graph for the Stamingo.

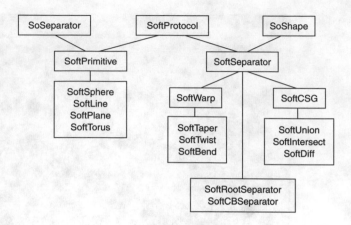

Figure 13.8 Class hierarchy for Inventor soft nodes.

Implicit Modelling Extensions to Open Inventor

The Open Inventor library is extended by subclassing the nodes and actions (Wernecke 1996b). To provide for the various implicit modelling primitives and operations described earlier, a new set of classes has been derived. All of the implicit surface nodes are derived from two base classes: *SoftPrimitive* for the primitives and *SoftSeparator* for the grouping operations. Figure 13.8 shows the class hierarchy for the new nodes in SoftEd. It can be seen that SoftPrimitive and SoftSeparator have been derived from SoShape and SoSeparator, respectively. The multiple inheritance feature of C++ has been used so that the two implicit nodes are also derived from SoftProtocol. SoftProtocol provides the interface to the implicit surface functions. SoShape and SoSeparator are Inventor nodes that allow the implicit surface nodes to be used as part of the Inventor scene graph. Most of the nodes have been described previously, however, two new nodes are *SoftRootSeparator,* which includes special children for the polygonized mesh and *SoftCBSeparator,* which is used for controlled blending (Guy and Wyvil 1995).

Realizing the Implicit Model

To polygonize or ray trace an implicit model the field value, the normal, and other values have to be calculated for a large number of points in space (see Wyvill and van Overveld 1996 and Wyvill et al. 1997) for details of polygonizing and the Blob Tree polygonization. The normal mode of using Inventor would be to traverse the scene graph to interrogate the model; however, the cost of this traversal is relatively high. To reduce this cost the scene graph is traversed once to produce a simpler tree that contains only implicit surface nodes.

This simplified tree is in fact a Blob Tree plus transformation and material information stored at every node. This tree is made up of SoftInformation objects that contain the transformation, material, a pointer to the Inventor implicit surface node, and a list of the child nodes. In the absence of warp nodes the transformations can be pushed down to the leaf nodes of the tree thus minimizing the number of matrix multiplications required to apply the transformations. If a warp node is present, the transformations must be applied before the warp operation takes place, in this case, the transformation stored at a node is from the warped local space to the local space of the node. One way this process can be optimized is that various information associated with the nodes can be precomputed. An example would be for the bend warp node in which various values are precomputed from the input parameters, these include sine and cosine calls.

Output Formats

The current output formats supported by the editor include the following:

- The polygon mesh produced can be output in various formats:

  A local mesh file format

  Open Inventor

  VRML

- Extended-rayshade file format for directly ray tracing the surface.

Extending VRML to Include Implicit Models

VRML (Hartman and Werecke 1996) is a method of communicating 3-D models across the World-Wide Web. To quote from the VRML 2.0 Handbook: "VRML is a two-year-old whirlwind that is still rapidly growing." The language defined in the handbook includes facilities for defining model geometry, including transformations, appearance, and animation. The geometry nodes of VRML include primitives such as *box, cone, cylinder, ElevationGrid, IndexedFaceSet, PointSet,* and *sphere.* Some modelling operations are also defined, such as extrusion, and animation operations, such as interpolation as well as collision detection.

Implicit surface models fit well into this scheme. Skeletal primitives can be defined (see following section), which have simple geometries. The specification of a complex object is reduced to a few primitives along with transformation, blending, warping, and field function information. A browser can then realize the models by invoking a polygonizer, using local compute power to produce the fully fleshed-out model as a mesh of polygons or by direct ray tracing.

Node Definitions

The nodes defined below constitute the basis of our proposed extension to the VRML standard. We have followed the conventions of Hartman and Werecke (1996) for our node definitions.

Group Nodes

The following nodes act like standard group nodes.

```
SoftSum {
  children [] # exposed field MFNode
}
```

The field values of the children are found and summed. The following groups have the same fields as `SoftSum`:

1. SoftUnion—Finds the union of its children.
2. SoftIntersect—Finds the intersection of its children.
3. SoftDifference—Finds $child_1$ - $union(child_2, child_3 ... child_n)$

The following nodes are a sample of possible warp nodes; see Barr (1984) for an explanation of the parameters.

```
SoftBend {
  ymin       # exposed field SFFloat
  ymax       # exposed field SFFloat
  bendRate   # exposed field SFFloat
  children []  # exposed field MFNode
}
SoftLinearTwist {
  twistRate # exposed field SFFloat
  children [] # exposed field MFNode
}
SoftLinearTaper {
  taperRate # exposed field SFFloat
  children [] # exposed field MFNode
}
```

Implicit primitives are defined by the SoftShape node:

```
SoftShape {
  skeleton         #exposed field SFNode
  fieldFunction    #exposed field SFNode
  weight           #exposed field SFFloat
  appearance       #exposed field SFNode
}
```

The *skeleton* field defines the geometry of the implicit skeleton and contains one of the nodes listed below. The *fieldFunction* contains a node that defines one of the many field functions described in the literature. Here we define the default function from Wyvill et al. (1986). The *weight* is a scalar multiple applied to the field value. The *appearance* node contains an Appearance node from Hartman and Werecke (1996).

The next set of nodes defines the actual implicit primitive shapes.

```
SoftSphere {
  radius #SFFloat
}
SoftRCylinder {
  radius #SFFloat
  height #SFFloat
}
```

The RCylinder has hemispherical ends.

```
SoftTorus {
  insideRadius #SFFloat
  outsideRadius #SFFloat
SoftPolygon {
  vertices[] #MFVEC3F
}
```

The polygon is actually an offset surface with hemicylindrical sides. The vertices must define a planar polygon, otherwise the effect is undefined.

```
SoftRCone {
  bottomRadius #SFFLOAT
  topRadius #SFFLOAT
  height #SFFLOAT
}
SoftPlane {
  normal     #MFVEC3F
  point  #MFVEC3F
}
```

The plane is in fact a half-space plane for use with the CSG nodes. Note that care must be taken to avoid the definition of an infinite model, which may cause problems with finite polygonizers.

Interaction with VRML Primitives

Blending and warping are characteristic of implicit models but are difficult to apply to other modelling primitives. However, ordinary VRML nodes also can be placed in the groups, but in our initial proposal these would not be subject to blending and warping.

Other operations, such as collision detection, can be done between implicit models using the method described in Gascuel (1993). We do not define collision detection between current VRML primitives and implicit models.

Conclusions and Future Work

In this paper we have presented a data structure for combining implicit primitives using blending, warping, and CSG operations. We have also presented some recommendations for extending VRML 2.0 to include implicit models.

The Silicon Graphics VRML browser, Cosmo player, is based on Open Inventor as is our own implementation of the Blob Tree. Conversions between VRML and Open Inventor, including the extension described in this paper, are relatively straightforward. A future project is to extend an existing VRML browser to include the proposed nodes. Further work is required to examine methods for more homogeneous interactions between existing VRML primitives and implicit models.

Acknowledgments

The authors would like to thank the following researchers: Kees van Overveld (Phillips Corp. and University of Eindhoven), Geoff Wyvill (University of Otago), and Jules Bloomenthal (Microsoft) for their very considerable contributions to this work over the years, and John Cleary (University of Waikato) for some helpful remarks. We would also like to thank Eric Galin of the University of Lyon for his collaborations on the development of the Blob Tree. This work is partially supported by grants from the Natural Sciences and Engineering Research Council of Canada.

References

A.H. Barr, "Global and Local Deformations of Solid Primitives," *Computer Graphics (Proc. SIGGRAPH '84),* Vol. 18, July 1984, pp. 21–30.

C. Blanc and C. Schlick, "Extended Field Functions for Soft Objects," *Implicit Surfaces '95,* Apr. 1995, pp. 21–32.

J. Blinn, "A Generalization of Algebraic Surface Drawing," *ACM Transactions on Graphics,* Vol. 1, p. 235, 1982.

J. Bloomenthal, "Polygonisation of Implicit Surfaces," *Computer Aided Geometric Design,* Vol. 4, No. 5, 1988, pp. 341–355.

J. Bloomenthal and B. Wyvill, "Interactive Techniques for Implicit Modeling," *Computer Graphics,* Vol. 24, No. 2, 1990, pp. 109–116.

M. Gascuel, "An Implicit Formulation for Precise Contact Modeling Between Flexible Solids," *Computer Graphics (Proc. SIGGRAPH 93),* Aug. 1993, pp. 313–320.

A. Guy and B. Wyvil, "Controlled Blending For Implicit Surfaces," *Implicit Surfaces '95,* 1995, pp. 107–112.

J.C. Hart, "Sphere Tracing: A Geometric Method for the Antialiased Ray Tracing of Implicit Surfaces," *Visual Computer* (in press), Dec. 1996.

J. Hartman and J. Werecke, *The VRML 2.0 Handbook,* Addison-Wesley, 1996.

D. Kalra and A. Barr, "Guaranteed Ray Intersections with Implicit Functions," *Computer Graphics (Proc. SIGGRAPH 89),* Vol. 23, No. 3, July 1989, pp. 297–306.

H. Nishimura et al., "Object Modelling by Distribution Function and a Method of Image Generation," *Journal of papers given at the Electronics Communication Conference '85,* Vol. J68-D, No. 4, 1985 (in Japanese).

A. Pasko et al., "Function Representation in Geometric Modeling: Concepts, Implementation and Applications," *Visual Computer,* Vol. 2, No. 8, 1995, pp. 429–446.

T. Sederberg and S. Parry, "Free Form Deformation of Solid Geometric Models," *Computer Graphics (Proc. SIGGRAPH 86),* Vol. 23, No. 3, Aug. 1986, pp. 151–160.

L. Velho, "Simple and Efficient Polygonization of Implicit Surfaces," *J. Graphics Tools,* Vol. 1, No. 2, 1996, pp. 5–24.

J. Wernecke, *The Inventor Mentor,* Addison-Wesley, 1996a.

J. Wernecke, *The Inventor Tool Maker,* Addison-Wesley, 1996b.

J. Woodwark and A. Bowyer, "Better and Faster Pictures from Solid Models," *Computer Aided Engineering Journal,* Vol. 3, No. 1, Feb. 1986, pp. 17–24.

G. Wyvill, C. McPheeters, and Brian Wyvill, "Data Structure for Soft Objects," *Visual Computer,* Vol. 2, No. 4, Feb. 1986, pp. 227–234.

B. Wyvill and K. van Overveld, "Polygonization of Implicit Surfaces with Constructive Solid Geometry," *J. Shape Modelling,* Vol. 2, No. 4, 1996, pp. 257–273.

B. Wyvill, E. Galin and A. Guy, "The Blob Tree, Warping, Blending and Boolean Operations in an Implicit Surface Modeling System," *Research Report No. 97/61/11,* July 1997.

## Chapter 14

Automatic Generation of Virtual Worlds for Electronic Commerce Applications on the Internet

Klaus M. Bauer

Computer Graphics Center (ZGDV e.V.)

. . . .

Abstract

The Internet has become an important media for commercial applications. Today, many vendors offer their goods with Internet-based technology, such as the World Wide Web. The work described here focuses on the advantages of 3-D worlds for shopping applications on the Internet and describes the prototype of a software tool that automatically generates 3-D models of virtual supermarkets. These supermarkets are distributed on the Internet or other media (e.g., CD-ROM). Based on the experiences with previously developed generators for Windows- and Java-based supermarkets, both design considerations and software aspects are discussed in detail. An important feature is the system's ability to generate individualized supermarkets for each customer's specific needs and demands. Finally, current and future work as well as new ideas for further development are presented.

Introduction

Artificial three-dimensional worlds are considered to be one of the most user-friendly metaphors for human-computer interaction. Today's virtual reality technology provides an impressive output quality. Suffering from high costs, powerful and 3-D-capable computers have not yet made their way to a larger number of home users. But this can be expected to change soon. Inexpensive 3-D graphics accelerator boards for PCs have been announced for the near future. It can be foreseen that a large user community of this technology will evolve.

An application area with a strong need for 3-D user-interfaces is electronic commerce, as soon as the interface becomes widely available for the customers. A large number of goods can be presented in a clear and attractive manner. Furthermore, the user can closely examine the model of each individual object. Additional features, such as posters, chat rooms, and many others, are easy to integrate. However, a big problem arises from the construction of these 3-D worlds. Traditionally, they are composed manually by highly skilled experts, using 3-D modelling and texturing tools. In the area of digital marketing these old methods are no longer sufficient. Requirements are minimal costs, fast updates, customer specific content, and the guaranty of total correctness. To fulfill these requirements, a new tool for the automatic generation of 3-D supermarket models (e.g., VRML) is under development at ZGDV. This paper describes the concept of this tool and its role within the entire system.

The Digital Marketing Support System Developed at ZGDV

To match the needs of both vendors and customers in the area of home shopping, the ZGDV has developed a prototypical Digital Marketing Support System (Bauer 1996). (Figure 14.6 gives an overview of the system's architecture.) Based on the information that is stored in a central database, a special tool selects goods and presentation styles for each individual customer according to his or her user profile. Afterwards, a generator produces a multimedia application that is suitable for the customer's platform. At the moment, HTML, Java and Windows-based supermarket applications can be generated. An authoring tool, a statistics tool, and several other components complete the system.

From the very beginning of this project it was a central requirement to produce comfortable, easy-to-use interfaces for the customers. The idea of a 3-D metaphor was implemented in a very early stage of the project. But, due to a lack of powerful 3-D computers at the customers' sites, 2-D technology was used to resemble 3-D appearance and functionality (Figure 14.1).

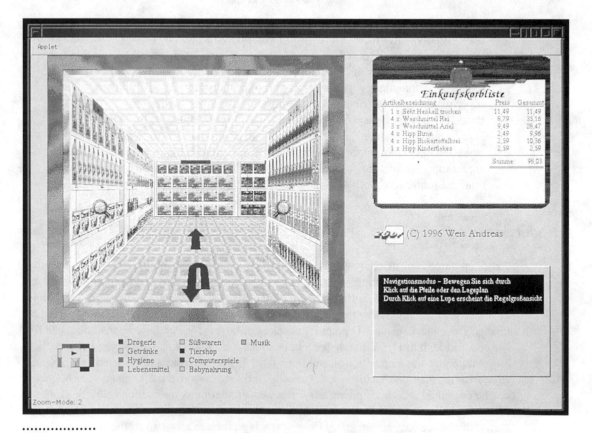

Figure 14.1 Automatically generated, Java-based supermarket with 3-D appearance (Bauer 1996). (*See color plate on page 327.*)

A Generator for Real 3-D Supermarkets

In the meanwhile, the time had come to extend the system with a generator for 3-D worlds (Encarnação et al. 1996). In a first step a 3-D model of a supermarket was designed manually to examine design options and technical requirements (Figure 14.2). Besides some rough performance estimations, it was mainly used to estimate the possible user acceptance for this style of presentation and interaction.

Analysis of the Requirements

At first, the needs and demands had to be surveyed both from the user's point of view as well as from the vendor's side. It is obvious that the users will not be satisfied if the only possible interactions are to walk around, to look at the goods, and to buy them. In this case, a 3-D world would simply be another display technique with no added value. To justify the increased effort, the capabilities of modern 3-D graphics and interaction technology have to be used to provide a number of advantages other technologies cannot offer:

• Give a virtual supermarket's customer the feeling of being really there. This includes the presentation of the complete exterior and interior equipment that is found at a real supermarket's site.

• Provide an unparalleled level of freedom for navigation within the shop and of interaction with all of its goods. Even stereo output devices and data gloves should be supported for a top-level VR presentation.

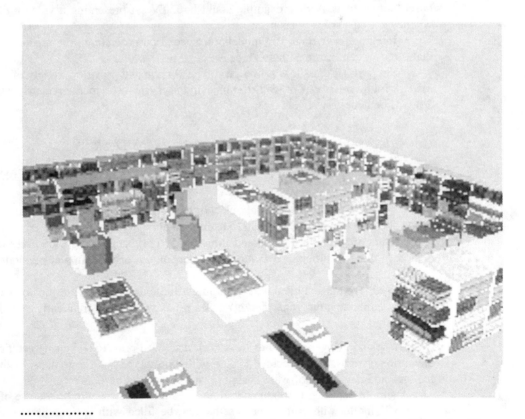

Figure 14.2 Manually produced prototype of a real 3-D supermarket. (*See color plate on page 327.*)

- Resemble the product's functionalities as perfect as possible. Give the user the ability to play with it (e.g., by pressing buttons of a simulated TV set to switch the channels).

Vendors that are willing to use 3-D technology for marketing and sales have a strong emphasis on several criteria:

- It must be possible to create a unique appearance of the supermarket and the products to make the characteristics of each brand clearly visible to the customers. In most cases they even wish to resemble the corporate design of their already existing shops (e.g., by using their logos and other design elements).
- Design schemes must be changeable with very little effort to be able to improve the design whenever it is considered appropriate. This is essential to respond to challenges from other vendors.
- Data input should be a very easy and schematic task to limit the efforts.
- 3-D models should be usable for planning tasks in the real world (e.g., for placement of shelves and products, as well as for the selection of colors and other design elements).

Based on these results, the technical requirements for the generator were defined:

- Provide a variety of quality levels for the output (e.g., different levels of detail for 3-D models or different resolutions for textures) to match different demands of users and different system capabilities.
- Create a realistic, high-quality model with textures, light sources, and a large number of details, if requested.
- Integrate the multimedia functionality of different systems, especially for WWW-based applications. A nice example could be a TV set presenting a video on its screen whenever a customer comes near.
- Integrate individual 3-D models for special purposes (e.g., the model of a camera that the user can examine closely).
- Support VRML 1.0 in a first prototype version and provide a variety of output formats in future releases. For well-designed and user-friendly supermarkets, at least VRML 2.0 will be necessary.

Considerations Regarding the Supermarket's Design

For the design of a 3-D world such as a supermarket, several aspects have to be considered carefully:

- What are the user's expectations? For example, does he/she want to see nice graphics or does the user prefer fast data transmission and interaction?
- What are the users' skills and limitations? For example, will the users be able to deal with the difficulties of 3-D navigation techniques or will simpler methods be more appropriate?
- What are the technical and economical limitations? What are the costs of setting up and maintaining the supermarket? Does it have to change frequently or is it a long-term design?

These and many more issues have been discussed prior to technical specifications of the entire systems. It was decided to start with a prototype that shows many of the desired features and that has a software design open to modifications.

A simple approach for the generation of a supermarket would be the use of a predefined 3-D model with empty shelves that can be filled with an individual selection of goods. This concept is easy to implement but lacks flexibility. Consequently, the size and the interior of the supermarket are computed separately each time a new supermarket is generated. This allows an optimal size and design of the market. In cooperation with sales experts, a number of design strategies have been defined:

• The supermarket is based on a rectangular layout. It should maintain a ratio of length to depth of approximately 4 units in length and 3 units in depth. In future versions of the generator, other shapes will be supported as well.

• Whenever additional shelves are added, the ratio 4 × 3 has to be kept. Gaps between shelves that are too wide to remain empty, are filled with baskets or other equipment. Larger gaps are used to place more shelves.

• Within one shelf only similar products may be placed. For example, bread and cakes are allowed to share one shelf, bread and soap, however, have to be stored on different shelves.

• All shelves have fixed sizes; there is no possibility to scale them. For packages, on the other hand, slight modifications of their size are allowed to fill the shelves properly and to avoid gaps. Of course, all modifications have to be small enough not to be noticed by the user.

• The number of checkouts correlates to the size of the supermarket.

It is important to note that the intention of all design principles described here is to achieve a nice and realistic appearance from the customers point of view. It is no CAD-like simulation of a real supermarket. In all cases, the user's perception of the supermarket is most important, not the absolute precision of all measures.

Database Design

It was found that only minor extensions to the multimedia content of the product database (see Figure 14.6) were necessary. For most of the products, a rough classification of their shapes (e.g., cubic, cylindrical, or spherical) is completely sufficient. The other data (e.g., pictures for texturing) can be reused from the already existing content.

The database had to be largely extended for the description of the supermarket's equipment. Generic description formats for shelves, baskets, fridges, and many other components have been integrated. Additionally, the user interfaces for their definition had to be developed (Figure 14.3). These interface are very important for the acceptance of the complete system by the supermarket companies. Only a few minutes should be spent to add a product to the database. The concept and design of the interface's functions ensure a high level of productivity. In a first step, several hundred products have to be added to the database. Finally, though, tens of thousands of products should be available.

Furthermore, a description format for the basic layout of supermarkets is under development. It is based on representations of different functional areas (e.g., shelves, walls, paths, and others) and on a set of rules for the combination of those areas. This ensures a layout functionality that is both flexible and realistic. Specific requirements can be added with little effort. Layout patterns are defined easily (Figure 14.4). Each pattern has a level of preference assigned (highly recommended, recommended, not recommended, forbidden).

The Texture and the Multimedia Archive

For the Digital Marketing Support System, many different types of multimedia data can be used. The basic data can be integrated to a variety of presentation styles for a large number of systems (HTML-based, VRML-based, etc.). Storage and retrieval of these data are

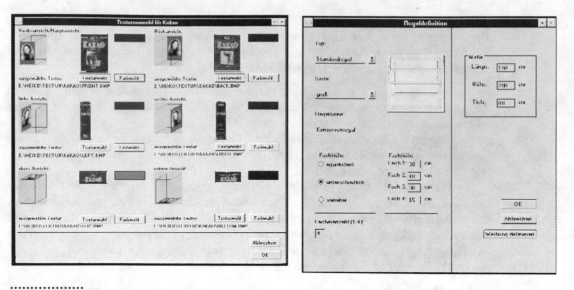

Figure 14.3   User interfaces for the definition of product textures (left) and shelves (right). (*See color plate on page 328.*)

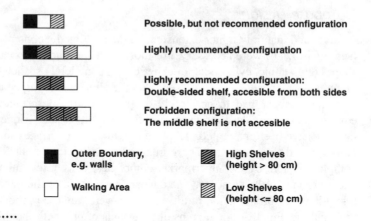

Figure 14.4   Examples for the layout scheme of a virtual supermarket.

crucial points for the effectiveness of the entire system. It is not the intention to define a fixed number of media or even media types for each product (e.g., two videos, one sound file). Instead, a maximum flexibility has to be ensured. With each product, a virtually unlimited number of media can be associated. For the classification of these media the MIME-type scheme is used.

Another problem arises from the fact that in Europe a large number of different languages should be supported. The database scheme was designed to provide maximum support for this by a specific description table and substitution rules. Not only the major languages (German, English, French, etc.) are supported. For example, in (German-speaking) Austria some names of products are different from their names in Germany, and in Switzerland even another name may be used. All media stored in the database are defined with attributes that allow the classification of the language(s) they belong to.

Obviously, textures play a major role for the realistic appearance of nearly all kinds of 3-D worlds. For the supermarket application described here, they are essential as well. For each cubic packaging, up to six pictures are needed. For irregular shapes, even more textures may be necessary. If other display techniques are used, additional pictures are required in some cases. The Java-based supermarket (see Figure 14.1), for example, needs up to ten pictures for each product to provide its full functionality. For Quicktime VR applications, dozens of pictures can be used to generate high-quality scenes. Therefore, it is necessary to store and to classify all the pictures by using a general scheme and finding an efficient way to minimize the storage capacity needed for them.

The pictures are not stored within the database itself, but references are. Either local path names (UNIX or DOS) or URLs can be used. This makes it possible to use one single copy of a file for both internal purposes on the server as well as for many different users with their clients.

A first naive classification approach would be to simply name all the pictures according to a specific scheme (e.g., "front," "top," etc.). However, this method lacks flexibility, and it would be difficult to implement tools to handle the pictures stored in this way. Therefore, another classification method was chosen. For each picture, the face normal of the (virtual) camera that took the picture is stored together with the reference to the picture file. A unique definition of each package's orientation has to be defined first and maintained for following work. For example, the front (main) view of a package is given by the vector (1,0,0), its top view by (0,0,1), and so on. This makes it possible for future systems (e.g., generators for other output format) to look for pictures that provide a specific view of the product or at least to find the best-fitting picture. Additional information about the resolution and the color depth is stored as well to simplify the searching.

The Generator's Functionality

The generator gets its input from the individualization tool (Kreutz 1996) that produces a customer-specific subset of the complete assortment of goods and gives some hints on the customer's preferred presentation style (e.g., whether or not decorations should be integrated into the market). Based on this information, the generator performs several steps:

- Start the marketing strategy tool to include additional products; for example, those that are on sale or are currently specially advertised.
- Select the shelves and other installations that are needed to display the products realistically. For example, frozen food should be found in a fridge, not on a normal shelf.
- Compute the required size of the supermarket depending on the number of products and the selected installation.
- Select and integrate additional components (e.g., posters on the walls and sound files for background music).
- Determine the supermarket's layout, and place the shelves and other equipment.
- Apply textures to the goods, and place them within the shelves.

- Choose an appropriate design for the supermarket's storefront and its environment to complete the 3-D model.
- Write system-specific code (e.g., VRML 1.0 or VRML 2.0, nff or something else) to a specified file in the WWW-server's data area.

Considerations on Software Design

Like all components of ZGDV's Digital Marketing Support System, the 3-D generators are designed to work as clearly separated modules. This makes maintenance work as well as future extensions quite easy. A major challenge arises from the design goal to support several different output formats but to avoid a complete software implementation for each of them. This is achieved by two strategies: All steps while assembling the complete supermarket are kept strictly neutral, and the output file is written when everything else is finished. Of course, not all formats support all features the supermarket generator can offer. For example, VRML 1.0 supports only static scenes (Figure 14.5), and VRML 2.0 supports objects with built-in functionality. It would neither be sufficient to simply leave the objects static nor would it be a good idea to try some kind of replacement strategy at a very late point

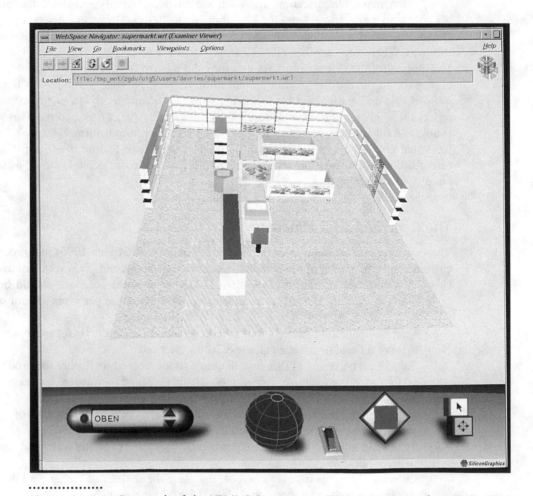

Figure 14.5 A first result of the VRML 1.0 generator. (The shelves are only partially filled for performance reasons.) (*See color plate on page 328.*)

of the generation process. To overcome this problem, a list of recommended features and a list of excluded features is kept for each output format that is supported by the generators. Furthermore, for each feature a list of possible alternatives is available. This makes it possible to avoid design problems or to offer some variations within the supermarket. Daily or seasonal changes of the supermarket's appearance can be achieved by this functionality.

Obviously, generators have to deal with data that describe output formats. It would not be a good idea to implement them hard-code directly in the modules. Three alternatives have been considered to store this information: (1) use a database to store the elements of each format, (2) use text files, or (3) use an abstract description such as a syntax tree. Of course, the third alternative is the most elegant, but for ease of use and a higher speed of development, the second one was chosen. For each output format, a set of text files that contains the specific syntax and some additional information for the generator is available. To achieve a better performance, these files are read once and kept internally for faster access during the generation process.

An important issue in the design phase of a new software is the choice of the programming language. Sometimes it can even become necessary to rethink an old decision when new aspects arise. In the case of the system described here, it was decided to implement it in Java (Bauer 1997) and to migrate the already existing tools from the first implementation in Visual Basic to Java as well. The most important reason for this is to achieve platform independence of the entire system. It can run on low-cost PC servers (with Windows NT) as well as on high-performance workstations. Not only from an academic point of view is this an important advantage. Electronic commerce is a very new field for research and development as well as for entrepreneurs. The platform independence of Java makes it possible to start a business with a relatively low budget (on a PC) and to migrate to more powerful computers without the need of changes to the software. One drawback of the Java programming language—its relatively low performance—has been overcome when compilers that produce native machine code became available. For a server-side application such as the generator described here, this is a very important point.

Current and Future Work at ZGDV

The functional requirements and the limitations for the 3-D world generator have been defined in the past. A prototype version of the generator with a limited functionality has been operational since December 1996 (Figure 14.6).

For further development, a number of tasks are planned and some ideas are currently being discussed:

• The appearance of the supermarket has to be improved significantly. The current prototype is sufficient for testing purposes, but for real applications, a much better quality of the graphics has to be reached (Figure 14.7).

• The generator for the storefront and its environment has not been implemented yet. Some ideas to achieve an attractive appearance are considered. One possible option is to take into account the different seasons as well as the actual time of day.

• The integration of human interaction and cooperation into virtual supermarkets seems to be very promising. This should even be possible for heterogeneous presentation techniques. Imagine several customers that use one type of a virtual supermarket that is displayed using different technologies (e.g., one in VRML, another in Java, a third one in MS-Windows). It seems to be an interesting task to develop avatars and interaction techniques for this purpose.

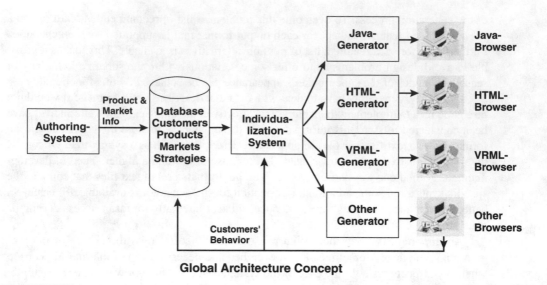

Global Architecture Concept

Figure 14.6 The VRML generator as a part of the reference architecture for the Digital Marketing Support System (Bauer 1996).

Figure 14.7 Two different views of the supermarket. Please note the open glass door of the fridge in the picture on the right. (*See color plate on page 328.*)

• The virtual 3-D worlds can be rendered to produce Quicktime VR applications that can be run on nearly all kinds of computers, providing a nice, yet not perfect, three dimensional feeling. This could lead to a partial reuse of software and a reduction of effort for the development of the complete system.

• Physically based simulations of objects can help to increase the user's understanding dramatically (Dai 1996; Dai and Reindl 1996). The user's ability to play with the virtual model of a camera and to request information that is specific for its current state can eliminate the need for salespeople in many cases. For both virtual reality applications and electronic commerce systems, this can demonstrate their usefulness.

• Other metaphors for the current virtual supermarket are also planned. Not only traditional approaches, such as virtual department stores and virtual exhibition grounds, are considered. New ideas for design concepts that do not have any counterpart in reality seem to be promising for innovative, application-oriented, and user-specific solutions.

• With minor changes of the design principles and some additional functions for interaction, the 3-D models of this system can also be used for planning purposes for real supermarkets. Interfaces for the integration of already existing layout data are under development.

Related Work

The work described in this short paper has tight relations to the developments in the area of electronic commerce (Janal; Werner and Stephan 1996), as well as to the work on virtual reality. Both are dynamically evolving research, development, and application fields. Several companies have published multimedia-based shopping applications. Some of them already use 3-D metaphors (e.g., the German company Konrad Elektronik with their CD-based system for PCs; but in fact, it is still 2-D technology). Operational 3-D supermarkets are not yet to be found. On the other hand, highly sophisticated VR techniques (Durlach and Mavor 1995; Burdea and Coiffet 1994) have been developed and are yet improved by both research institutions and commercial companies. It can be assumed, though, that home and office applications will be limited to medium-quality rendering and simple input equipment, such as 2-D and—in rare cases—3-D mouselike devices.

Conclusions

The ongoing work shows clearly that it is both useful and possible to create 3-D models of virtual supermarkets. Very few extensions to the existing data model were sufficient to enable the automatic generation of 3-D supermarkets. The results that were achieved are very encouraging. The further development of the current work will surely demonstrate that with this approach, individualized, high-quality, 3-D worlds can be produced at very low costs. Furthermore, additional functionality can be integrated easily to enhance both the appearance and the functionality.

Our performance tests showed that the major drawback of 3-D supermarkets is the lack of cheap, yet powerful, computers. The 3-D model of a full-scale supermarket that offers several hundreds of different products consists of thousands of polygons—many of them are even textured. Today, a model of this size can be displayed for real-time interaction on high-performance workstations only. The consequence is that for the next two or three years, these 3-D applications will be used mainly for special purposes and as a demonstration for the state-of-the-art techniques of computer graphics and virtual reality. But in the long run, it is very likely to be an attractive and useful interface metaphor for many home and business users.

References

K.M. Bauer, "Java-Anwendungen für Home-und Teleshopping-Systeme," *Business Online,* Vol. 4, Oct./Nov. 1996, pp. 68–70.

K.M. Bauer, "Kundenspezifisch individualisierte Virtuelle Supermärkte unter Anwendung von Internet-technologie und Java," *Proc. Europäische Congressmesse Online '97,* Hamburg, Germany, February 3–7, 1997.

G. Burdea and P. Coiffet, *Virtual Reality Technology,* John Wiley & Sons, New York, 1994.

F. Dai, *Lebendige Virtuelle Welten—Physikalisch-basierte Modelle in Computeranimation und Virtueller Realität,* Springer-Verlag Heidelberg 1996 (in print).

F. Dai and P. Reindl, "Enabling Digital Mock-Up with Virtual Reality Techniques—Vision, Concept, Demonstrator," *Proc. ASME Design for Manufacturing Conf.,* Irvine, Calif., August 18–22, 1996.

N.I. Durlach and A.S. Mavor, eds., *Virtual Reality: Scientific and Technological Challenges,* National Academy Press, Washington DC, 1995.

J. Encarnação et al., "New User Interface Aspects to the Information Highway—Applying Java and VRML," *Int'l J. Information Technology (IJIT),* Vol. 2, No. 1, 1996, pp. 79–95.

D.S. Janal, *Online Marketing Handbook,* http://www.thomson.com/vnr/omh.html

A. Kreutz, *Konzepte und Werkzeuge für das Individualisierte Marketing im World Wide Web,* diploma thesis, Darmsstadt Technical University, FG Graphisch-Interaktive Systeme, 1996.

A. Werner and R. Stephan, *Marketing-Instrument Internet,* 1996.

## Chapter 15

A Case Study in the Use of VRML 2.0 for Marketing a Product

Nick D. Burton, Alistair C. Kilgour, and Hamish Taylor

Department of Computing & Electrical Engineering
Heriot-Watt University

• •

Abstract

The addition of behaviors, animation, and sound to Virtual Reality Modelling Language has created interesting new opportunities to market products on the Internet. This paper investigates these opportunities and illustrates their potential through a case study in promoting electronic bagpipes.[1] It also reports on an evaluation into 3-D navigation in VRML 2.0 that exposes problems users have with current navigational interfaces in VRML 2.0 browsers and suggests a way in which such problems may be overcome.

Introduction

Recently there has been a marked increase in the use of the Internet for many business activities. One of the main reasons for this is the ease with which multimedia documents can be constructed for the World-Wide Web (Web, or WWW). It is now possible to buy anything over the Web from military equipment to Italian sausages. Estate agents have access to a world-wide customer base, trade associations are using the Internet to keep their members informed with up-to-date information, and so on. The richness of Web page content has improved dramatically. It was not long ago that it was possible to display only text and monochrome pictures. Now that it is possible to include color pictures, sound, movies, 3-D animation, and interactive programs, there is much greater potential for marketing products. One of the more interesting recent developments in Web technology has been version 2.0 of the Virtual Reality Modelling Language. VRML 2.0 is a scene description language that enables the construction of interactive, animated 3-D objects and environments on a Web page.

....

[1]http://calligrafix.co.uk/demo/bagpipes/index.html
nickb@cee.hw.ac.uk, ack@cee.hw.ac.uk, hamish@eee.hw.ac.uk

This paper investigates the possible use of VRML 2.0 to enhance the Web-based marketing of manufactured objects using interactive and animated solid models (Burton 1995–1996). The work was carried out in conjunction with Calligrafix,[2] a multimedia and Internet service provider in the Scottish Borders. The intention was to develop a Web-based virtual shopping mall within which products could be displayed for purchase over the Internet. It soon became apparent that the limitations of VRML 2.0 would restrict the attractiveness of such a virtual shopping mall, and it was decided to investigate the marketing potential of VRML 2.0 for producing Web-based 3-D models of a product instead. Rather than choosing a well-known product, it was decided to focus on something novel that would be amenable to 3-D modelling. The promotion of an electronic bagpipe seemed suitable as a case study since its shape was regular and the requirement for sound added extra interest.

Any technology for marketing a product on the Web must

- Attract and maintain a potential customer's attention
- Be easy to use by inexperienced computer users
- Be capable of providing useful information about the product
- Be cost-effective to employ

Hypertext mark-up language (html) is the technology most commonly used to construct Web pages. Text and 2-D pictures can convey a great deal of information about a product, but both are static and are limited to what the Web page author chooses to present. It would be more useful when marketing a manufactured product, if a potential buyer were able to interact with an animated 3-D model of the product. This would better engage the buyer's attention and would convey information about the product that could otherwise be gained only from real experience of it. It is just such an enhancement that is expected from the use of VRML 2.0.

There are a number of Web technologies that offer 3-D and interaction, and these are discussed briefly in the next section. Then we give an overview of the aspects of VRML 2.0 that were pertinent to this case study and discuss the construction of the 3-D models used. The strengths and weaknesses of VRML 2.0 for marketing a product on the Web are also discussed. Later in the paper, we describe an experiment that was conducted in order to gauge users' reaction to the problems of 3-D navigation. The final section discusses problems with the user interface controls on some of the VRML 2.0 browsers and describes a prototype custom interface for object manipulation in VRML 2.0 that attempts to rectify some of these problems.

Existing Technologies

At present there are a number of technologies that promise 3-D and animation on the Web. Of these the most interesting are

- Java
- Macromedia Shockwave
- QuickTime VR
- Virtual Reality Modelling Language

---

[2]http://calligrafix.co.uk/

Java

Java is a general purpose, object-oriented programming language developed by James Gosling of Sun Microsystems in 1990. It is (almost) platform independent due to the fact that the source code is compiled into "byte code," which is then interpreted by a Java virtual machine on the host computer. Java can be used to produce stand-alone applications. It can also be embedded in Web pages, where it is referred to as an *applet*. Applets provide executable content for Web pages. Since May 1995, when Java was first "loosed on the Internet," (Semich and Fisco 1996) numerous applets have appeared, which have added animation, both 2-D and 3-D, to Web pages. Some of these, such as the "ticker tape" applet, quickly become irritating. A more interesting example of 3-D Java animation is the *Virtual Rubik's Cube*," which can be found at the "Applet Arcade." An example of an interactive 3-D environment written in Java is the game *Dungeon Disaster*, which has some similarities to the well known computer game *Doom*. The two previous examples demonstrate that Java is quite capable of providing both 3-D and interaction in a Web environment, but the amount of development required to produce such applets (as evidenced by the available source code) would not make it cost-effective for marketing products. Another problem is that applets, which are more than just trivial eye-catchers, can take a long time to load into a Web browser, and this is not attractive for business use.

Shockwave

Macromedia Shockwave is a *plug-in* for the Netscape Web browser. A plug-in is a piece of third-party software that is used in conjunction with the Netscape Web browser to add functionality. The Shockwave plug-in allows Netscape to run Macromedia Director multimedia presentations. These are interactive presentations with clickable buttons, video, and sound. A disadvantage with Shockwave is that it doesn't allow the user to explore a 3-D environment by navigating in and around it.

Quicktime VR

Quicktime VR is software that displays an interactive panorama on the Web. The panorama is derived from photographs of real scenes, and users can navigate around the panorama by clicking on the particular point that they wish to visit. Quicktime VR is good for architectural applications such as displaying the interior of buildings, and a nice example of this can be seen at the Calligrafix Web site. Navigation in Quicktime VR is restricted to jumping between prearranged viewpoints. It does not have the flexibility required to allow potential customers to investigate a product as they choose.

VRML 2.0

VRML, as its name suggests, is a modelling language that allows unrestricted virtual environments to be built. In other words, the geometry of a virtual world in VRML is not constrained to be orthogonal (as it is, for example, in the popular game *Doom*). The viewer of a VRML virtual environment, usually described as an *avatar*, is similarly unrestricted in the direction of travel through the environment. This makes VRML rather flexible. VRML 2.0 added interaction, animation, and sound to the solid modelling and navigation capabilities of the original VRML, making it more promising for marketing solid manufactured objects on the Web. The case study suggests that although VRML 2.0 has considerable potential for marketing, there are several problems with its use at the moment, preventing it from being acceptable in a real business environment.

VRML 2.0 Concepts

It is important to understand key concepts of VRML 2.0 to appreciate how it can be used as a marketing aid. This section explains relevant VRML 2.0 concepts.

Overview

VRML is a 3-D scene description language in ASCII file format. A Web browser requires a VRML 2.0 plug-in or helper application to render the 3-D scene described by a VRML 2.0 file. A user may navigate in and around a VRML 2.0 scene and may interact with it, causing it to display certain animated behaviors. A VRML 2.0 scene can contain links to other VRML 2.0 scenes and to html-based Web pages.

Basics

Conceptually a VRML world consists of a scene graph in which each entity is a node. Each node has a type name, fields, and in some cases, events that it can receive or send. The events carry information among nodes to allow animation and interactivity within a VRML 2.0 world. A scene graph consists of group nodes and leaf nodes in a hierarchical structure. Group nodes are containers for child nodes, which can be group or leaf nodes. Transform nodes are the most useful group nodes. They allow their children to be transformed or rotated in the coordinate system of their parent. If a transform node is at the top of the scene graph (i.e., it has no parent), then its children will be transformed or rotated with respect to the world's coordinate system. Leaf nodes describe the components that make up a 3-D world; for example, shape, lighting, color, sound, and 2-D text. VRML 2.0 has the cube, cone, cylinder, and sphere as its elementary shapes. Complex shapes can be built up from a combination of elementary shapes, or they can be described by *indexed face sets*. An indexed face set is defined by a number of points in 3-D space. These points are joined to form polygons (usually triangles), which combine to make up the complex surface or shape. Figure 15.1 shows a simple example of an indexed face set.

VRML 2.0 allows authors to define viewpoints, or locations in 3-D space from which a user may view the world in a preset direction. A number of viewpoints may be defined and then made available for a user to select using a browser interface control. Extensive use of viewpoints makes it easy for a user to investigate a 3-D world without having to grapple with the browser's navigation controls.

Interaction

VRML 2.0 provides a number of different ways in which users can interact with a scene. A user can investigate a 3-D world by moving among its viewpoints or by navigating into it using the VRML 2.0 browser's interface controls. A user can also traverse hyperlinks that connect parts of the scene to other VRML 2.0 scenes or to html documents. Finally, a user can initiate behaviors or animation in a VRML 2.0 scene using the mouse while the

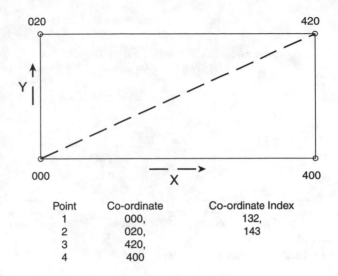

Point	Co-ordinate	Co-ordinate Index
1	000,	132,
2	020,	143
3	420,	
4	400	

Figure 15.1   A simple indexed face set.

cursor is over a variety of sensors. Sensors are not visible but are used to detect user input. Geometry sensors generate events based on user actions. The TouchSensor generates an event when the cursor is over an object and the user clicks the mouse button. Three of the geometry sensors; CylinderSensor, PlaneSensor, and SphereSensor generate events if the cursor is clicked and dragged when over an object. The dragging movement is translated into a translation of the object (in the case of the PlaneSensor) or a rotation (in the case of the other two sensors). Time sensors generate events depending upon the relative values of their start-time and stop-time fields. Time sensors are used in conjunction with geometry sensors to sequence key-frame animation.

Animation

VRML 2.0 supports key-frame animation. When a user triggers a geometry sensor, it generates an event that causes a time sensor to start. The time sensor then generates a stream of (time) events at regular intervals, and these are sent to an interpolator. The interpolator calculates data values for each moment in time and sends these data values to the node containing the object(s) to be animated. Depending on the type of data received, the objects will be translated, rotated, or change color, size, or shape. Animations can be programmed to run once only or to repeat themselves. Figure 15.2 illustrates the process.

Key-frame animation is suitable for simple linear animation. If more complex animation is required, for example, opening a door if it is shut and closing it if open, then scripting is required. Scripting opens up VRML 2.0 to the computational power of languages such as JavaScript and Java.

Authoring the 3-D Models

At the time that the case study was conducted, VRML 2.0 authoring tools were not yet available. Models could be authored in VRML 1.0 tools and then translated to VRML 2.0, or they could be authored from scratch by hand. The electronic bagpipe (or mini-pipe), is

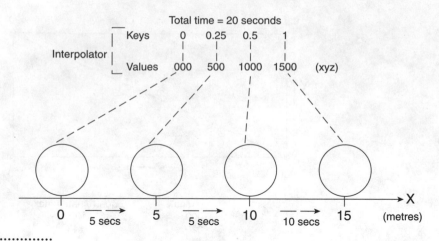

Figure 15.2 Key-frame animation.

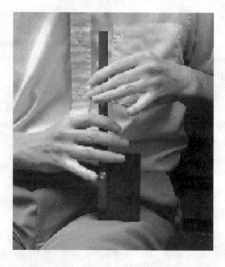

Figure 15.3 The mini-pipe in use. (*See color plate on page 329.*)

played by touching a number of finger pads that are set into a hollow tube known as a chanter (Figure 15.3). Authoring the model by the first approach required the use of sophisticated software such as *Alias Animator* to craft the complex shape of the holes in the chanter, and this presumed familiarity with the use of such software. In addition, the translation software was in beta release, the VRML mailing list reported that it suffered from bugs, produced huge files, and still required handcrafting in order to implement VRML 2.0 specific features such as animation. The second approach (authoring the models by hand) required the tedious calculation of many points in 3-D space to define indexed face sets for shapes that could not be represented by VRML 2.0 primitives. The second approach was judged to be the quickest and was adopted. Three objects were modelled: the mini-pipe itself, a pp3 battery, and a set of earphones. Figures 15.4 through 15.7 show the 3-D models.

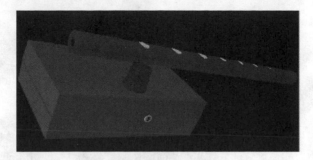

······················
Figure 15.4 VRML 2.0 model of mini-pipe. (*See color plate on page 329.*)

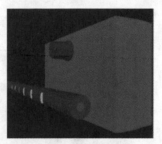

······················
Figure 15.5 VRML 2.0 model of mini-pipe. (*See color plate on page 329.*)

······················
Figure 15.6 VRML 2.0 model of pp3 9V battery. (*See color plate on page 329.*)

Complex Shapes

More than a thousand points were used to model the chanter section of the mini-pipe. Figure 15.8 shows a close-up of one of the finger pads set onto a hole in the chanter tube. These cut-out holes were the most difficult part of the mini-pipe to model. The difficulty was caused by having to represent the material thickness of the chanter tube. Solid models in computer graphics are represented by closed surfaces and are not actually solid. This is true whether the objects are modelled using polygon mesh surfaces (Foley et al. 1991a), as in VRML, or constructive solid geometry (CSG) (Foley et al. 1991b) (which is used in some ray-tracing systems). It is possible to represent a hole in a nonblack plane surface of

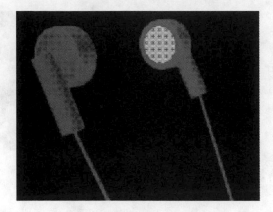

Figure 15.7 VRML 2.0 model of earphones. (*See color plate on page 329.*)

Figure 15.8 Close-up of finger pad on chanter. (*See color plate on page 330.*)

an object by positioning a black object on that surface. This technique was used to model the earphone socket hole and the hollow parts of the battery terminals, but it is not very realistic as the holes do not appear to have any depth. It was not possible to use this technique to model the finger-pad holes in the chanter as the finger pads had to be recessed below the outer surface of the chanter tube as can be seen in Figure 15.8.

Animation

Two components of the mini-pipe were animated: the lid of the battery compartment and the volume control knob. The battery and earphones were also animated. In all cases it was necessary to control the animation using scripts. The battery compartment lid opens and closes, and the battery will go into the battery compartment or come out of it only if the lid is open. The volume control knob turns up or down, and the earphones plug themselves into the ear phone socket. The models were developed for the Cosmo Player browser, which meant that the scripting language was VRMLScript (a subset of JavaScript). The biggest problem with the animation was working out 3-D coordinates for objects that were simultaneously rotating and translating. The local coordinate system rotates with each incremental rotation of the object, which means that the direction of translation along a particular axis also changes each time. The problem is compounded if rotation (with respect to the world coordinates) is required about more than one axis. A compromise solution is to complete any rotation before starting a translation, and this was the approach used for the battery compartment lid and earphone animations. The battery animation used simultaneous rotation and translation and required a great deal of trial and error to determine the path of the battery. Hopefully,

VRML 2.0 authoring tools that simplify this process will become available soon. Animation is acknowledged to be one of the most difficult aspects of VR authoring.

Strengths and Weaknesses of VRML 2.0 for Product Marketing

Strengths

Web Based

Nowadays, Web browsers are available on virtually all platforms and are used by many people with a broad range of computer expertise. Obviously, information about a product that can be displayed on Web browsers will reach a wide audience with the minimum of distribution effort. There is no real alternative to VRML on the Web for the widespread dissemination of interactive 3-D information.

3-D

With so much information available on the Web, one of the main concerns of Web authors is to attract users to their page. People find 3-D fascinating, and if there is sufficient user interaction built-in to a 3-D environment, this will not only attract interest but maintain it as well.

Relatively Easy to Create

VRML 2.0 is a scripting language. It is not necessary to know anything about computer graphics in order to produce interesting 3-D scenes. This makes VRML 2.0 attractive to content producers who will be coming more from an artistic than a computer science background.

Demonstration of Product Behavior

It is possible to demonstrate the functionality of a manufactured product using video embedded in a Web page. As far as the viewer is concerned, this is a passive activity. Engaging a user's attention through interactive involvement with the 3-D environment makes VRML 2.0 more attractive. The user and not the computer is in control!

Weaknesses

More Than One Browser Required

Additional software in the form of a plug-in or helper application is required in order to view VRML 2.0 scenes. VRML 2.0 was supposed to be a standard, but unfortunately, during the negotiation of the scripting part of the VRML 2.0 specification, there was a language war involving Javascript and Java. The result is that some browsers support only Javascript and other browsers support only Java. VRML 2.0 content authors tend to develop for a particular browser, and their files will not work in all of the VRML 2.0 browsers. It is to be expected that browsers of the future will support both Javascript and Java, but at the moment it is a real problem that most do not.

A Web browser will display a VRML 2.0 scene in its own window. In order to allow VRML 2.0 scenes to be embedded within, and controlled from an html file, the Java External Authoring Interface (EAI), was developed. Unfortunately, not all the browsers support the EAI, so it is quite common to download a VRML 2.0 file and find that it doesn't work in a particular browser. The lack of portability across VRML 2.0 browsers dramatically reduces the potential audience that authors can expect to reach and makes VRML 2.0 much less attractive for the marketing of products.

Unacceptable File Download Time

A number of factors contribute to the size of VRML 2.0 files. First, it is an ASCII file format. Second, complicated shapes may have a high polygon count. Third, textures tend to require a large amount of specification. Fourth, sound files tend to be large. The result is that if you have packaged your complicated product in a virtual environment with realistic textures, enhanced with sound, the user may have to wait 20 minutes for the scene to load. If your VRML 2.0 scene contains many links to Web pages or VRML 2.0 scenes that also have to be downloaded, your potential customer is going to lose interest quickly.

Slow Rendering of Complex Scenes

VRML 2.0 scenes are rendered in real time by the VRML plug-in or helper application. This can result in unsatisfactory animation due to the current limitations in the processing power of typical end-user computing vehicles. One of the worst effects is the loss of textures that can occur if a scene is changing rapidly. If a scene relies heavily on texture mapping to convey photo-realism, it is unacceptable for these textures to be replaced by minimal wire-frame outlines while the scene is moving. It is hardly surprising that the rendering capability of VRML 2.0 is limited when you consider that each frame in the computer-generated film *Toy Story* took up to 15 hours to render. Nevertheless, people's expectations are high, largely due to the excellent graphics quality in modern computer games. For users who do not have an appreciation of the technical difficulty of rendering 3-D scenes in real time, the results seem to be crude and not state of the art.

Time to Author Complex Shapes and Behaviors

Simple shapes that can be built up from the elemental shapes can be authored quickly and efficiently. More complex shapes, requiring indexed face sets, are time-consuming to author by hand. Authoring packages are available now that make the task of modelling complex shapes easier, but there are still areas where a certain amount of hand tweaking is required. One of the most difficult shapes to model is a hole punched through a solid object. Simple key-frame animation is likewise relatively straightforward to achieve and is available on at least one of the VRML 2.0 authoring packages. Complex animation on the other hand, requires much handcrafting. For inexpensive products, the cost of authoring a marketing aid using VRML 2.0 is going to be prohibitive for some time to come.

Wide Color Variation across Platforms

The rendering of color in VRML 2.0 scenes varies so much from platform to platform that a working group has been set up to try and address the problem. In some cases (e.g., 3-D games), obtaining the correct absolute colors may not be critical, although the aesthetic appeal of a scene may be adversely affected. For marketing products such as pictures, textiles, or clothing, it is essential that the colors created by the VRML 2.0 author are reproduced faithfully in the browsers of potential customers.

Awkward 3-D Navigation

The main reason why VRML 2.0 should be such an effective marketing aid is that it promises potential customers the ability to explore a virtual product. The interface controls on VRML 2.0 browsers are based on the concept of a viewer (or avatar), who navigates in and around a virtual environment, but the controls are so awkward to operate that navigating a VRML 2.0 environment is frustrating—not at all the situation that will encourage users to stay at a site. In addition there is no built-in interface that will allow users to manipulate individual objects within the 3-D scene. This functionality has to be programmed into the VRML 2.0 file. The next section reports on a study that was carried out to gauge users' reactions to VRML 2.0 navigation (Burton 1995–1996).

3-D Navigation Experiment

The purpose of this study was to ascertain the attitude of users towards 3-D navigation in VRML 2.0. An experiment was conducted in which subjects were timed as they navigated a 3-D road circuit (Figure 15.9). The main objective of the experiment was to ensure that each subject had a similar experience of 3-D navigation in VRML 2.0 prior to forming a judgment about its suitability for navigating virtual shopping malls. Also of interest was the degree of improvement (if any) between a subject's first and second lap of the course. Following the experiment, the subjects were asked a few questions relating to the system they had just used. They were also asked to choose a preferred environment for navigating a virtual shopping mall: VRML 2.0, a 2-D map, or a list of shops or products.

Subjects

Ten volunteers took part in the study, seven male and three female, ranging in age from 15 years to 47 years old. Four of the subjects were of school age (15 to 17 years old), three were between 20 and 40 years old, and three were over forty years old. Of the ten subjects, only two

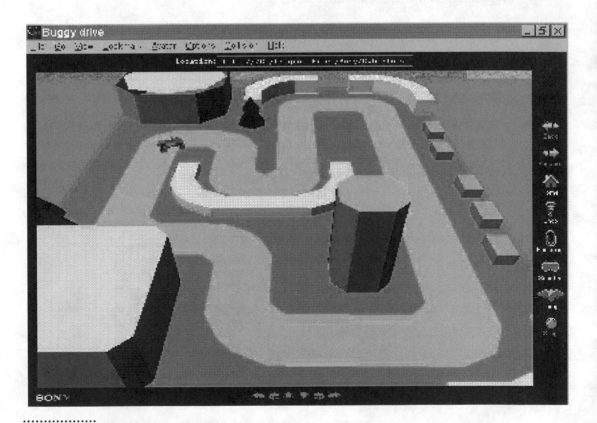

Figure 15.9   The course for the navigation experiment.

used a computer less than once a week, three used a computer on average once a week, and five used a computer every day. Two of the subjects were not British but spoke fluent English.

The Experiment

Every subject was given the same task. They were asked to navigate around a VRML 2.0 road circuit, and then after a pause of about one minute, they were asked to navigate around it again. The time taken for each complete navigation of the road circuit was recorded. No subject had experience of the 3-D course or VRML 2.0 prior to the experiment.

Materials

Elonex 486/66 PC with 16Mb RAM, running Windows 95\newline

Cyber Passage V2 beta 2 (Sony's VRML 2.0 Browser)\newline

Netscape 2.02\newline

VRML 2.0 file 'drive/main.wrl' from Sony Pictureworks

Results

Figures 15.10 and 15.11 summarize the preferred navigation method. Results for the navigation experiment are summarized in Figure 15.12.

Assessment of Results, 3-D Navigation

Figure 15.10 shows that a majority of subjects rejected 3-D navigation. This was not a statistically significant result as the chi-square values in Figure 15.11 show. The observed chi-square was 2.6, which is much less than the calculated value of 4.61 (for an alpha of 0.1). It is necessary to go to an alpha of 0.5 to get a calculated chi-square less than the observed value. This can be explained by the small number of people in the sample (for example, if the ratio of preferences was kept at 1:4:5, but 100 people had been sampled instead of 10, the observed chi-square would have been about 26).

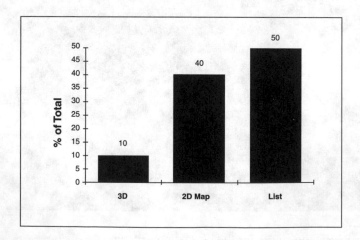

Figure 15.10 Summary of navigation preference for shopping mall.

Sample Size	10
Degrees of Freedom	2
Chi-square (observed)	2.6
Chi-square (tabulated, alpha = 0.1)	4.61
Chi-square (tabulated, alpha = 0.5)	1.39

Figure 15.11   Chi-square for navigation preference.

	Lap 1	Lap 2	Combined
Sample Size	9	9	18
Mean Lap Time (mins)	3.98	3.1	-
Sum of Squares	14.86	4.22	-
Variance	1.86	0.53	1.19
Standard Deviation	1.36	0.73	-
S.e.d.	-	-	0.52
t (measured)	-	-	1.71
t (tabulated, alpha = 0.1)	-	-	1.75

Figure 15.12   Summary of navigation experiment results.

The object of the study was to get a feel for people's attitudes rather than to prove a theory, so the statistical significance of the result was not a prime consideration. Initially, two subjects chose 3-D as their preferred navigation method, both were in the youngest age group and did so because they thought that it was fun. One changed after reflecting on the question (they hadn't initially related it to shopping). It is interesting that the only subject who stayed with 3-D as a preferred navigation "because it was fun" declined the invitation to do a second lap. Maybe it wasn't that much fun!

The results for improvement of lap time from the first to the second lap were also not significant. Figure 15.12 shows that the measured t value was 1.71 compared to a calculated value of 1.75 (with an alpha of 0.1). A possible reason for this is that some subjects tried to use the In-screen controls on the first lap and went wildly off the course (one started going the wrong way round the track). These subjects reverted to the safer Arrow controls for the second lap. The standard deviation for the first lap was 1.36, which is high compared to a mean lap time of 3.98 minutes. The second lap had a much more acceptable standard deviation of 0.73 with a mean lap time of 3.1 minutes.

Conceptually, the navigation controls were straightforward, which is why Cyber Passage was used in preference to Cosmo Player. In practice, the controls proved to be awkward and frustrating to use as is illustrated by the following selection of comments:

Subject 2:   "I certainly couldn't use the controls on the screen; I would get lost. I was glad there were two types of control. It's too slow for shopping malls."

Subject 3:   "This is difficult" (twice). "I wouldn't have the patience; I would find it very irritating."

Subject 5:   "It's not doing what I wanted it to do; I don't know where I am. Oh, here we are." (This subject went completely off the road a couple of times.)

Subject 9:   "Your concentration is on the navigation and not where you are."

The comments and results just discussed reinforce the belief that building a virtual shopping mall in VRML 2.0 would not be beneficial, at least without a major improvement in the navigation interface(s) offered by VRML 2.0 browsers. The situation will improve when the appropriate rendering software is implemented in hardware. A VRML 2.0 binary file format is being developed, and this should improve download times. It is not just speed though which needs to be addressed; the basic usability of the controls has to be improved. For marketing products it is desirable that potential customers can manipulate objects directly rather than navigating around them. The next section describes a prototype interface for this purpose.

VRML 2.0 Browser Interfaces

There is a need for a VRML 2.0 interface aimed at object manipulation rather than 3-D navigation. A prototype object manipulation interface is described.

Design of Custom Interface

The interface consists of four icons that represent four modes of object manipulation: translation in the plane of the computer screen, translation in a vertical plane going into the computer screen, rolling about an arbitrary point, and spinning about a single axis. In addition to the iconic representation of each mode, a textual description appears when the cursor is passed over the icon and disappears when the cursor is moved off again. An icon is highlighted to show that it has been selected. Once an icon has been selected, objects may be manipulated by clicking the mouse when the cursor is over the object and then dragging the mouse. The dragging movement of the mouse is converted into movement of the object according to whichever mode is current. The standard browser interface is still available if required, or it can be switched off. Figures 15.13 and 15.14 depict the custom interface within a sample VRML 2.0 file.

Ideally the architecture of a VRML 2.0 system should allow an interface implementation to be separated from the file that is being viewed. At the time of the project, not enough

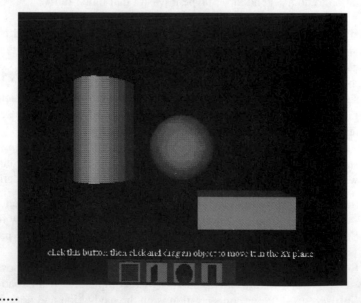

Figure 15.13 Object manipulation interface, mouse over first icon. (*See color plate on page 330.*)

information was available to allow a separate interface to be developed. The interface was therefore programmed into a VRML 2.0 world. It is possible to define an interface as a VRML 2.0 prototype that can be reused and adapted. This would be an improvement over the present situation, but it would still be an interface in a file.

The icons used in the interface are themselves VRML 2.0 primitive objects, and in order to get them to stay stationary with respect to the viewer, it is necessary to position them within the front half of the volume occupied by the viewer's imaginary avatar (i.e., at the current viewpoint). The icons must be in the front half of the avatar to ensure that they are not clipped when the browser renders the scene. In order to keep the icons tracking the viewpoint (or avatar position), a proximity sensor must be attached to the icons. The proximity sensor should be large enough to cover the entire virtual world so that the viewpoint will always be inside it. As the viewpoint moves around inside the proximity sensor, the sensor sends position information to the icons' transform node to enable it to move the icons in synchronization with the viewpoint. Figure 15.15 illustrates this concept.

The icon design attempts to convey the type of movement associated with each mode. The translation icons depict the plane in which movement occurs, and the rotation icons depict an object whose rotational behavior is similar to the rotation of the objects (e.g., a ball and a cylinder).

A small red bar appears at the top of an icon to show that it has been selected. Selecting an icon is achieved by clicking on it with the mouse.

In order for objects to respond in the appropriate manner, they must have a range of geometry sensors attached to them. The translation behaviors of the first two modes are achieved with a PlaneSensor. Sideways movement of the mouse is converted into sideways object movement for the first mode and object movement in or out of the screen for the second mode. A script node detects which mode is current and routes the translation information to the X or Z coordinate of the objects transform node accordingly. Up or down movement of the mouse is converted into up or down movement of the object in

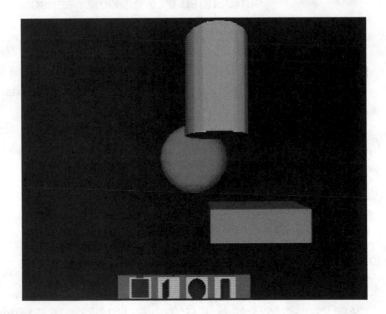

··················
Figure 15.14   Object manipulation interface, cylinder moved. (*See color plate on page 330.*)

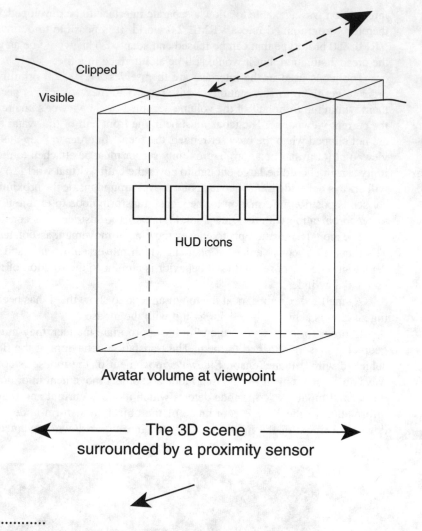

Clipped

Visible

HUD icons

Avatar volume at viewpoint

The 3D scene
surrounded by a proximity sensor

Figure 15.15 Object manipulation controls inside imaginary avatar volume.

both cases. For the rotational modes, movement of the mouse is converted into rotation of the object. In the case of the ROLL mode, the rotation may be about an arbitrary axis. The axis of the SPIN mode is fixed.

A switch node activates the appropriate icon and deactivates all others. A further switch node causes an appropriate text message to be displayed when the cursor is over a particular icon. When the cursor moves off the icon, the text message is removed.

Conclusions

A case study has shown that VRML 2.0 has the potential to enhance the marketing of products on the Web; however, at the present time there are too many problems for its use to be of real benefit. The main advantage that VRML 2.0 has over rival technologies is that it allows the exploration of a 3-D environment with movement in all degrees of freedom.

This gives a potential customer the sense of being in control. In addition the interaction and animation capabilities of VRML 2.0 can convey much information about a product's functionality. What limits VRML 2.0's usefulness as a marketing aid is that it is not cost-effective for inexpensive products, it is nonstandard across VRML 2.0 browsers and platforms, the rendering of color varies unacceptably, it is slow, and the browser interfaces are not user friendly.

References

"Applet Arcade," http://members.aol.com/shadows125/arcade.htm

N. Burton, *Interactive Solid Models for Enhancing Business on the Internet,* master's thesis, Heriot-Watt University, Edinburgh, UK, 1995–1996.

"Calligrafix," http://www.calligrafix.co.uk/horizons/tour.html

"Dungeon Disaster," http://www.igd.fhg.de/wiener/java/dungeon/doc/index.html

J. Foley, A.F. van Dam and J. Hughes, *Computer Graphics, Principles and Practice,* 2d ed., ch. 11.1, Addison-Wesley, 1991a.

J. Foley, A.F. van Dam, and J. Hughes, *Computer Graphics, Principles and Practice,* 2d ed., ch. 12.7, Addison-Wesley, 1991b.

"Java External Authoring Interface for VRML 2.0. http://vrml.sgi.com/moving-worlds/spec/ExternalInterface.html

W. Semich and D. Fisco, "Java: Internet Toy or Enterprise Tool," Datamation, March 1996.

"Virtual Rubik's Cube," http://people.taligent.com/~songli/cube/cube.html

"VRML 2.0 Binary File Format," http://www.research.ibm.com/vrml/binary/

"The VRML 2.0 Specification," http://vag.vrml.org/VRML 2.0/FINAL/

## Chapter 16

A Virtual Environment for Collaborative Administration

D. England

Connect Centre, University of Liverpool

W. Prinz

GMD FIT-CSCW Institute

K. Simsarian and O. Ståhl,

Swedish Institute for Computer Science

• •

Abstract

In this paper we describe a virtual reality interface to a distributed cooperative system for government workers. Our aim is to improve the levels of mutual awareness among colleagues by reinforcing the concepts of shared documents and showing some measures of sharing. The work is implemented using the DIVE virtual reality system using the VRML language. We also point out some limitations for our approach and suggest some future directions.

Introduction—Virtual Organization

The aim of a virtual organization is to provide support for members of one or more actual organizations who are physically distributed or otherwise unable to meet. While it cannot replace face-to-face meetings, it can assist in the coordination of tasks, such as shared document processing. In this paper we take the POLITeam system and see how a virtual reality interface might provide additional facilities for cooperation and coordination. The actual mechanism of support is achieved by mutual awareness of the activities of others in the virtual organization.

Another aim of the virtual organization is to blur the distinction between synchronous and asynchronous means of human-to-human interaction. In CSCW these two forms of interaction are usually classified separately and do not overlap. Thus, we have delayed, asynchronous means of interaction such as fax and e-mail, and synchronous means of interaction such as video conferencing and on-line chat systems. CSCW systems have also been classified as to whether they support interaction taking place in the same or a different location. Again, the idea of a virtual organization is to blur this distinction so that issues of location are transparent to the users.

Awareness in the real world consists of those cues which, though often not consciously realized, make coordination possible. Thus, speakers are aware of the alertness, or otherwise, of their audience, and controllers in the London underground (Heath and Luff 1991) are aware of the state of action of their colleagues. These real-world examples of awareness demonstrate the need for such a feature in computational environments. Indeed, the lack of the ability to monitor and communicate awareness is a factor leading to the relative bareness of virtual worlds compared with their real-world counterparts. Some classes of application positively discourage mutual awareness. For example, the concurrent users of database systems are unaware of each other's existence.

In order to counter the bareness of on-line worlds, models of awareness have been proposed. Existing models can be classified in three ways: models based on current user actions, those based on event notification, and those based on the structure of the application graph. What the proposals lack is a firm understanding of how users might interact with any awareness mechanism based on their models.

In the first model, exemplified by Mariani and Prinz (1993), users are operating in a database. When they begin to perform similar operations on related data items, they are given an indication by a bar chart that other users are also acting on those data items. The drawback with the implementation of this approach was the overhead of providing the monitoring.

Fuchs' et al. model (1995) uses event notification to inform users when an event happens on an object of interest. Thus, users can declare which objects and which events they are interested in. Notification may occur synchronously or asynchronously. The interaction question here is, How do we allow users to manage declarations of interest? For example, at some point in time the number of events of interest may be very large and overwhelm the user. What further filtering do users require on events notification?

Rodden (1994) proposes a general spatial model of awareness. The geometric spatial model (Benford and Fahlén 1993) arose from work in the DIVE virtual reality system and allowed users to control such factors as their mutual visibility and audibility to other users and objects. Thus, a user could set a narrow focus and hear and see objects within only a certain range in the same world. This was generalized by Rodden so that the measure of *distance* was not just a geometric measure of separation but could also mean the distance that separated users and objects as measured by their position in the graph of the application.

In other work, Bray (1996) looks at links between Web sites in terms of their mutual links and creates a 3-D map where the distance between sites indicates the degree of commonality. However, the links are measured on a site-by-site basis, which cannot be ascribed to any individual at those sites. In our work we are looking at how individuals can collaborate.

In our work the medium of awareness is the landscape of a virtual world, formed from the shared data from a group of collaborators. The hypothesis is that users gain an awareness of the activities of the group from the shape of the landscape and the placement of shared items within it. The shared data comes from the POLITeam project, which we describe in the following. We also examine some of the current problems of collaboration via POLITeam and suggest how our approach may solve some of those problems. Our interface work is implemented in the DIVE shared virtual environment, and we outline the architecture we devised for bringing together POLITeam, VRML, and DIVE. Finally, we point out some limitations for our current approach and discuss some future directions.

POLITeam Groupware in a Government Ministry

In 1991, the German government voted to move the capital from Bonn to Berlin. Not all federal agencies and ministries will move at once, which poses new challenges for the application of telecooperation technologies between the two cities. The POLIKOM research program was launched in 1993 by the German ministry for education, science, research, and technology to develop the necessary telecooperation infrastructure and tools. The POLITeam project is one of the projects within this larger research effort. The main goal of POLITeam is the development and introduction of a system supporting cooperation in large, geographically distributed organizations. In this paper, we describe our experiences with the introduction of the POLITeam system: the system, its method of introduction, the cooperative design process, and some early user reactions.

The Application Domain, a German Ministry

The POLITeam system is being developed for the support of individual office work and cooperative processes in a German ministry. To provide an understanding for our application environment, this section considers the scenario of preparing a speech for the minister. The scenario is suitable since it introduces the different hierarchical levels and the cooperation patterns applied within and between the levels.

The request for the speech is issued by the minister's office. Together with the request, additional background information that is useful for the preparation of the speech is collected in a circulation folder. The minister's office addresses this folder to the head of the department responsible for the speech topic. It is also common practice that the minister's office designates beforehand the sequence of addressees down to the unit level of the ministry.

Figure 16.1 illustrates that the processing of this request involves several hierarchical levels. The managers at the department and subdepartment level are each supported by a secretary, who performs additional tasks on incoming or outgoing documents. For example, this includes the sorting of incoming folders according to their priority, which is indicated by three different colors of the circulation folder. The manager at each level acknowledges the receipt of the folder (i.e., the speech request) and additionally provides comments or advice for further processing. Even though the subsequent addresses of the circulation folder are already predefined by the minister's office, the manager may reroute the folder or inform subsequent recipients about the process as the situation demands. Before the circulation folder is received by the units, its content is registered by the registry.

At the unit level of the ministry the speech is prepared in a cooperative process between different people of the same unit or different units. This process is not determined by any particular order. The person who initiates the cooperative process decides who should or must be involved in the production of a document or whether contributions to the document are collected in a sequential order or in parallel. Members of the unit level are supported by a writing office that types the handwritten or audiotaped drafts they produce. The cooperation process is characterized by a frequent exchange of documents between the members and the writing office if the modification of a draft requires the production of new versions.

After the creation of a typed draft version, it is sequentially processed by the managers of the unit, subdepartment, and department. All managers review the speech and additionally annotate the document with their own comments to the proposal. Before the proposal is forwarded to the next higher level, the managers approve the cover note that is attached to the proposal with their signature. A peculiarity of the system is that the comments and the signature are performed using a colored pen. The color corresponds to the role and

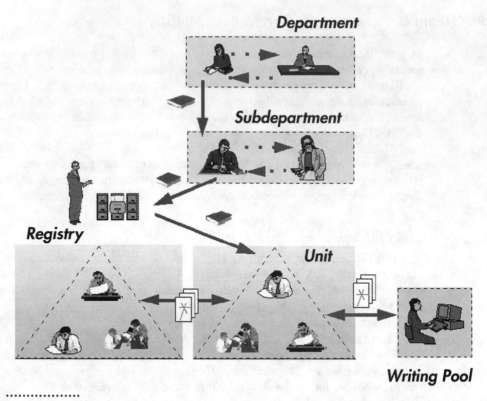

Figure 16.1   The flow of a circulation folder in a ministry. (*See color plate on page 331.*)

position of the manager in the hierarchy. If necessary, the manager may return the proposal to the preceding recipients, but more often questions or comments to the proposal are discussed by phone.

After the proposal for the speech has been finally commented and approved by the minister or his/her assistants, it contains a lot of handwritten comments. This document is then retyped by the writing office. Often its members use the previously stored draft version of the text and modify this according to the handwritten annotations on the paper document.

This scenario briefly discussed the major steps that are involved in the preparation of a speech for a minister. Although this describes just a special process out of a large number of different processes that are handled by a ministry, it is typical in the sense that most processes are treated in the same way: a top-down processing of documents through the organizational hierarchy, a cooperative production of a document at the unit level, and then a bottom-up approval process through the organizational hierarchy.

The POLITeam System

This section describes the base system and the two basic components of the POLITeam system.

The Base System: LinkWorks

The POLITeam developments are based on and integrated into the groupware platform LinkWorks by Digital (1995). Two reasons led to this decision: first, to start the development from scratch is not reasonable since commercially available groupware systems already provide a sufficient basic functionality; and second, the introduction of a research prototype into a ministry for the support of work processes is actually not appropriate. The decision to choose LinkWorks is determined by the fact that LinkWorks already supports the basic POLITeam functionality. Furthermore, it is extendible, which allows us to integrate additional functionality components, and it provides APIs that allow external applications access to the LinkWorks functionality and data. The current POLITeam system is based on the following modifications and extensions to the base system:

- *Tailoring* of the user interface: terminology, metaphors, available tools and menus
- *Configuration* of the overall system: document types, access rights
- *Adaptation* to the two POLITeam cooperation tools: electronic circulation folders and shared workspaces
- *Extension* by newly developed applications: electronic signatures, administration functionality for documents, workspaces, and circulation folders

The following sections describe in more detail the two basic POLITeam cooperation tools that have been realized with LinkWorks.

Electronic Circulation Folders

Initial interviews with the application partners indicated that a work-flow functionality that is similar to the circulation folder that is used in most organizations would satisfy the basic cooperation support needs. More sophisticated work-flow coordination mechanisms are not applicable in this organizational environment. Therefore, POLITeam provides electronic circulation folders as a flexible work flow coordination medium. It corresponds to the circulation folder that is described in the previous application scenario as the document transport medium (Figure 16.2). With that cooperation tool, we aim to support the cooperative processes in the ministry by the provision of a user-tailorable cooperation medium and not by the provision of predefined cooperation mechanism (Bently and Dourish 1995; Grudin 1994).

The electronic circulation folder is capable of transporting arbitrary electronic documents (i.e., spreadsheets, presentations, audio, or video objects). It is fully integrated into

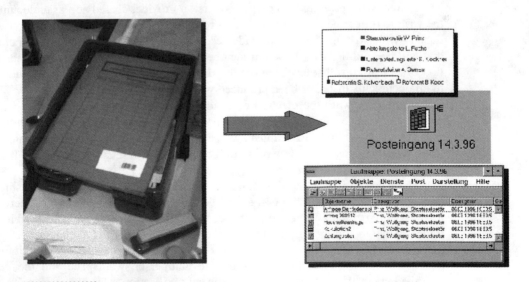

Figure 16.2   Real and electronic circulation folders. (*See color plate on page 331.*)

the POLITeam electronic desktop. Users add or remove documents by simple drag and drop operations between the folder, the desktop, other folders, and containers. Access to the documents is controlled by access rights that determine the operations that users can perform on the folder (e.g., the removal or inclusion of new documents) and its contents (e.g., reading or modification of documents).

The route of the circulation folder is described with a configurable circulation slip. It sequentially lists all recipients of the circulation folder. If a specific work flow requires parallel processing of a document, this can also be expressed. The contents of the circulation slip can be modified easily by the recipients of the circulation folder. This allows users to react flexibly on situations that require a change to the described route of the circulation folder.

Shared Workspaces

Whereas electronic circulation folders are mainly used to support the structured cooperation between the different levels of the hierarchy, shared workspaces (Agostini et al. 1995) support the cooperation among members of a unit or different units and the writing office. The POLITeam workspace provides its members with a virtual space for cooperation and communication. According to the cooperation patterns studied in the application domain, we differentiate between a person responsible for the workspace and "ordinary" members. For initialization, the user creating the workspace is responsible for it, but may delegate this task to another member of the workspace. Only the responsible person is allowed to invite new members to the workspace or to exclude an existing member by withdrawing the share. The decision to leave a workspace is left to each member.

Since the coordination of the activities within a workspace is not predetermined, but left to its members, it is important that other means of coordination support are integrated into the workspace. The POLITeam notification and information service provides its members with shared feedback about ongoing and past activities of their cooperation partners as far as cooperation issues are concerned (Bray 1996; Sohlenkamp and Chwelos 1994). This information allows the members to react on the activities of each other.

Some Problems with Collaboration

The shared workspace has two purposes: First, it provides shared access to documents for the writing office and the unit members for the following: to exchange a text initially produced by the writing office; to provide access to a text for the production of a clear copy that will be modified later by a unit member; and to allow access to a text for the production of a clear copy that has been produced by a unit member. Second, it provides access to all documents that have been produced within the unit to the unit leader.

In the case when the electronic version of the document is modified and then forwarded on paper for further processing (the management levels are not yet equipped with POLITeam), the writing office must incorporate the annotated paper changes into the electronic version. Therefore, the writing office is required to retain access to the latest electronic version of the document.

The solution in practice is as follows. Documents are exchanged by all using shared workspaces. For each member of the unit, a workspace exists, which is shared between this member and members of the writing office. All these workspaces are contained in another folder, called the unit folder. This allows the writing of organized documents according to the units and members of a unit. Whenever a document is produced by the writing office, this document is placed in the appropriate unit/person workspace and the person is informed about the new document by an event notification. This convention for how the shared workspaces are organized is logical for the writing office members' work process.

It is determined by the name of the user (unit member) for whom it was typed and the date of production. We refer to this as the *order* view. The sorting criteria for the documents in a workspace is the date when the document was typed.

This solution seemed appropriate at the beginning of the POLITeam use. But after some time of practical experience using this file organization, the following problem occurred. The unit members wanted to organize their documents according to their work processes (i.e., they wanted to collect documents produced by the writing office in a task or process-specific folder, rather than according to unit members). Their sorting criteria for documents in a workspace depends not on the date of production, but rather on the content of a document (e.g., a speech on a transit problem, or on an economic issue). We refer to this as the *work-process* view. With the current configuration only one view, the order view, is supported.

Thus, we found that users prefer to structure their information using their own pattern. We also found that the different views of the users corresponded to different level structures: the typists prefer to use a relatively flat information structure, and some of the unit leaders prefer a deep multilevel structure. The problem for the group arises when the different users collaborate in a shared workspace, which requires one common information structure for the groups' documents. Some users often could not find documents among the vast array of information in the shared workspace because the system supported the order view, which was not their view.

A second example of a convention is one set via a workshop: work must occur in the shared folder (i.e., a document must not be removed). However, in practice, the unit member would drag the document out of the folder shared with the writing office into their task specific folder. After that operation, the writing office no longer had access to the electronic document, which was crucial for changes. Thus, the unit member violated the conventions that were established for the use of the shared folder. Even documents that are produced only by the unit members must be stored in the folders shared with the writing office to provide an electronic version for the clear copy production.

The Difficulty of Setting Conventions

The groupware environment lacks cues for conventions. One may walk into a conference room and know at once from observing the environment how one should behave. Cues, such as type of dress, posture, tone of speaking, and so forth, immediately provide information on whether the group is formal or informal, what type of language is appropriate, and whether the group might be following strict rules of procedure or not. A groupware environment lacks such cues, making it difficult for people to know what is appropriate behavior to use with the system. For example, a novice e-mail user might include a top-level manager on a distribution list, not realizing that the manager would then become bombarded with messages.

Technological Support

Technological support can be applied to target the difficulties identified. It can facilitate convention use in three ways. First, support can be given in the form of awareness. Second, for many routines, default conventions can be implemented into the system. Third, support can enable users to retain individual conventions.

Awareness and Convention Use

A groupware system, such as a shared workspace, is a social environment. Here, people's actions are not independent. For example, one may make changes to a shared document, which are visible to all others. Not only do people's actions interact with one another, but one's actions can have dire consequences on others. If a document is removed from a shared workspace, then others cannot find it. If someone rearranges files within a folder, creating subdirectories, then others will not readily find them. If someone renames a document, then another user will have serious difficulties.

Many groupware systems act like single-user systems; we can see our own actions, but we are unaware of the actions and their effects from other users. A multiuser database is an example of such a system. We can develop individual conventions, such as search strategies, but group conventions are not needed to work with the system. However, a shared workspace is an example of a groupware system where each one's actions and their effects are experienced by those who are cooperating together.

Providing Cues for Conventions

The system should also provide cues about the appropriate behavior in a shared workspace to support conventions. Four levels of support can be distinguished: signalling conventions by *visualization; notification,* and *provision of feedback* about action, to support the active learning of conventions; and *automatic enforcement* of conventions, to avoid inappropriate user behavior.

The visualization of conventions through different workspace appearances is an implicit way of signalling conventions in a shared workspace. This can be achieved by the application of appropriate interface metaphors (e.g., rooms). Thus, whenever a user enters a shared workspace, the appearance of that workspace signals its type and the conventions that apply to that type of workspace. This makes it easier for the users to distinguish between different workspace types and to identify immediately that they have entered a shared workspace.

It is interesting to note that the metaphors applied by our groupware platform LinkWorks all stem from single-user objects as shown in Figure 16.3 (e.g., cabinets, folders, drawers). These are appropriate for the organization of a single-user environment, but do not fit for the organization of a shared environment. The semantics of these metaphors are applied to govern and control how containers are nested. They have no implication for their use as shared workspaces, nor do they change their behavior when they are shared. The only difference in the appearance between shared and nonshared workspace is the use of a different color for the name. Our experiences show that this is not sufficient, since often users were not aware whether the workspace they are using is shared or not.

The proposal to investigate a 3-D interface for POLITeam aims to support the convention of shared and nonshared workspaces. This is done by providing public and private spaces. Documents placed in the public areas of the 3-D landscape are available to all members of that landscape. The idea is to provide a mechanism for users whereby they can mark their documents as shared or not and thus overcome some of the problems described. This can be taken further as with earlier work in the DIVE system where different barrier objects were used to indicate the accessibility of objects. So, for example, an object behind a virtual pane of glass can be seen but not selected, and if an object is behind frosted glass, the user is aware of its existence but cannot get further details.

Supporting Individual Conventions

When individual conventions may be difficult for users to give up, a technical solution is to enable users to retain their individual conventions. Not only does this offer the large advantage that users may keep conventions that might logically relate to their task, but also when users must assume other roles, such as substitutes, they can manage other users' files in a shared workspace using their own familiar routines.

In our case, different views to the shared documents were required by members of the writing office and unit members. The writing office prefers the order view (i.e., the

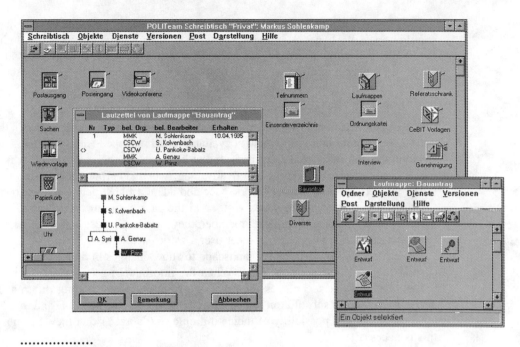

....
Figure 16.3 The POLITeam desktop.

organization of documents according to the employee). The unit members prefer the process view (i.e., the organization of documents according to their type of work process (see Figure 16.1). The order view requires a two-level structure of the documents according to the orders. The process view requires a deeper, fine-grained structure. It requires that documents are further organized into folders, where each folder represents a distinct process. However, this structure is not acceptable for the writing office since they have no knowledge about the structure. Therefore, a directed browsing through the structure is not possible. The provision of a search facility is also not acceptable since a document is remembered by its location, and the name of the document is not exactly known.

The provision of different views in the 2-D interface is possible by the creation of additional aliases. By placing the alias of a document in a task-specific folder, the unit members can organize documents according to their individual views. However, subsequent changes to the document are still accessible for the writing office. Thus, the alias provides a translation between the different views. Aliases can similarly be provided in a 3-D environment.

We must now consider some problems that are raised when users have different views on shared information. The major concern is that although a document is shared within a group, everybody has a different view to the document without a common reference (i.e., without a common group view). For example, a member of the unit and a typist may refer to the same document differently, because they see the document from different views (i.e., different folders or workspaces). We must provide functions that map the individual views to a common group view. Two aspects must be considered for the common reference: the identity and locality of shared objects in the individual views.

• *Object identity:* Users must be able to confirm whether two objects they are referring to are identical. This can be achieved either by a unique object share that is accessible by the users, or by a function that tells the users whether the object they both have currently selected is the same.

• *Object locality:* To become aware of the structuring preferences of other users it should be possible to get information about the locality of a shared object on another users' desk. This will allow a better understanding of other users with whom they collaborate. However, a user might not want to publish certain information (e.g., if a user placed a shared document in the wastebasket of his/her electronic desktop). Therefore, it must be possible to assign appropriate access rights.

Thus, users can retain their own information structure and still collaborate in the shared workspace. A common group view is available that is independent of the number of, and configuration of, individual user views. Individuals can use their own views, which is cognitively easier for them. This is a large advantage when (1) group membership is dynamic (i.e., members can join a shared workspace and adapt a file structure to their own needs), (2) when the information becomes vast, as in the case with our users, and (3) when one member must substitute for another member, they can simply access their own information structure and retrieve the file in another user's workspace.

We can imagine adapting the landscape to show changes in membership, thus broadcasting these changes to other users. When the information grows too great in the landscape, we must use techniques for partitioning the word and hiding different levels of detail. Membership substitution can be carried out the same way in 3-D and in 2-D, but we have the additional possibility of indicating on a user's avatar that they are substituted if this is necessary.

DIVELink: System Overview

DIVE is a multiuser virtual environment system that enables geographically dispersed users to meet and interact inside a 3-D graphical world. The 3-D world is constituted of objects of various types that are stored in a replicated database, one copy for each DIVE process. Initial object descriptions can be in the file formats VRML and/or VR (a custom file format). Each process has the ability to manipulate and change the database and the objects and may also in this way control how the objects behave in the virtual world. All manipulations to the database are done by one process and are immediately distributed to the others, so that the database is always kept synchronized.

The database objects may have scripts associated with them. These scripts are stored in the database together with the objects and executed when certain events occur. For instance, scripts may be run when a user interacts with an object or when two objects collide. Since scripts typically involve operations that change object attributes in different ways, scripting is another way of controlling object behavior. The advantage of using scripts to control objects, compared to using a process, is that they can be defined in the object data files (i.e., the data files that define the objects graphical representations). This means that the data file contains both the visual and behavioral description of the object and any process parsing the file will be able to execute the object's correct behavior.

In the POLITeam/DIVE project, DIVE is used to visualize the 3-D representation of the 2-D POLITeam desktop. The workspace representation is a VRML model of a desk and file cabinets, complete with drawers containing documents. One goal of the project is to add behaviors to the desktop objects so that it would become possible for users to open a drawer, take out a document, read the document, and put it back again. Furthermore, operations performed on POLITeam objects in the 3-D world were to be directly visible on the 2-D desktop, and vice versa. This meant that if a DIVE user opened a 3-D drawer, a POLITeam user would see the opening of the corresponding 2-D folder, and the other way around.

To handle the behavior control of the 3-D POLITeam objects and also communicate events back and forth between DIVE and POLITeam, the DIVElink application was developed. DIVElink is a DIVE application that when started, finds the objects in the DIVE database that represent the POLITeam desktop objects and then assigns scripts to each one of them. The scripts define how these objects will behave when users interact with them in the 3-D world or when POLITeam users interact with the corresponding objects in the 2-D desktop world. There are different types of scripts for the different types of objects (e.g., drawer and document), and each script implements different operations that may be performed on that particular object type. For instance, the drawer script implements the behavior of an open and a close operation, and the document script also adds a delete behavior.

DIVElink also sets up a connection to the POLITeam system, thus enabling messages to be sent in either direction whenever DIVE or POLITeam detected a user interaction. A protocol was developed to allow the two systems to exchange information. The protocol defined a number of message types, one for each possible operation, and the unique 65-character-length POLITeam object identifier string was used to identify objects in both environments. DIVElink has access to these identifier strings since they were part of the initial VRML model parsed into DIVE. This unique identification makes it easy to synchronize the operations that are performed on an object.

User interactions in DIVE are mostly performed by clicking on 3-D objects with the mouse pointer. In some cases interaction also involves mouse clicks to select the object and then menu selections to perform an operation on the selected object. The clicks trigger the execution of associated object behavior scripts, which initiate the object behaviors as well as make DIVElink aware of the initiate operation. This results in a message being composed and sent off to POLITeam where the appropriate visualization of the event is performed. In POLITeam, interactions are also performed by mouse clicks, for instance, the click on a folder to open it. In similar ways, such interactions lead to a message being sent from POLITeam to DIVElink. Whenever DIVElink receives such an event message, it uses the POLITeam identification string in the message to find the corresponding DIVE object in the database. This explicitly triggers the object behavior script and allows the object to react in accordance with the specific operation.

Shaping the Landscape for Groups

Our initial virtual office took a simple approach of taking objects from a user's POLITeam desktop. These were translated into VRML objects such as a desktop, mail boxes, documents, folders, drawers, and cabinets as shown in Figure 16.4. The VRML is in a form suitable for DIVELink and enables interaction on the objects so that drawers and folders can be opened and the documents accessed. Additionally, the user can access the document from the LinkWorks database by selecting the URL of the representative object in the 3-D world. The generation of this simple world is controlled via a CGI form where the user is allowed a limited amount of tailoring over the positioning of the cabinet objects inside the office.

In our next virtual world, the pattern of the landscape was formed by a graph of the shared workspaces, where the location of each workspace from its neighbors is decided by some factor of similarity. Each POLITeam object has a value for its creator and later editors. From this attribute we can synthesize a set of members for container objects such as cabinets, drawers, and folders. We can use shared membership among container objects as

Figure 16.4   An initial virtual POLITeam office. (*See color plate on page 331.*)

a factor of similarity. Other factors that can be used include dates of some activity (updating, reading, deletions, adding/deleting members, versioning), the type of some activity, and so forth. Thus, in Figure 16.5, the pattern shows a group of workspaces located according to the number of members they have in common.

Another user coming into this world and seeing this arrangement would see the workspaces he or she was interested in and others with similar membership. For example, in Figure 16.5, workspaces A and B have a greater shared membership, indicated by the distance D, than that between the other workspaces. The user might decide it is worthwhile to examine those workspaces.

Limitations

There are some drawbacks with our approach that limit its current usefulness. For example, we are taking data that users have originally organized in two dimensions and rearranging it in three. First, the original 2-D arrangement is usually chaotic and direct conversion to 3-D layouts can produce messy displays. We have chosen our demonstration data to avoid this. Second, the users may have difficulty in relating their original objects to their 3-D counterparts.

Also, the size of a typical desktop of one of the POLITeam development members was two to three thousand objects. Directly mapping all these objects into 3-D produces unmanageable scenes. Thus, we have to employ some mechanisms for hiding details and partitioning the worlds.

The degree of interaction with the 3-D objects is limited. This is partially an indication of the state of our implementation. Only a limited number of the underlying POLITeam functions are currently available in the DIVE interface. It is fairly straightforward to map some of the POLITeam operations into 3-D interactions: actions such as deletion and copying and moving of objects. However, actions requiring text interaction have to be done with the original documents or via pop-up dialogue boxes.

A further concern is the tension between the requirement for consistency in an interface and the evolving nature of the landscape. We might imagine that the landscapes drawn

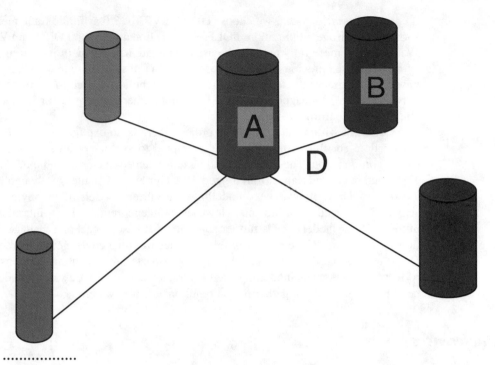

Figure 16.5   Workspaces positioned by a common factor.

from POLITeam may evolve slowly and thus users can cope with the pace of evolution of the landscape. However, for other classes of application the evolution may be more rapid and even dramatic. In rapidly evolving landscapes, the emphasis on landscape layout needs to be altered to focus on the indication of change rather than on measures of similarity as we are using at present.

Finally, the distribution of workspaces by similarity factors is not in itself sufficient for making a navigable landscape. We still need other techniques (Ingram and Bedford 1995) in order to annotate the landscape and highlight features and areas of interest.

Future Directions

We can first deal with the current limitations. The tension between 2-D and 3-D working is one that has faced interaction designers in a number of situations. We need to prove the benefits of the 3-D interface to users in order that users might choose that mode of interaction where appropriate. For example, we could provide an opening 3-D world that gives workers an overview of the current situation in their unit. They are then free to navigate this world and switch to a more detailed textual view when necessary. This would also help manage problems of scale. Currently, due to limitations in the coverage of VRML, we are not able to take full advantage of its detail-hiding features. Similarly, we can make more use of world partitioning to reduce the size of individual worlds.

Currently, we can output VRML 1.0 and VRML 2.0, but we have not yet made full use of the behavioral features of VRML 2.0. Here behaviors are written in Java rather than TCL/TK, and supporting browsers are now available for the class of machine used by current POLITeam users.

Regardless of whether we choose DIVE or VRML 2.0 as future implementation plat-forms, we envisage adding more POLITeam functionality to the POLITeam VR interface. We could also more fully explore the use of animation as a way of indicating features.

As we stated previously, our approach is not a full solution to making VR worlds more navigable. It tries only to add features of awareness in order to increase cooperation between on-line colleagues. We need to explore other techniques for legible landscapes (Ingram and Benford 1995).

Currently, within the POLITeam project, there is no explicit awareness logging and notification. For this project we have attempted to synthesize awareness data from the existing information. The next step will be to implement the proposed model (Fuchs et al. 1995) and build landscapes from logged data. Our biggest challenge, though, comes from the state of the art of awareness modelling. The current models fill an obvious gap in pre-sent on-line shared environments. However, problems remain. First, interaction with the output of these models needs further study: How do users react and manage the informa-tion they are given? The second problem comes from the coverage of awareness models. Questions still need to be asked about what aspects of application domains need to be reflected in awareness modelling. Perhaps rather than explicit awareness modelling, we need better methods for gathering the requirements for awareness.

References

A. Agostini et al., "Contexts, Work Processes, and Workspaces," *Proc. COOP'95,* INRIA, ed., Le Chesnay Cedex, 1995, pp. 219–238.

S.D. Benford and L.E. Fahlén, "A Spatial Model of Interaction in Large Virtual Environments," *Proc. 3d European Conf. CSCW (ECSCW'93),* Milano, Italy, Kluwer Academic Pub., Dordrecht, 1993.

R. Bentley and P. Dourish, "Medium Versus Mechanism: Supporting Collaboration through Customisation," *Proc. ECSCW'95,* Sept. 11–15, Stockholm, Sweden, Kluwer Academic Pub., Dordrecht, 1995, pp. 133–148.

T. Bray, "Measuring the Web," *Fifth Int'l World-Wide Web Conference,* Paris, France, 1996.

L. Fuchs, U. Pankoke-Babatz, and W. Prinz, "Supporting Cooperative Awareness with Local Event Mechanisms: The GroupDesk System," *Proc. ECSCW'95,* Sept. 11–15, Stockholm, Sweden, Kluwer Academic Pub., Dordrecht, 1995, pp. 247–262.

J. Grudin, J., Groupware and Social Dynamics: Eight Challenges for Developers," *Communications of the ACM,* Vol. 37, 1994, pp. 92–105.

C. Heath and P. Luff P, "Collaborative Activity and Technological Design: Task Coordination in London Underground Control Rooms," *Proc. ECSCW 91,* L. Bannon, M. Robinson, K. Schmidt, eds., Kluwer Academic Pub., Dordrecht, 1991.

R. Ingram and S. Benford, "Improving the Legibility of Virtual Environments," *Proc. 2d Eurographics Conf. Virtual Environments,* Monte Carlo, Jan./Feb., 1995.

LinkWorks User Manual, Digital, http://www.digital.com/info/linkworks, 1995.

J.A. Mariani and W. Prinz, "From Multi-User to Shared Object Systems: Awareness about Coworkers in Cooperation Support Object Databases," *Informatik, Wirtshaft, Gesselschaft,* H. Reichel (Hrsg.), Springer-Verlag, 1993.

T. Rodden, "Objects in Space, the Spatial Model, and Shared Graphs," *COMIC Document 4.2, The Shared Object Service and Prototypes,* Oct. 1994.

M. Sohlenkamp and G. Chwelos, "Integrating Communication, Cooperation, and Awareness: The DIVA Virtual Office Environment," *Proc. CSCW'94,* Oct. 22–26, Chapel Hill, N.C., 1994, pp. 331–343.

• About the Authors •

David England is a lecturer at the University of Liverpool, UK, based in the Connect Internet Centre. He was previously a research fellow at the GMD FIT.CSCW Institute in Sankt Augustin, Germany. His PhD, from Lancaster University, is in the area of graphical support for the specification of interaction.

Wolfgang Prinz is a project leader at the GMD FIT.CSCW Institute in Sankt Augustin, Germany. He is currently leading the work by GMD on the POLITeam project. His PhD from Nottingham University, is entitled, *A Framework for Organizational Information Support in Cooperative Environments*.

Kristian Simsarian is a researcher in the DCE group at the Swedish Institute of Computer Science. He is also a doctoral candidate in the Computer Vision and Active Perception group at the Royal Institute of Technology in Stockholm, Sweden.

Olov Ståhl is a researcher at the Swedish Institute of Computer Science, where he is involved in the development of the DIVE system. His research interests include design and implementation of distributed virtual reality and simulation systems, behavior modelling of virtual worlds, and computer graphics.

## Chapter 17

Collaborative Theatre Set Design across Networks

Ian J. Palmer and Carlton M. Reeve

EIMCU, University of Bradford

. . . .

Abstract

This paper examines collaborative design of theatre sets over the World-Wide Web (WWW).

Historically, theatre sets required paper plans or low-tech models for implementation. However, these plans are hard to visualize, and the models are difficult to make and awkward to alter. Neither design method can accurately recreate the desired lighting conditions or give any feeling of movement or "sightlines" within a theatre space. Also, with the plans drawn many weeks before the actual construction of a set, the performers struggle during rehearsals because of a vague understanding of their ultimate surroundings.

The Internet and its associated technologies of WWW and VRML enable set designers to create fully working models. These virtual environments are accessible from anywhere on the computer network, and VRML 2.0 scripting enables the switching between scenes. Not only does this allow different lighting states, it also offers the opportunity to view scene changes in context. These virtual sets allow the designer, director, and producer equal access to the model and provide a simplified mechanism for its development. Once constructed, these designs enable a cast to appreciate their scenery before its physical construction, thereby aiding characterization and confidence.

The Theatre in the Mill is a typical small-scale regional theatre. It is part of Bradford University's campus. The "Mill" is home to an active drama group that comprises professionals, members of the local community, and students. Each theatre season hosts visiting performance companies and three productions by the resident drama group. With all such venues, visiting companies need detailed information about performance space. Traditionally, this information is paper based and usually requires a physical visit. This is the project's second target.

The nature of VRML facilitates the placing of particular sets within an existing model. This ability greatly increases efficiency by reducing familiarization time for a space and identifying potential problems before a company arrives.

Figure 17.1 The Theatre in the Mill.

This interactive VRML model of the Theatre in the Mill and its recent use in a production of A Midsummer Night's Dream significantly enhances the production process and initiates a system that eases the installation of future visiting shows.

Introduction

As the name suggests, Theatre in the Mill (Figure 17.1) was once part of a textile mill. The stage occupies the upper level of the two story building. It is, in essence, a "black box" performance space: a studio theatre. It does not have a proscenium arch, and all the fixtures and furniture within the space are movable. This means that, potentially, the whole of the 13.7 m × 12.2 m area is available for performance. In general, however, with a conventional seating arrangement for an audience of 100, the available space is approximately 6 m × 11 m. To accommodate the maximum 140 people, the portable seating rigs reduce this space still further. The Mill's performance area has four entrances: two are internal, giving access to the ground floor, and the others open to the outside. The main internal entrance is the audience door to the auditorium. The second is a spiral staircase coming up from the dressing rooms. The external openings are a fire exit on the east side and an old loading door from its days as a mill on the west.

Traditionally, paper plans of the Mill's performance space have been used as a starting point for set design and adaptation. The standard plan used for this purpose is shown in Figure 17.2. If a local theatre group is designing the set, typically they will have access to the Mill and can visit it to help the design process. However, if the Mill is already being used for another performance (as is often the case), companies cannot test actual sets until the current show has ended its run. This may lead to tight deadlines if unforeseen problems necessitate redesigns.

The problems are magnified when the theatre group is not local. The paper plan, possibly together with interior photographs, may be all that a set designer has to work with initially. For a travelling company to effectively use the performance space, a visit to the Mill prior to decisions about the set is virtually essential, but this may be difficult to arrange, and multiple visits may be impractical due to work schedules or the distances involved.

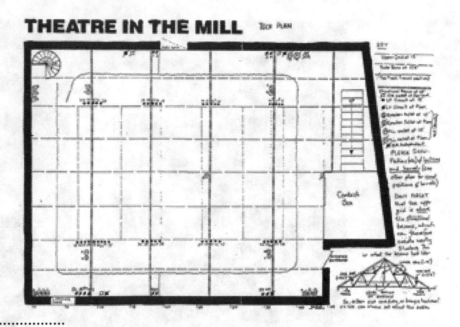

Figure 17.2   The interior plan.

In both these scenarios, what is required is that set designers have access to the performance space and are able to try out ideas before committing to physical set construction. It is with this aim in mind that the work described in this paper was carried out.

The Model of the Mill

It was decided to model the Mill as a virtual environment as this would allow maximum interaction with the space and the ability to visualize the interior. The structure of the theatre has both advantages and disadvantages for the modelling process. VRML 2.0 was chosen as the implementation language for the following reasons:

- The implementation was to be made available over the WWW, and VRML was designed from the outset with this in mind
- The addition of scripting in VRML 2.0 (using VRMLScript) allowed the interaction required for the model

The 3-D model was constructed jointly in Alias/Wavefront *Power Animator* and SGI *Cosmo Worlds*. *Power Animator* allows exporting to VRML-enhanced inventor files and was used because of its powerful modelling facilities, while *Cosmo Worlds* produces VRML 2.0 files directly and was used to add the interaction.

The Main Area and Seating

With few fixed objects within the theatre, the shell is essentially an open-top cuboid. Besides using the model to visualize the interior, it was essential that the set designer can access data about various aspects of the space. The VRML model has all the building

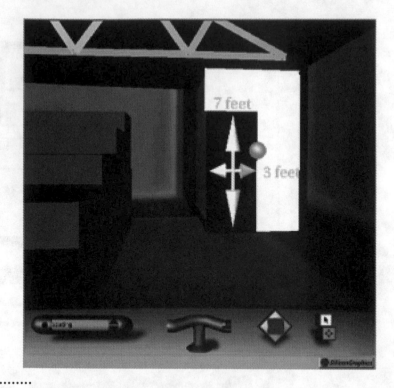

Figure 17.3 View of doorway showing dimension display. (*See color plate on page 332.*)

measurements available in it. Dimensions are available at any time by clicking on the appropriate button. Figure 17.3 shows the horizontal dimensions of one of the doorways being displayed. All the entrances have their dimensions attached so that incoming companies can decide if the Mill can accommodate their sets. (It is worth noting, however, that the very nature of touring theatre shows requires easily transportable, modular sets and properties.)

Due to the Mill's extreme flexibility in housing shows of all shapes and sizes, the model reflects the ease by which the layout can alter. All four seating rigs are independent objects that can move around the space. The designer can open them out so they can be used or retract them to save space. In this way the Mill is suitable for all kinds of performance, from traditional fourth-wall theatre to promenade. Examples of the seating arrangements are available within the model.

The Roof Space and Lighting

Since it is a converted building, the roof presents a number of problems. It is an A-frame construction, a little over 3 m above the floor (Figure 17.4). Not only is this quite low for a performance space—restricting the height of the set—the complicated nature of the supports makes rigging the lights difficult because of shadows. It is the most intricate part of the model. The lighting racks and barrels are mainly among the five A-frames, so it was important to have all the struts and supports accurately represented. Figure 17.5 shows these as they are modelled in the VRML model.

Figure 17.4   Roof A-frames.

Figure 17.5   General view of model showing A-frames. (*See color plate on page 332.*)

Controlling the Lighting Model

One of the key elements and strengths of the model is the ability to manipulate the lighting of the stage. Lighting is key to the overall feel of a performance. Lighting "dresses" the set and gives it a sense of emotion. Obviously, the type of lighting depends largely on the style of show. Naturalistic shows require realistic lighting—the inside of a sunlit room during the afternoon for example. More abstract pieces use harsh, surreal lighting techniques where particular colors might represent certain characters or feelings. It is crucial that the light condition is sympathetic to the tone of the scene and the portrayal of the script by the performers. For example, bright blue lights make a stage appear very cold. Blue light is inappropriate for a loving tender moment but improves a scene of rejection.

Although at present the lighting model is a subset of the full rig, ultimately the model will offer complete control over the position, direction, color and strength of all the lights. With the complicated and unconventional ceiling construction, shadows are a key consideration. In most purpose-built theatres, the lighting rig is unobstructed, thereby allowing the maximum flexibility in lighting design. In the Theatre in the Mill, the A-frames can cause some problems. This issue of shadows significantly reduces the freedom of a designer and is very important for visiting companies who might require very distinctive and tight lighting. With the working model, designers can position particular lights in specific places, arrange their direction, and set their intensity.

Color Control

Regarding the color of the lights, there is a standard but extensive set of lighting gels used for coloring white light. The model can represent these colors accurately but not need a detailed knowledge of the gel numbers. This has two advantages. The first is for visiting lighting designs with known colors. The designer can quickly and easily reproduce the necessary lighting conditions for a mock-up of the space. The second advantage is for new designs. The technical and artistic directors can quickly and arbitrarily select colors for the design. The model will identify the appropriate gel numbers for the physical construction at a later stage. In this way the designers can try numerous combinations without the laborious task of physically replacing and plotting the light. Figure 17.6 shows the model lit for different effects in different scenes.

Lighting of Scenes

In our working model the lights are interactive. They have limited movement along the roof frames, a set color, and the ability to switch them on or off. For the test set model of *A Midsummer Night's Dream,* separate switches enable the viewer to see lighting and set for a particular scene. It is rare, after all, that a play has a single lighting state or a lone set for its duration. The lighting complements the set in portraying different locations, times, and emotions. Distinct scenes commonly indicate these changes. Figure 17.7 shows a view of the stage area with three lights in position.

Using the Set for Assisting Rehearsals

The ability to view the set for individual scenes is particularly important for the performers. It is unusual for actors to have the opportunity to explore their theatrical environment before the physical construction of the set and its placement within the theatre. This often does not

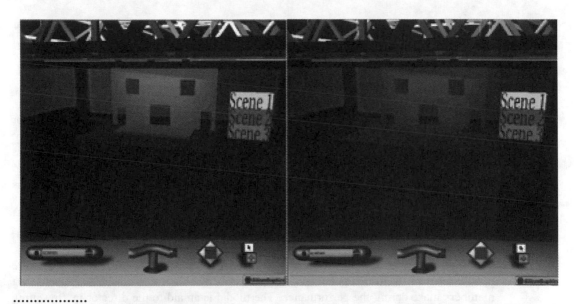

Figure 17.6   Different lighting moods. (*See color plate on page 333.*)

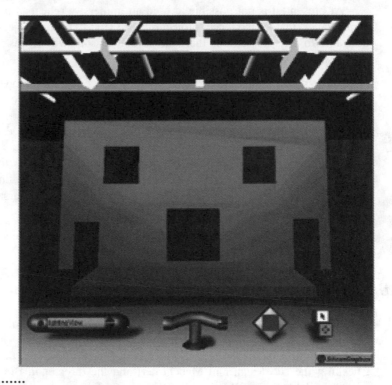

Figure 17.7   View of stage area with three lights. (*See color plate on page 333.*)

take place until a few days before the actual performances. If the cast does not have a clear idea about the set, it is more difficult for them to develop characterization. If actors do not really appreciate the style of show or location of properties, it is harder for them to create believable characters. In the worst cases, an unexplained set or one not integrated into the rehearsal process disorientates the actors during the critical last days before a performance.

The virtual set offers a solution to these problems. The virtue of a scale model that one can explore is that it gives a feel for the ultimate environment. The model allows an actor or director to "see" a representative human figure within the proposed set. One important yet obvious consideration is whether the audience can actually see all the action. Sometimes set or other characters can obscure what is going on. The virtual theatre permits the user to "sit" anywhere within the auditorium and look at the stage. With this capability, set designs are easily modified to take account of sight lines and access.

The model can help develop an understanding of the spatial relationships between characters and properties. An audience "reads" a great deal about the intercharacter dynamics from the physical spacing between them. The director can place multiple human figures, representing the characters, on stage at any time. This will help in deciding when and where people move. It identifies potential problems with blocking, that is, the movement cast members make during the performance. The model is an aid to the director in preventing actors from bunching up or not using the available space properly. Inexperienced directors often overuse a good idea until it becomes tired and predictable. The virtual set offers the potential to see early on in the rehearsal process if an area of the stage is overused.

At present, it is not possible to hold an immersive rehearsal in the virtual set in the same way as that for the recent project by IBM (IBM Theatre Projects). While it would be desirable to offer this option, there are several reasons why it was not felt appropriate in this case:

• The project was designed to allow people access to the model using existing technology. Typically, theatre groups will not have access to anything other than basic PC-type technology.

• At the moment, the technology for immersive VR (i.e., headsets and body suits) is prohibitive for theatrical performances and far too expensive.

• The model of the Mill was not designed to *replace* access to the actual space for activities such as rehearsal. It was designed to make sure that the limited time in the Mill that was available was used effectively (i.e., that it could be rehearsal time rather than set design/redesign time).

Incorporating immersive technology is a possibility for future work, but it is not seen as a priority.

Using the Model for A Midsummer Night's Dream

The first use of the model was for the set design of a production of *A Midsummer Night's Dream*. The members of the production team, although from the Theatre, were separated across the north of England for much of the rehearsal schedule: the technical director in Cheshire; the set designer in Manchester; and the director in Bradford. With the basic model available on the Web, each individual could assess the design and suggest improvements. General communication took place via e-mail, but the model had world read permission so that everyone could copy the master, modify the design, and store the result. In this way each of the team could develop their ideas, then share the data with the others.

The general process was one of refinement. The director started the design by visualizing his ideal set, including entrances and exits. Initially this was a crude model, very much an impression. With a basic outline, the designer took the model, added detail, and enhanced the overall look. During this stage, the technical director could see whether the design was physically possible within the space. As the set took shape they began a basic lighting design that was finalized when the model was agreed.

Throughout the design period, stewardship of the model passed along the team, each adding his unique contribution, identifying difficulties and suggesting alternatives. At every stage, all the team could inspect the work. In this way they had a common master model of the Mill that did not preclude individual exploratory work on copies.

This working practice demonstrated a number of benefits over traditional methods. The production team and cast shared a common vision of the ultimate set. This removed any ambiguity by description or personal interpretation. Consequently, no one was disorientated by the eventual arrival of the physical set within the space. With the designs complete and tested within the space, the construction of the set started much earlier than normal. Its installation and the lighting rig were completed much quicker, saving precious time within production week.

Conclusions and Future Work

The use of VRML to model the Theatre in the Mill has proved extremely valuable. It has been used to assist the design of sets in a real production and has greatly reduced problems with set design performed by a number of the production team in remote locations.

A project with a similar aim of using VR technology to aid the production process of a performance event has since been carried out by IBM (IBM Theatre Projects). This was a larger scale project that allowed virtual immersive rehearsal as opposed to collaborative set design. Such projects imply that work in the area will increase in the future as it is a valuable and effective use of VR techniques.

Problems with the current system mainly center on the relatively simple interaction supported by VRMLScript. Complex operations (such as large scale movement of lighting rigs, redesign of scenery, etc.) have required users to edit the VRML model itself. This naturally requires some knowledge of VRML 2.0. An alternative, as used in this case studied here, is to communicate the changes to someone who could carry out the modifications. This brings the process much nearer to traditional modes of working but with the advantage of a shared visualization of the changes to confirm their correctness.

To overcome this, the model is currently being rewritten to use Java as the scripting language, allowing much more flexibility (Java Programming Language; Java Scripting in VRML). This will allow more natural interaction and a greater range of information to be provided, such as pop-up windows with a menu scheme for selecting lights and gels, seating configuration, and so on. The experience gained with the production of *A Midsummer Night's Dream* will feed into the construction of this model, and the enhanced version is scheduled to be more extensive throughout 1997.

References

Cosmo Worlds, http://vrml.sgi.com/cosmoplayer/
"IBM Theatre Projects," http://www.ibm.com/sfasp/theatre.htm
"Java Programming Language," http://java.sun.com/
"Java Scripting in VRML," http://vrml.sgi.com/moving-worlds/spec/part1/java.html
Power Animator, Alias/Wavefront, http://www.alias.com/
"VRMLScript Specification," http://vrml.sgi.com/moving-worlds/spec/vrmlscript.html
"VRML 2.0 Specification," http://vrml.sgi.com/moving-worlds/

Moving the Museum onto the Internet: The Use of Virtual Environments in Education about Ancient Egypt

William L. Mitchell

Department of Computing,
Manchester Metropolitan University

· ·

Today's museum faces various challenges to the role it has traditionally performed. Virtual environments provide the means of meeting these challenges. Two projects are described. The first project resulted in the construction of an Internet-based virtual environment model of the tomb of the Egyptian noble Menna, using VRML. The second project describes ongoing work into the development of a virtual environment representing the pyramid builders' town of Kahun. This VE has the specific aim of providing support for primary-school visitors to Manchester Museum.

Key Terms

virtual heritage   museum informatics   education

The Challenges Facing Today's Museum

The museum is an institution that has traditionally performed two main roles. First, it provides an area for displaying artifacts and information about them to the general public. This has usually been done by putting these artifacts into glass cases with accompanying information panels. The second main role of the museum is to act as a center for knowledge about a particular subject area. Keepers of collections and research staff help in the conservation and storage of the artifacts and build up information about the artifacts such as when and where they were recovered.

Today's museum faces various challenges (Vergo 1989; Bennett 1995, MacDonald and Fyfe 1996). There is a tension between those who see a visit to the museum as an educational experience and those who see it as a leisure experience. The boundaries between entertainment and information have become blurred. The way in which heritage is viewed is undergoing a change. The increasing exposure of the general public to technology means that their expectations of display techniques have changed. The increasing spread of the Internet has changed expectations of how and when people are to access information.

One opportunity that has presented itself to museums is the increase in funding due to the National Lottery. Lord Rotschild, chairman of the Heritage Lottery fund, has described this as "a once-in-a-lifetime opportunity to bring about a much needed renaissance for museums of the United Kingdom" (Glaister 1997). It has been recognized that new technology can play a key role in the work of the modern museum (Department of Heritage Report 1997).

Virtual Environments

Virtual environment technology provides a means of broadening the role of the museum. Several subfields such as museum informatics and virtual heritage have developed in response. Virtual environments (VEs) can provide several advantages:

- Simulation of the physical layout of the museum and the artifacts
- Removal of time and distance constraints on access to displays
- Transformation of abstract data into a virtual artifact
- Ability to interact with artifacts

The majority of work in creating virtual museums has focused on simulating the museum on computer. Such a VE typically provides a map or plan of the museum's galleries that replicates the physical layout of the museum. In addition, the VE will also contain simulations of the artifacts (such as paintings or other objects) on display in the museum.

Visitors are limited by opening hours as to when they can view artifacts in the museum. A virtual museum can be visited at any time, particularly if it is made accessible on the Internet. Internet access also ensures that visitors can view artifacts in the museum without having to travel great distances.

A VE can also be used to create virtual artifacts that are not direct simulations of real-world objects. Visualization techniques can be used to transform information into a virtual artifact that can be easily viewed and understood by the user.

A VE also provides a means for the user to interact with an artifact. For example, a virtual artifact can be created that is an exact copy of the real artifact. This virtual artifact can be handled by the user without fear of breakage or loss.

The rest of the paper describes two projects that are investigating some of these issues.

Objectives of the Tomb of Menna Project

The first project investigated how information about an Egyptian tomb could be displayed in a more intuitive fashion (Mitchell and Pendlebury 1996).

Two main objectives were identified:

- To investigate the process of creating a virtual artifact from existing information sources
- To provide alternative ways for visitors to access the virtual artifact

One of the main criteria used in deciding on a possible artifact was that the information should require as little interpretation as possible to be turned into an artifact. This would provide a means of demonstrating how existing information sources can be quickly and simply used. At a more practical level, this was a pilot project the subject experts were unable to commit time and resources to interpreting the information into a form suitable for use in a virtual environment.

An Egyptian tomb was chosen due to its simple geometry. This would help to simplify the design and construction process. In order to model a tomb accurately, information is needed about the plan of the tomb (to base the geometry of the artifact upon) and photographic data would be needed for every surface of the tomb. Unfortunately, very few Egyptian tombs have been completely documented.

The tomb that was chosen was the Tomb of Menna (TT. 69). The Tomb of Menna is located at Sheikh Abd el-Qurna, the most central area of the Theban necropolis and where most of the tombs of officials are located. Menna lived in the Eighteenth Dynasty either during the reign of Tuthmosis IV or that of his successor, Amenophis III. His title (Scribe of the fields of the Lord of the Two Lands) suggests that his official duties were connected with agriculture, and this is reflected in one of the best known scenes from his tomb. He had a wife called Henuttawi, two sons, and four daughters, all of whom are pictured in his tomb.

It was decided to use the Tomb of Menna because of the completeness and ready availability of data about it. The data was gathered in a survey conducted by Robert Mond between 1914 and 1916 (Malek). Mond measured and created a plan of the tomb (Figure 18.1). He also photographed the interior ceiling and wall and floor coverings, covering 90 percent of the tomb's surfaces. Data from the Mond survey is held at the Griffith Institute,

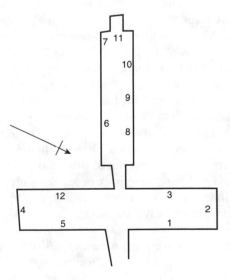

....

Figure 18.1 Mond's plan of the Tomb of Menna.

Figure 18.2   Mond's model of Tomb 85 (not Menna).

Ashmolean Museum, Oxford. The Griffith Institute kindly gave permission to make use of this information in the project.

Mond used a sophisticated set of photographic techniques that have parallels with those used today to capture images for 3-D modelling purposes (Mond 1933). He created a framework mounted on rails around the interior of the tomb. To this he attached a simple camera and lighting equipment. He then photographed each surface by taking a sequence of overlapping photographs from a distance of 18–20 inches away. These photographs were then developed, and the prints were cut and arranged (like a mosaic) onto cardboard. The resulting mosaic was then retouched by hand to hide the joins between the prints and then photographed again. Mond himself realized the problems of displaying this information to the general public. His solution was to enlarge the photographs and paste these onto a cardboard reconstruction of the Tomb (Figure 18.2).

The data available suggested that the virtual artifact could take the form of a virtual environment representing the interior of the tomb that the user could walk through and inspect. There is thus a parallel between the problems and display techniques adopted by Mond and those faced by the current project 80 years later.

To meet the second objective of ensuring wide access to the virtual artifact, it was decided to make it available on the Internet. In order to ensure as wide a range of access as possible, it was important not to place too many constraints on the target platform. Factors considered were the processor speed, availability of memory, and speed of communication links. With these factors in mind, the target platform was assumed to be a reasonably modest, commonly available PC (at the time, a 486 66 MHz machine with 16 Mb of memory). Communication links were assumed to be as limited as a 14.4 K modem. It was also decided not to assume the availability of special hardware devices such as head-mounted displays or six degrees of freedom controllers.

The system was targeted initially at the general Internet user. This allowed the level of background information to be kept reasonably simple and allowed the project to concentrate on issues surrounding the virtual walk-through.

Figure 18.3 Virtual walk-through of the tomb.

Development of the Tomb of Menna System

The virtual walk-through of the tomb was constructed using VRML (Virtual Reality Modelling Language) version 1.0 (Figure 18.3). As VRML was being developed as an industry standard, it was felt that this would allow the tomb to be viewed across as wide a range of platforms as possible.

Development of the model was carried out with the cooperation of Silicon Graphics who provided access to a suitable authoring platform. The VRML authoring was carried out on a Silicon Graphics Indigo2 Extreme (192 Mb, 2 Gb, 200 MHz, R 4400). The construction of the model consisted of several stages: creation of the geometric model, the application of textures, adding details to the world, and creation of supplementary material.

The plan of the tomb was used to create the underlying geometry using the Medit software package. The lack of detail in the source plan meant that the geometry of the model of the tomb was fairly simple. Each wall was modelled as a separate object to allow a hyperlink to be associated with it. The walls were then combined into a scene within Medit, and the final model was exported in the Open Inventor 2.0 format.

In some cases, parts of the tomb were not included in the geometrical model. For example, a small recess at the back of the tomb (wall 11) was not modelled geometrically but was instead represented via photographic means (Figure 18.4). This improved system performance without harming the user's experience of the tomb.

The second major stage involved texture mapping of images onto the geometrical model using Medit. Prior to this mapping, the textures had to undergo image processing, which accounted for the majority of the work in the project. As mentioned previously, Mond's photographic techniques were quite advanced for the time and are actually fairly compatible with the notion of 3-D modelling. However, there were still areas of distortion that had to be removed by applying various image processing techniques. These included rotating images, joining together of images, blurring of the resulting joins, and cutting and pasting areas from other images. These techniques also had to be used to remove distortions introduced when scanning the Mond photographs.

Figure 18.4   A recess containing a statue of Menna (wall 11).

One problem encountered was that for some areas of the tomb Mond's photographs were incomplete. As Mond was more interested in the main content of the pictures (such as the figures or hieroglyphs), details of the borders were sometimes missing. This was solved by copying similar pieces of detail from neighboring regions. Several small sections of wall also did not have any Mond data, namely, the walls of the small opening facing the observer when standing in the entrance. These were left completely blank with the underlying wall material showing. This appears as a gray color on most VRML browsers. There were also no photographic data available for the floor, so it was given a mottled rough texture to simulate an earth floor (see Figure 18.3). The Mond photograph of the ceiling texture was of only a small section of the pattern. This had to be retouched to allow a small texture map to tessellate and represent the pattern on the ceiling.

The textures initially created would have been of the order of 16 Mb. This would have placed unreasonable memory requirements on the target platforms and perhaps have taken quite some time to download due to limitations on communication links. It was thus necessary to reduce the size of the texture maps. Two versions of the tomb were created: a PC version containing smaller textures and a workstation version. Textures were reduced by rescaling and applying a box filter.

WebSpace Author (an SGI product) was used to create the VRML model by handling issues such as viewpoints and lighting. A series of viewpoints were defined to assist the user's navigation around the model. This was important due to the lack of experience of 3-D navigation in the target user group.

Supplementary material was created in HTML (Hyper Text Mark-up Language) (Figure 18.5). Hyperlinks were made to individual pages, detailing walls and showing a higher quality image than was otherwise possible in the 3-D model due to memory constraints on the target machine. Descriptive text was added to simulate a guided tour. This text was provided by the Griffith Institute.

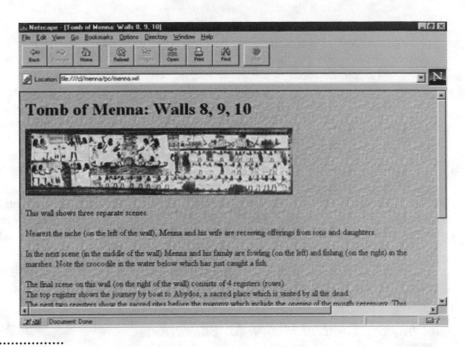

Figure 18.5   Details of walls 8,9,10.

Assessing the Tomb of Menna Project

The first version of the Tomb of Menna was placed on the Internet in March 1996 and evaluated by means of feedback forms. The form was designed to collect simple technical information about the user's computer (what generation it was, whether it possessed 3-D graphics acceleration) as well as their reasons for visiting the site (whether Egyptological, educational, out of an interest in VRML, or just out of curiosity). The form also provided an area for free text comments.

While the site registered 200 visits over the first 10-day period, the number that responded to the form was somewhat disappointing as only 12 responses were collected. These included a range of users from historians and archaeologists to general visitors. The feedback was positive, showing strong interest in the area of virtual heritage. Informal feedback from the on-line Egyptological community suggested much interest in the prototype. Most users had a 486 computer with no 3-D acceleration. Most reported problems navigating around the model. The major problem users encountered seemed to be that of getting their VRML browsers to work correctly on their own machines.

One of the advantages of an Internet-based application is that evaluation can take place continuously. This has proved to be the case with the Tomb of Menna. Subsequent feedback has confirmed the results found in the initial evaluation period. The amount of feedback has varied over time as the Tomb of Menna site has been linked to more Web sites and has been publicized in the press (Woolley 1997). One refreshing development has been the increasing amount of queries about the Tomb of Menna itself. This shows a shift from users interested in the technology to users who are interested in the subject of Egyptology. The site can be visited from the author's home page:

```
http://www.doc.mmu.ac.uk/STAFF/B.Mitchell
```

Overall, it was felt that the project successfully met the first objective by showing that existing information sources could be used with little further interpretation by subject experts to create a virtual artifact.

The project partially met the second objective. The Griffith Institute was satisfied with the site and has put a link from its own Web site to the Tomb of Menna Web site. The Tomb of Menna site is thus seen as an artifact in its own right that is made available to a different set of (Internet) visitors. One failing was the problem encountered by visitors in ensuring that their particular viewing platform was suitable for the VRML model.

The virtual walk-through of the tomb provided a good context for the information about the tomb. Due to limitations on target platforms, more detail could not be included in the VRML model itself. This had to be accessed by the user clicking on relevant areas of the tomb.

There were two main issues that the Tomb of Menna project did not address. First, the system was aimed at the general user and did not aim to meet specific educational aims. Second, the system was primarily a means of displaying information. It did not allow the user any real interaction aside from the ability to walk through the tomb and click for more information. To investigate these issues, it was decided to carry out further development and evaluation of the Tomb of Menna system.

Developing the Tomb of Menna for Education

It was decided to focus on the educational potential of the Tomb of Menna and, in particular, its application to children of primary-school age. Manchester Museum agreed to act as an evaluation site.

There are several problems with evaluating an Internet-based system via feedback forms. It is difficult to get subjects to respond. It is difficult to get a representative cross section of subjects. The anonymous nature of the Internet means that the evaluator is reliant on the subjects responding honestly about their background. These issues are compounded when the subjects are of primary-school age.

To deal with these problems it was decided to evaluate the Tomb of Menna during the museum's activity week. This is organized by the museum's Education Service to coincide with the half-term holidays. Parents and children are invited to take part in various activities that reflect the work of the museum. These include storytelling, inspecting specimens through a microscope, and writing in Egyptian hieroglyphs. The Tomb of Menna system was added to the list of activities.

The system was rewritten and reorganized to reflect the abilities of the new target user group. A stand-alone version was created as there was no Internet access from the Education Service's rooms. The original version of the Tomb of Menna provided a limited range of interaction (mainly navigating around the tomb and clicking). To test issues of interactivity it was decided to create a small interactive application and incorporate it into the Tomb of Menna site. This took the form of a simple Java application that allowed a child to write words in English and translate to hieroglyphs, and vice versa (Figures 18.6 and 18.7). Words were created by dragging symbols from the top section of the screen to the writing area below. This application mirrored one of the other activities provided for the children where they used crayons and pencils to do the writing.

Evaluation techniques such as on-line feedback forms or written questionnaires would not have been suitable for the children. These are quite formal techniques that might be suitable in a classroom environment but not in an activity week during the holidays. Interviewing also proved difficult with children as they tried to provide "correct" answers. The technique used was a mixture of observation and interview.

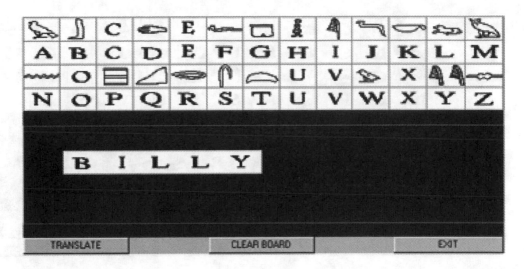

Figure 18.6   A name in English.

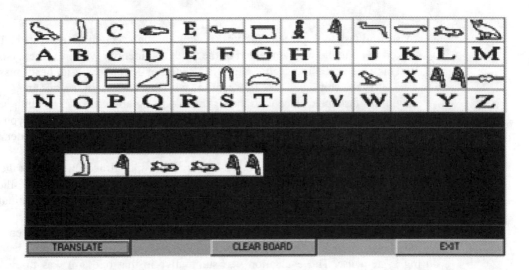

Figure 18.7   The name translated to hieroglyphs.

Each child was given a short introduction to the system and informally asked about their background (age, whether they knew about ancient Egypt, their experience with computers). The child was then asked to carry out a series of tasks and was observed. The child was allowed to make mistakes, though if they persisted, they were given help before they became frustrated. After a few minutes of use the child would become more comfortable with the evaluator. At this point the child was more receptive to interviewing and in some cases would volunteer information about the system.

Three tasks were developed for the child to carry out. The first task was to explain what was going on in a particular scene (walls 8,9,10, which contained a depiction of Menna fishing and fowling [see Figure 18.5]). The child was encouraged to identify the crocodile in the scene. This task involved the child clicking to view a larger, more detailed version of the wall photograph and then panning over it.

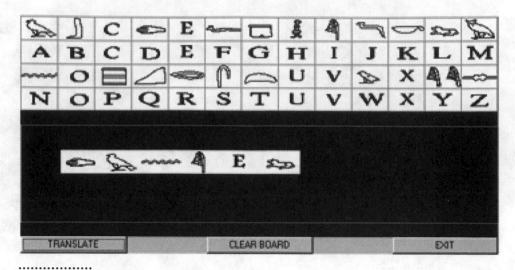

Figure 18.8   Daniel's missing hieroglyph.

A second task took place in the virtual walk-through section of the system. The child was asked to go to the wall at the end of the tomb and identify what was there (see Figure 18.3). This was wall 11, which contained a statue of Menna with its head missing (see Figure 18.4). The child was asked for reasons why the head might be missing.

The third task took place with the Egyptian writing application. The child was asked to write his/her name and then translate it. In some cases there was no hieroglyph that corresponded to a letter (Figure 18.8). The child was then asked what other letters (if any) might be substituted.

Evaluations took place over four 2-hour periods spread over 3 days. As the periods were not formally structured, the number of children and the time each used the system varied. On average eight children per period used the system for more than 5 minutes each. The children ranged in age from five to ten years old.

Conducting the evaluations in this way rather than via on-line feedback ensured that the number of responses was maximized. It also meant that the child could be observed during system use. The evaluation was summative in that one aim was to evaluate the Tomb of Menna system. However, it was also formative in that another aim was to investigate general issues of VE interaction for subsequent projects. Thus, the information that was being gathered was of a mainly qualitative nature, so measurements (such as the time taken to complete a task) were not taken. In any case, the unstructured and informal evaluation situation would have made such measurements very difficult.

Results of the Evaluation

The evaluation uncovered several general interface and interaction issues. It was found that children had to be too precise in lining the symbols up in the hieroglyph writing task. A more sensitive mechanism for having symbols snap into place would have been useful. Also, some children clicked on symbols rather than dragging them. One unexpected but, with hindsight, obvious finding was that very young children had trouble selecting symbols. This was not because they were not capable of spelling (albeit by sounding the letters rather than naming them) but because the mouse was physically too large for them to move and hold down a button at the same time.

Another problem was the granularity of the information presented. It was not always immediately apparent to some children that they needed to scroll down a page to see more information. For example, for the pages in the site that contained information about the walls, it was necessary to scroll down past the photograph to read the description (see Figure 18.5). For some children the scrolling operation was quite difficult to perform due to the small size of the scroll bars presented by the browser used (Netscape). It would have been preferable to design each page so that it matched the screen size, thus removing the need for scrolling. However, this conflicts with the normal rule of designing Web pages of larger grain to minimize the number of pages that need to be downloaded. This is a classic example of the trade-off between performance and usability considerations.

The problem of the separation of a photograph and its description could be overcome by making the photographs sensitive so that when a child clicked on an area, the description would be displayed. This would remove the need for scrolling. In some cases, scrolling would still be necessary. For example, due to its large size a detailed version of a wall photograph cannot be displayed on the screen all at once. In this case, scrolling needs to be made easier for the child by providing landmarks that indicate where in the photograph the child has panned to and by making the scrolling controls more explicit.

The children found the virtual walk-through of the tomb quite easy to use. Some commented on the similarity between the walk-through and games they had used before, such as *Doom* (a mazelike game). Some children found it quite disorienting being able to go through the walls of the tomb into darkness. This was due to the lack of collision detection in the model. The ability to return to a fixed viewpoint was thus found to be an important feature. Some children also commented on the lack of color in the photographs.

The findings pointed to a wider problem with designing interfaces for Internet-based applications. The actual form of the interface is dependent on the user and the particular platform and browser software they might be using. For example, the scroll bars may have been too small with the particular browser used (Netscape) but may well have been adequate with another. Also, the particular VRML browser used (in the evaluation this was the Live3-D plug-in from Netscape) determined the way in which controls were presented to the user (see Figure 18.3). These parameters could have been controlled in the evaluation as the evaluator was responsible for setting up the platform. There would be less control in a "real" on-line situation. Interface design in this case is thus not just about designing one ideal interface but becomes much more challenging.

In addition to feedback from children, feedback was also obtained from parents. In some cases, parents were primary-school teachers who were also on their half-term holidays. Both children and parents appreciated the variety of the tasks that the Tomb of Menna system supported. This is an important issue in the use of VEs for education. Of equal if not more importance than the visual quality of the environment is the provision of a range of meaningful tasks and guidance in performing them.

To investigate the wider issue of the use of VEs as a means of providing education about existing artifacts, it was decided to set up a second project at Manchester Museum.

Objectives of the Kahun Project

The aim of the second project was to investigate how VEs could be used as an educational resource to support the work of Manchester Museum. The resource would support the work of the Museum's Education Service, which caters for visits by groups of primary-school children.

Manchester Museum contains galleries ranging from Botany, Mammals, and Ethnology to Egyptology. The two main Egyptology galleries concentrate on funerary beliefs and customs (the mummies) and daily life. Egyptology is seen as one of the particular strengths of the museum. With this in mind and the background of the Tomb of Menna project, it was decided to continue working in the Egyptology area.

One problem encountered at the outset of the project was the lack of familiarity of the Education Service staff and teachers with virtual environments. This meant that it was not possible for them to supply a suitable set of requirements for the resource. This can be seen as a problem of opportunity identification. The education staff were unable to identify areas in the museum's collection that would be suitable for use as virtual environments. This is a problem that has been identified in the area of soft systems methodologies (Checkland 1990). To get around this problem, the project was structured into a series of stages:

- Development of a "rich picture" of the domain area
- Proposal of possible virtual environment resources
- Development and evaluation of prototypes with users

The first step in developing a rich picture of the area was to identify the various stakeholders in the resource:

- Educationalists
- Egyptologists
- The Museum
- Technologists
- Children

Educationalists include the Museum's Education Service staff as well as the teachers. These stakeholders bring to the project topics to concentrate on and teaching methods. Teachers also have to ensure that a resource fits in with demands made by the national curriculum. The involvement of the Keeper of the Collection and other Egyptologists is important to ensure the authenticity of any resource developed. The museum itself has a stake in the design as any resource developed can help to attract more visitors to the galleries. Technologists not only provide the means for implementing a resource but more importantly provide knowledge of the capabilities of virtual environments. The children are the end users of the resource.

The second step was to perform a task analysis of a typical school visit. This involved examining materials used in the visit as well as observing actual visits.

In advance of a visit, each teacher is sent an education pack containing notes about the exhibits and diagrams of various artifacts. In addition, the teacher fills in a questionnaire indicating the background of the children, such as their age range and the purpose of the visit.

A typical visit consists of three main stages. In the first stage, Education Service staff give a short talk of about 20 minutes to the children to provide a general introduction to ancient Egypt and the museum's collections. In the second stage, children get the opportunity to handle actual artifacts, related materials (e.g., papyrus), and models of some of the more valuable artifacts. In the final stage, the children go around the Funerary Beliefs and Daily Life Galleries, which are open to the public. The exhibits consist of artifacts in glass cases with supporting information panels. During this stage, the children complete work sheets based on the exhibits.

Observations are focused on the final stage of the visit when the children are carrying out work activities in the public galleries. The work activities are centered around a set of five work sheets covering burial customs, daily life, writing, jewelry, and the pharaohs.

There are difficulties in trying to observe 20 children moving around a gallery. The work sheets proved useful as they guide the children to particular displays and set specific tasks. The work sheets thus provided a good means of structuring the observations.

The activities mainly involved drawing artifacts on display and answering questions about them. This was meant to help the children to understand what each of the artifacts was used for. Short interviews were carried out with the children to gauge whether the activities helped in their understanding of the artifacts.

The Funerary Beliefs and Customs Gallery contains a range of information from the details of the mummification process to mummies themselves. The children seemed well motivated to learn about and understand the displays in this gallery. This was due to the subject material itself (they seemed in particular to have a morbid fascination with the canopic jars that were used to "contain the insides of the dead person" when they were being mummified).

The Daily Life Gallery contains information about activities and the objects that were used in them. These include work activities such as agriculture, metal working, stone masonry, and carpentry as well as leisure activities such as writing and literature, children's games, and cosmetics. The children had much more difficulty becoming interested in this subject.

Children seemed to have a great deal of trouble understanding what the various objects on display were and what they might be used for. One problem was that the limits on the time available for each visit meant that the worksheet activities sometimes weren't completed. More seriously, children seemed capable of completing the associated task (drawing or answering questions about the objects) without developing any real understanding. In one particular case, one girl drew a perfect copy of a brick mould (Figure 18.9) and was able to correctly identify what it was. When asked if she knew what it was for, she smiled, said she didn't know, and then skipped off to do the next task.

Interviews were also conducted with teachers accompanying the parties and Museum Education Service staff. These indicated that teaching about ancient Egypt was done for a wide variety of reasons, from history and geography to design (the jewelry on display).

With these problems in mind, three main aims were identified for the resource:

• To teach children about everyday life in ancient Egypt. This is a topic area that the children are not well motivated to learn about. It therefore poses a strong challenge for developing alternative means of displaying the information.

• To provide children with a context for the objects in the museum's collection. The children found it difficult to relate to objects removed from their context and placed in glass cases.

• To show how objects were used in various activities. Ideally this would be by handling the objects or seeing them in use. The risk of breakage or loss prevented this option.

Figure 18.9   Brick mould.

The Proposed Kahun Resource

To determine the form that the resource should take, a study was undertaken of the museum's collections. One particular identified strength of the Egyptology department was the collection of artifacts recovered from the pyramid builders' town of Kahun (David 1986). This site differs from many other sites in that the objects recovered were from the everyday life of the pyramid builders rather than the more common artifacts of distinguished persons recovered from tombs. These artifacts make up the majority of the displays of daily life at the museum.

Work on situated cognition and constructivism has shown the need for providing appropriate contexts for education (Lave and Wenger 1991). It was proposed that the objects in the museum could be provided with an appropriate context by placing them in a virtual environment. This virtual environment would represent Kahun, from which the objects were actually recovered. The VE would contain the various buildings that made up the town. The environment would be populated by various objects from the museum's collection. Children would be able to interact with these objects in order to understand their functions. Such an environment would provide two main benefits. First, it would allow the children to see and learn about what the town of Kahun might have looked like. Second, it would also provide a powerful means of placing an object in its correct context as well as allowing a child to learn about an object by interacting with it.

It was also proposed that this resource should be based on the Internet. This would allow it to be used both before and after a school visit, thus increasing the value of the actual visit to the museum. Use of the system before a visit would mainly be concerned with providing a context for the actual visit. Use of the system after the visit would provide support for follow-up project work. It should be noted that the resource was not seen as acting as a substitute for an actual visit, so it was not being designed to merely replicate the experience of visiting the galleries.

The project is currently entering its first phase of prototype development. The work can be divided into two main areas: the construction of a virtual model of the town and the provision of objects that the children can interact with in this environment.

A small section of the town has been identified for reconstruction. As the town itself is covered in sand and not visible, this will involve using measurements and written observations that already exist about the buildings in order to develop a virtual model. Children will be provided with three views of the town: a bird's-eye view (an overview of a plan of the whole town), a street-level view (containing facades of buildings and details of the streets), and a view within individual dwellings.

Initial activities identified for inclusion in the virtual environment are

- Brick making (children showed difficulty in understanding the function of the brick mould [see Figure 18.9]).
- Starting fires (the Keeper of the Collection indicated that the fire stick [Figure 18.10] in the museum is unique).

Certain sections of the town will be set aside for each of these activities. Children will learn about these activities by interacting with the objects associated with them or viewing simulations of them in use.

Implementation of the prototype is being carried out in a mixture of VRML (to construct Kahun) and Java (to handle the activities).

One fault of many virtual environments is the lack of things to do once the user has done a walk-through. This becomes even more of a problem when the users are children

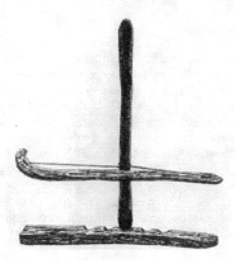

Figure 18.10 Fire stick.

who have comparatively high expectations of such environments (possibly due to their exposure to applications such as computer games). Evaluating a limited system with a group of children might not result in the formative information required.

With this in mind the initial prototype will be evaluated in two stages. First, the prototype will be evaluated with teachers who can make allowances for the limited nature of the system. This will be done as part of one of the INSET (in-service training) days run by the Education Service at Manchester Museum later in the year. Due to the time pressures on teachers, getting their opinions has proved difficult. The time during a visit to the museum is limited, and the teacher is occupied with the children. Time is also limited at school for interviews. Evaluating the prototype in a workshop context will allow access to a group of teachers interested in the subject area who are not in a pressured classroom environment. Second, the prototype will be evaluated by Egyptologists at Manchester Museum to check the authenticity of buildings and objects.

Conclusions

Initial findings from the two projects suggest four key points:

- Virtual environments can play a crucial part in museum work
- Internet-based resources pose particular problems for interface design and evaluation
- Children are a special category of end users and place different demands on system design
- There is a distinct lack of methodologies that can be readily applied to the development of virtual environments for museums

The Kahun project has begun to outline a methodology for developing VE resources. Once a prototype has been developed, work can begin on evaluation methods for determining the educational worth of such virtual environments.

Acknowledgments

I gratefully acknowledge the contributions of the following:

Matthew Pendlebury for developing the Tomb of Menna system.

The Griffith Institute for granting permission to use data from the Mond Survey. The illustrations used in Figures 18.1, 18.2, and 18.4 are copyright of the Griffith Institute.

Nigel John at Silicon Graphics for providing facilities for developing the Tomb of Menna system.

Prof. Lianfang Liu and Mr. Yujun Wu from the New Software Technology Laboratory of Guangxi, Guangxi Computing Centre, China, who worked on the hieroglyph writing application.

Mike Holme for implementing the first Kahun prototype.

Manchester Museum and, in particular, the Education Service for cooperating on the Kahun project.

References

T. Bennett, *The Birth of the Museum,* Routledge & Kegan Paul, London, 1995.

P. Checkland and J. Scholes, *Soft Systems Methodology in Action,* John Wiley, Chichester, UK, 1990.

R. David, *The Pyramid Builders of Ancient Egypt,* Routledge & Kegan Paul, London, 1986.

Department of Heritage Report, *A Common Wealth: Museums and Learning in the UK,* 1997.

D. Glaister, "Tate Gallery Scoops £18m Jackpot in Big Lottery Payout," *The Guardian,* Feb. 21, 1997, p. 9.

J. Lave and E. Wenger, *Situated Learning—Legitimate Peripheral Participation,* Cambridge Univ. Press, 1991.

S. MacDonald and G. Fyfe, *Theorizing Museums: Representing Identity and Diversity in a Changing World,* Blackwell, Oxford, UK, 1996.

J. Malek, *The Private Tombs of Thebes, The Photographic Survey by Sir Robert Mond 1914–1916,* Griffith Institute, Ashmolean Museum, Oxford, UK.

W.L. Mitchell and M. Pendlebury, "Reconstruction of the Egyptian Tomb of Menna using VRML," *Short Paper Proc. 3d UK Virtual Reality Special Interest Group Conf.,* De Montfort University, Leicester, July 3, 1996, pp. 67–73.

R. Mond, "A Method of Photographing Mural Decorations," *The Photographic Journal,* Jan. 1933.

P. Vergo, *The New Museology,* Reaktion Books, London, 1989.

B. Woolley, "All in the Past," *Personal Computer World,* Vol. 20, No. 3, Mar. 1997, pp. 306–307.

• About the Author •

William Mitchell is a lecturer in Computing at Manchester Metropolitan University. He received his PhD in 1993 from MMU for a thesis on the use of models of human discourse in human-computer dialogue. His main interest is human-computer interaction, in particular, the use of virtual environments for educational applications. He has also done work in the design of systems for users with special needs.

Chapter 19

The Virtual Reality Responsive Workbench: Applications and Experiences

Lawrence Rosenblum, James Durbin, Robert Doyle, and David Tate

Virtual Reality Lab/Information Technology Division
Naval Research Laboratory

Introduction

Virtual reality (VR) is a complex and challenging field (Earnshaw and Rosenblum 1995; Rosenblum and Cross 1997), and several distinct types of systems have been developed for displaying and interacting with virtual environments. One of the newest is the Virtual Reality Responsive Workbench (Kreuger and Froehlich 1994; Kreuger et al. 1995; Rosenblum, Bryson, and Feiner 1995). The Workbench is an interactive VR environment designed to support a team of end users such as military and civilian command and control specialists, designers, engineers, and doctors. The Workbench creates a match for the "real" work environment of persons who would typically stand over a table or a workbench as part of their professional routine. For example, the Workbench could be used to represent fluid flow over a ship's hull while supporting a design team in interactive visualization. Perhaps the greatest strength of the VR Responsive Workbench is the ease of natural interaction with virtual objects. Current interactive methods emphasize gesture recognition, speech recognition, and a simulated "laser" pointer to identify and manipulate objects.

This paper classifies VR systems into three categories: immersive head-mounted displays (HMDs), immersive non-HMD systems, and partially immersive tabletop systems. We discuss the utility of each classification. Several applications that we have developed in the Virtual Reality Laboratory of the Information Technology Division (ITD), Naval Research Laboratory (NRL), are examined, and we discuss our experiences with VR Responsive Workbench interfaces and software architecture.

Systems for VR

There is no accepted definition for VR. One important reference, the U.S. National Research Council report *Virtual Reality: Scientific and Technical Challenges* (Durlach and Mavor 1995), does not attempt a definition. Rather, characteristics of a virtual environment are given. These include a man-machine interface between human and computer, 3-D objects, objects having a spatial presence independent of the user's position, and the user manipulating objects using a variety of motor channels. Virtual reality can be subdivided in many different ways; here we will categorize based upon the visual channel.

Head-Mounted Displays/BOOMs

HMDs, which typically also include earphones for the auditory channel as well as devices for measuring the position and orientation of the user, have been the primary VR visual device for much of the 1990s. Using CRT or LCD technology, HMDs provide two imaging screens, one for each eye. Thus, given sufficient computer power, stereographic images are generated. Typically, the user is completely immersed in the scene, although HMDs for augmented reality overlay the computer-generated image onto the view of the real world. Low-end HMDs can be obtained for less than $10,000. These suffer from information loss (resolutions of approximately 400 × 300 pixels; typical field of views are between 40° to 75°). Extremely low-end "glasses" cost only hundreds of dollars, but these systems are not yet usable for serious applications and find their role in system testing and in research. High-end HMDs overcome these limitations at very high costs and thus are utilized for only a limited number of applications such as military flight training. In addition, ergonomic limitations such as weight, fit, and isolation from the real environment make it unlikely that users will accept HMD-based immersion for more than short time periods until advances in material science produce lightweight, eyeglass-size HMDs. They are, however, more portable than are other VR systems.

An alternative to HMDs is the BOOM (Binocular Omni-Orientation Monitor). Two high-resolution CRTs are mounted inside a package against which the user places his eyes. By counterbalancing the CRT packaging on a free-standing platform, the display unit allows the user six-degree-of-freedom movement while placing no weight on the user's head. The original version of the BOOM had the user navigating through the virtual world by grasping and moving two handles and turning the head display much as one would manipulate a pair of binoculars. Buttons on the handgrip are available for user input. A more recent desktop version (the Fakespace PushBOOM) allows the user to navigate by pushing his/her head against a spring-loaded system.

HMDs and BOOMs are similar devices in that the user is fully immersed in the virtual environment and does not see his/her actual surroundings. The BOOM solves several of the limitations of the HMD (e.g., resolution, weight, field of view), but at the expense of reducing the sense of immersion by requiring the user to stand or sit in a fixed position. This loses the freedom of movement associated with HMDs where users typically take steps and turn their body to determine direction (the BOOM also restricts the user's hands).

Immersive Rooms

Immersion does not necessarily require the use of the head-mounted displays that are the most common method for presenting the visual channel in a virtual environment. The CAVE™ (Cave Automatic Virtual Environment), a type of immersive room facility developed at the University of Illinois, Chicago, accomplishes immersion by projecting on two or three walls and a floor and allowing the user to interactively explore a virtual environment (Cruz-Neira et al. 1993). An immersive room is typically about $10' \times 10' \times 13'$ (height), allowing a half-dozen or more users to examine the virtual world being generated within the space. Computer-generated stereographic images are produced by calculating right and left eye images and using stereographic shuttered glasses to synchronize these alternating images. To determine the view, the group leader's head position is tracked using magnetic sensors to determine position and orientation. Both by walking within the immersive room and by utilizing an interactive device called a *wand*, which has a second tracker for position identification and buttons for issuing commands, the group leader navigates through the data. All users see the same image; thus, other team members view the scene from an incorrect perspective with the resulting distortion depending upon differences in location within the immersive room. Since the stereographic shuttered glasses are see-through, all users see each other. This facilitates group discussion and data analysis.

While HMDs require that users interact in virtual spaces (they cannot see each other in their "real" environment), the immersive room offers the significant advantage of permitting user interaction, discussion, and analysis in the real world. However, the computational cost of generating scenes within an immersive room are very high. Two images must be generated at high refresh rates for each wall in the immersive room. In addition, each wall requires a high-quality projector, and since back projection is used, a large allocation of space is required for projection length. Costing over one-half million dollars, immersive rooms exist only in a handful of large research organizations and corporations.

The VR Responsive Workbench

The two paradigms discussed above are both fully immersive. However, there are many applications for which full immersion is not desirable. A doctor performing presurgical planning has no reason to wish to be fully immersed in a virtual room and with virtual equipment. Rather, he would like a virtual patient lying on an operating table in a real room. He would like to reach out and interactively examine the virtual patient and, perhaps, practice the operation. Similar remarks apply to engineering design, military and civilian command and control, architectural layout, and a host of other applications that would typically be performed on a desktop, table, or workbench. These applications are categorized by not requiring navigation through complex virtual environments but rather by demanding a fine-granularity visualization and interaction with virtual objects and scenes. Thus, the Workbench supports VR for a large class of applications that are substantially different from the fully immersed, navigation-oriented applications supported by HMDs and immersive rooms.

The VR Responsive Workbench operates by projecting a computer-generated, stereoscopic image off a mirror and then onto a table (i.e., workbench) surface that is viewed by a group of users around the table (Figure 19.1). Using stereoscopic shuttered glasses (just as is done in the immersive room, users observe a 3-D image displayed above the tabletop. By tracking the group leader's head and hand movements using magnetic sensors, the Workbench permits changing the view angle and interacting with the 3-D scene. Other group members observe the scene as manipulated by the group leader, facilitating easy communication between observers about the scene and defining future actions by the group leader. Interaction is performed using speech recognition, a pinch glove for gesture recognition, and a simulated laser pointer. Figure 19.1 shows a schematic of the Workbench.

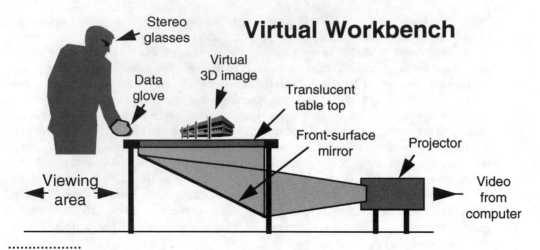

Figure 19.1   A schematic diagram of the VR Responsive Workbench

Table 19.1 presents trade-offs between these systems. Table 19.2 indicates the strengths of each type of system.

In 1994, the NRL/ITD VR Lab designed and fabricated the first Virtual Reality Responsive Workbench in the United States (Rosenblum, Bryson, and Feiner 1995) based in part upon earlier work at the German National Research Center for Information Sciences (GMD). The remainder of this paper discusses our interactive methods for the Workbench as well as two applications: medicine and situational awareness. A third application that we have developed, not discussed in this paper, is engineering design where we show how the Workbench was used to find new information about a preliminary version of a ship design (Rosenblum et al. 1996).

Interactive Techniques

This section discusses three types of interactive techniques that we employ on the Workbench: (1) direct manipulation using a pinch-glovelike system, (2) voice recognition, and (3) a simulated laser pointer (wand).

Direct Manipulation via a Glove

For direct manipulation, a user places an instrumented hand into the virtual environment and attempts to interact with virtual objects as if they were physically present. For example, in our medical application a user can reach into the skeleton and select and grab bone groups or internal organs. Grabbing is accomplished via a pinch-glovelike system. The user makes a pinching gesture with his/her index finger and thumb to indicate a desire to grab the currently selected item. To let the user know that a specific object is selectable, we change the color of the object when the user's hand is close to the object. Once grabbed, the object is attached to the hand. The user manipulates it as he/she would a real object; in/out and sideways hand movements zoom and pan while hand rotations rotate the object.

The strength of this metaphor is that it is simple and intuitive. The user interacts exactly as he/she would with a real-life object, reaching out, grabbing it, and manipulating it. There is no artificiality, and the need for a learning process for the interaction ("mouse-

Table 19.1 **System Characteristics**

	HMD (mid-range)	PushBOOM	Immersive Room	Workbench
Immersion	Full	Full	Full	Partial
Resolution	Low	High	Medium	Medium
User acceptance[3]	Fair	Fair	Fair/good	Good
Detailed interaction	Low	Low	Low/medium	High
Group interactions	Low	Low	High	High
Portability	High	High	None	Low
Cost (device only)	$10K[1]	$35K[1]	$150K[2]	$60K[1]

[1]Requires high-end graphics workstation for most applications, cost approx. $150,000; some applications can be performed using less expensive computational engines.
[2]Requires multipipe, high-end graphics workstation, cost approx. $400,000.
[3]Refers to the willingness of users to stay within the virtual environment for extended periods.

Table 19.2 **System Usage**

	HMD/BOOM	Immersive Room	Workbench
Strengths	Navigation	Navigation collaboration	Detailed interaction collaboration
Sample applications	Architecture walk-through Single-user mission rehearsal	Scientific visualization Information visualization Multiuser mission rehearsal Engineering design	Medicine Engineering design Mission planning Scientific visualization Data mining

ology") is eliminated. We have found that users adapt to the metaphor in seconds. It largely fulfills the goal of VR, producing a natural environment where objects have presence (i.e., they are treated and act exactly as real objects would).

While this method is natural and intuitive, issues remain to be solved. The most obvious is that the user must physically wear the glove and that the glove is attached by wires to a control box. Ideally, one would like to recognize gestures without using a glove. Investigations into natural gesture recognition are taking place using techniques such as optical flow to identify hand motion. However, better algorithms and faster computers are required before the glove can be replaced by a camera-based system. Our glove metaphors are limited to natural actions such as grabbing and touching. Glove-based systems that require users to learn, memorize, and remember unnatural and nonintuitive gesture combinations defeat the purpose of VR.

Another issue in glove-based gesture recognition is that the instrumented hand blocks some of the imagery. To perceive stereo images correctly on the Workbench, the user's visual system focuses on the imaging surface (tabletop). However, the eyes also must converge at a point in space necessary to correctly perceive the depth of a specific object. For example, the eyes converge above the table for objects that appear above the imaging surface and converge below the table for objects below the imaging surface. This is not how the human visual system usually works and thus takes a user some time to adjust. Eye strain can result. Introducing the user's hand into the projected space causes the eyes to attempt to both focus on imaging surface and converge elsewhere and focus and converge on real physical object. Usually the human visual system will default to what is normal: it will focus and converge on the real object and lose the correct perception of the projected object. This often causes a user difficulty in selecting objects as he/she loses depth cues when the hand is near the projected virtual object.

NRL has also developed a two-handed glove system (Obeysekare et al. 1996). In this system, the user wears both a right-handed and left-handed pinch glove and can pick up objects in each hand. The two-handed system has been used to examine molecular manipulation and related applications.

Speech Recognition

The ability to issue verbal commands to a computer and have the computer understand and respond has long been a desired goal of the human-computer interface community. Ideally, the computer would understand conversational English, including the correct handling of pronouns, context between sentences, and continuous speech. Researchers are beginning to produce systems that move toward this goal. However, systems available today have less capability.

Speech input is ideal because speech is our primary communication channel. Many humans even talk to inanimate objects such as their cars and toaster ovens, although those inanimate objects have no speech recognition capacity whatsoever. Humans are simply comfortable with speaking to or at things in an attempt to communicate.

A well-designed system will support common commands that a user might want to execute. These commands are often highly dependent on the task that the application was designed to solve. Verbal commands are often much easier for a human to remember than a contrived keyboard key combination or button combination on an input device. Associating a short phrase describing the functionality of a button combination is one way to memorize what each combination does. Allowing the user to directly speak this short phrase removes the step of associating with a button, thus making speech more intuitive than buttons.

However, there still is the problem of knowing or learning what commands are available for a given environment in a given state. This is often referred to as the *habitability* problem. The richness of the English language allows a speaker to convey the exact same meaning by using completely different words. It is impossible to prepare actions for every possible human phrase that a user may utter during a session. The ability to make the user aware of what commands are supported in a given situation is presently a very large unsolved problem.

Today's technology still often requires slow, deliberate speech. This is changing as more research and commercial speech recognition groups perfect their software and hardware. The goal in many peoples' minds is to achieve the ubiquitous voice-activated computer in *Star Trek*.

We have used a commercial system as a voice recognition tool on the Workbench. The system has worked well in terms of understanding different accents; foreign speakers for whom English is not a native language have been understood. We have limited the number of commands, thus limiting the user's learning curve. However, the fundamental limitation of voice recognition remains. Users still need to whisper to us, Well, what do I say? since a user can't know the precise phraseology required. Progress toward natural language recognition is a step toward removing this limitation.

Simulated Laser Pointer (Wand)

Our third interface method is a simulated laser pointer (wand). To create this, we modified a PC flight stick and programmed a virtual laser beam to appear to emanate from the wand. The wand's motion is detected by the addition of an internal magnetic tracker, and the position of the laser is adjusted accordingly.

When the laser beam intersects an object, a sphere encircles it, indicating its selection. Movement of a selected object is enabled by pulling the wand trigger. The object is "grabbed" by the laser beam and may be moved to any location on the Workbench. Releasing the trigger caused the object to drop and fall to the surface. While the object is grabbed, it may be rotated by pushing the leftmost button, and it may be tracked in or out (with alternate button pushes) by pushing the rightmost button. The latter is convenient for moving objects long distances.

The wand is also used for applications with terrain to provide a convenient and intuitive method for object scaling, rotation, and translation. Button actions combined with user hand movements permit zoom, translation, and rotation. Once the correct button combinations are pressed, moving the hand up/down generates a zoom. Moving the hand left/right or forward/back generates a pan, and rotating the wand generates a rotation of the terrain. In addition to object intersection and movement and terrain motion, the wand may be employed as a query device.

The wand was designed to simplify the use of the Workbench. All of the previously mentioned functions are enabled by movement of the wand or combinations of trigger pulls and button pushes involving only one trigger and two buttons. However, the wand does not fulfill the long-term goal of VR of fully natural interaction. In addition, the wand requires the user to learn a sequence of unnatural interaction techniques: combinations of buttons and triggers are required for each interaction. This is reasonable when the number of interactions remains small. However, having to insert into the virtual environment a menu of required button/trigger interactions would not, in our view, be an effective interactive method.

Conclusions about Interactive Methods

We have found all three interactive methods discussed to be effective, although each has limitations. The next sections discuss two of our Workbench applications. For the first, we use a combination of glove and voice, while the second uses the wand. We plan to perform user evaluation studies to determine which combinations of interfaces produce the most effective interactions.

A Medical Application

The first application we developed on the Workbench as an early proof of concept was to display a human skeleton on the Workbench and investigate interactive methods for manipulating body parts. We selected this because of the very natural paradigm involved. A doctor standing over a patient on an examining (or operating) table knows the procedures he/she will undertake. Even nonpractical demonstrations, such as removing an organ, examining it, and replacing it, are in some sense "natural," whereby it is clear what needs to be done. This application uses

- Direct manipulation
- Simple pinch gesture recognition
- Speech recognition
- Tracked stereographic projection

The model used for this application is a commercially purchased human adult skeleton with many of the major internal organs. An articulated glove model was used as an avatar for the user. The purpose of this application was to experiment with direct manipulation, simple pinch gesture recognition, and speech recognition. Head-tracked, off-axis

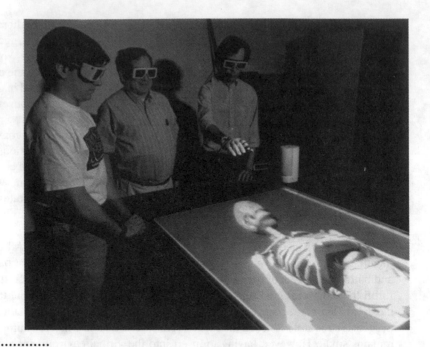

Figure 19.2   A medical application of the Workbench. Both gesture recognition and speech recognition are used to interact with the virtual patient. (*See color plate on page 333.*)

stereo projection was used to provide the user with the best possible perspective into the virtual environment. Figure 19.2 shows the medical application on the Workbench.

Interaction for the Medical Application

A user had two methods of interacting with the model. The first was using a direct manipulation method; the user literally translated his hand into the projected skeleton model. The intersection of the hot spot of the glove avatar with a bone group or organ caused the object to be highlighted. This indicated that the user had selected that bone group or organ. The completion of a pinching gesture when a bone group or organ was highlighted caused the selected item to become attached to the glove avatar. In effect, the user had grabbed that item. The user can then manipulate the item as if he/she was really holding it.

Direct manipulation worked well for large bone groups and coarse movements of selected items. However, there were problems selecting small groups due to the precision movements required that were not possible with the tracker system and in the real world highly dependent on tactile feedback. This is also true for replacing bone groups. Without modeling collisions with other bone groups and many other aspects of the simulation, a user could not place the bone groups back exactly where they belonged. Manipulation of the entire model also became an issue. The interface could have been constructed using keyboard key sequences, additional pinch gestures, or complicated combinations of both solutions. However, we felt that this would increase the burden on the user. We wanted to make the interface as intuitive as possible. Thus, we experimented with speech recognition to address some of these concerns.

The second method of interaction was speech recognition. We used the commercially available HARK system from BBN. This system is a speaker-independent system that does not have to be trained to each user's voice. It requires a predefined grammar and thus has limited recognition capabilities. For this application, our grammar consisted of very simple and short commands.

The user could select (highlight) any bone group or organ that he/she could name (e.g., "select clavicle"). The user could also "grab" an already selected (highlighted) item or could directly request a specific bone group or organ to be grabbed (e.g., "grab heart"). The named item would move up to the glove avatar. Using speech recognition, a user can select any bone group or organ on or off screen, regardless of how difficult it may have been to select that item using direct manipulation.

Finally, the user had very basic manipulation control over the entire human model. The user could "rotate" the model, and he/she could "scroll" the model left, right, up, and down. These operations could be done only through the speech recognition system.

While primitive, this proof of concept provided a starting point for conversations with medical professionals and other researchers. Most of the medical professionals that viewed this application immediately saw the potential for educational use. However, they saw the system to be even more applicable to medical procedure visualization and planning. They would like to have the ability to visualize real-patient 3-D X ray, CAT, MRI, or other datasets in a near real-time manner. This would allow a doctor to visualize and plan a medical procedure with the actual anatomy of the patient.

Situational Awareness Using the Workbench

In this section we describe a situational awareness Workbench system that was fielded in March 1997 (Rosenblum et al. 1997).

Overall Requirements

Our task is to provide situational awareness for the complex logistical task of directing the movement of U.S. Marines and material over rugged terrain, day and night, in uncertain weather conditions. This difficulty is multiplied by the well-known dangers of amphibious assault, long considered the most difficult problem in warfare.

Even with the advent of computers and sophisticated decision-making software in Marine Corps Combat Operation Centers (COC), command and control are predominantly undertaken with paper maps and acetate overlays. This is a cumbersome, time-consuming process. In addition, detailed maps and overlays can take several hours to print and distribute. There currently exists no overall picture of the battle space that provides a commander with a dynamic range of resolution sufficient to track units ranging from aircraft carriers to six-marine fire teams. Furthermore, a mechanism is needed to deliver information, on demand, concerning the status of any unit of interest (fuel supply, ammunition, casualties, etc.). The resolution and bandwidth requirements to deliver this "big picture" in 3-D is beyond the capabilities of the PCs and low-end workstations typically found in a COC. The Workbench is one item being demonstrated for possible use in an Enhanced COC (ECOC). The goal of this preliminary demonstration is to show the Workbench's capability to represent a large area terrain on the Workbench at a resolution comparable to maps used in the field and to utilize a selection of the interactive techniques previously discussed to manipulate icons representing forces and objects on that terrain. Figure 19.3 shows a mission-planning application on the Workbench, while Figure 19.4 shows several marines being trained on the system discussed in this section.

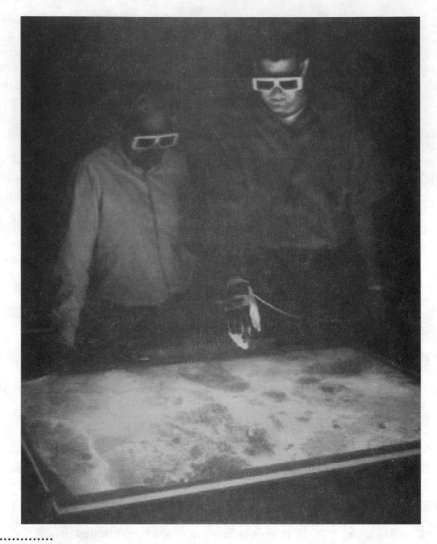

Figure 19.3 NRL's VR lab personnel study battlespace terrain for an ONR-sponsored Command and Control research project. (*See color plate on page 334.*)

Terrain and Texture

Reasonable 3-D terrain resolution of an area the size of the training base (a 62 × 72 km area) requires a minimum of 20,000 vertices for high resolution. Complicating the construction of the terrain was the requirement for a virtual "ocean" outside a road network bordering the training area. To this end it made sense to utilize a commercial modeling and terrain package that could both automatically construct terrain from raw Defense Mapping Agency (DMA) data and provide the tools to create a reasonably realistic ocean. DMA data used was a height field on a 100 m grid for the area of interest.

Commercial software was used to read DMA data directly from CD-ROM, select a usable resolution, and employ Delauney triangulation to calculate vertices and place them efficiently. The construction of the ocean required that the texture map be applied to the terrain so that the borderline road network could be seen. Vertices lying outside the desired border could be selected in groups and their elevation decreased to zero. It was important

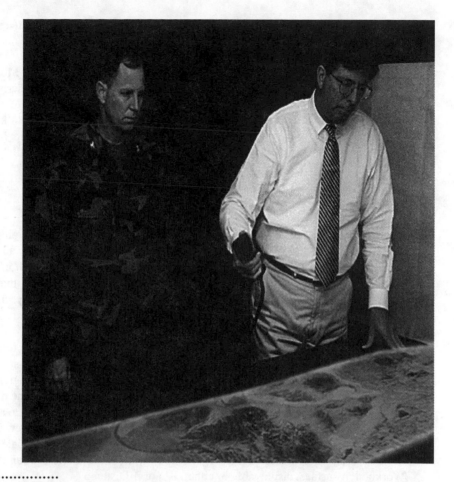

Figure 19.4   Special Purpose Marine Air Ground Task Force (Experimental) Commanding Officer Col. Tom O'Leary receives training on the Workbench at Enhanced Combat Operations Center in preparation for the Hunter Warrior exercise for Sea Dragon. (*See color plate on page 334.*)

to select only vertices that were not part of an edge that crossed into the land area. Otherwise, the terrain contours of the exercise area would have been distorted. Thus, a trade-off was made that left some ocean vertices at an elevation of greater than zero. These were concealed using an alpha channel in the terrain texture.

Even more demanding are the requirements for texture-map resolution because, in addition to aerial and satellite photos, the terrain must be textured with line-drawing maps with a geographic resolution approaching 1:250,000. This requires a minimum of an 8 k × 8 k pixel image that will occupy 125 MB of storage in raw 16 bit format. Reduction of line drawings to more manageable sizes is impracticable because of the unavoidable loss of contours, grids, and text legibility (although a 2 k × 2 k map was used for ocean creation). This is beyond the capacity of the maximum texture memory for an Onyx (64 MB). Thus, we used a clip mapping approach to terrain texturing. Clip mapping hardware is standard on the Onyx Infinite Reality.

A large map image (8 k × 8 k) of the exercise area was extracted from a DMA Compressed ARC Digitized Raster Graphics CD-ROM. This was too large to work with directly and our "ocean" had to be drawn on a 2 k × 2 k copy using Adobe Photoshop.

The image was then scaled to 8 k × 8 k, and the monochrome ocean area was selectively copied and ultimately composited over the original 8 k × 8 k image of our map using the San Diego Supercomputer Center's Imaging Tools.

An alpha channel was added to render the ocean area transparent. Modulation of the texture over the terrain resulted in the transparency of the polygons underlying the ocean. To create the clip-map texture, the final image was cut into a pyramid of 1 k × 1 k tiles; for example, 64 tiles from the 8 k × 8 k image, 16 tiles from a 4 k × 4 k image, and so on, using various Silicon Graphics imaging tools. The clip mapping hardware in an Onyx/Infinite Reality Engine manages the paging of the 1 k × 1 k tiles between disk, RAM, and texture memory.

Models

Approximately half of the military models used were of commercial origin and were, typically, models of complex equipment such as tanks, ships, helicopters, and so forth. The simpler models were constructed at our laboratory. Large units (battalion size or larger) were represented by flags bearing the name and seal associated with that unit. Smaller units (platoons, squads, and fire teams, for example) were represented by simple cubes textured on all sides with standard military symbols such as an *X* for infantry and a sideways *E* for engineers. These are easily recognizable by the users.

Placement of units on the terrain can be achieved in two ways. The first uses a simple token scheme in a designated status file. Each token is followed by data such as unit name, lat/lon, altitude, and so forth. The file is read automatically by the program when a time change is detected. An output file of the same format may be saved at the discretion of the user. The second method uses on-the-fly electronic messaging. A separate application, initiated manually, downloads the e-mail containing status updates. The new icons are introduced upon receipt of the e-mail. The user maintains control of the updates, because they are delivered on demand. Figures 19.5 and 19.6 illustrate the terrain and models on the Workbench surface, displayable in either 3-D or 3-D stereo.

Interaction

An overall view of the battle space, including hundreds of operational units, is useful in itself for command decisions; however, more detailed information is needed to prosecute an action. A simulated laser pointer was used for interrogation of icons, and presentation of their underlying data was done on a heads-up display (HUD) on the right side of the workbench surface. The HUD displays all the information previously discussed and automatically disappears 5 seconds after a unit is deselected or immediately when a new unit is chosen.

User controlled movement of an intersected icon is enabled by pulling the flight stick trigger. The object is "grabbed" by the laser beam and may be moved to any location on the terrain. Releasing the trigger causes the icon to drop and fall to the terrain surface. Aircraft are encoded to remain at whatever altitude they are released. While the icon is grabbed, it may be rotated by rotating the wand about the laser's axis, and it may be tracked in or out (again with alternate button pushes) by pushing the right-most button. The latter is convenient for moving icons long distances across the terrain.

The flight stick provides a convenient and intuitive method for scaling and translating the terrain. The following actions are always enabled with the trigger pulled in, the terrain intersected, and a button pushed. Pressing the left-most button and raising the flight stick causes the terrain to uniformly scale up. Moving the flight stick left or right

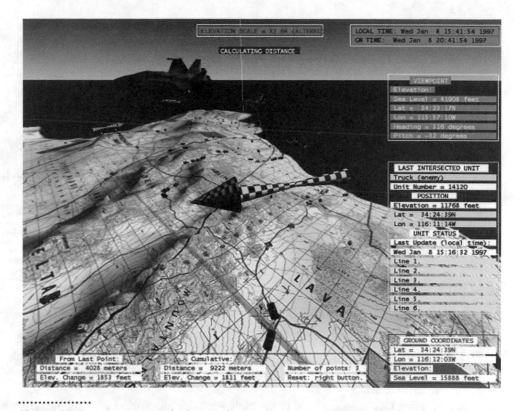

Figure 19.5   Icons are overlaid atop the terrain along with textural information. The image seen appears in 3-D on the Workbench. (*See color plate on page 335.*)

and forward or back, with the same button pressed, moves the terrain in the same direction. Pressing the right-most button and rotating the flight stick around the Z-axis causes the terrain to rotate in the same direction around the Z-axis. Rotating the flight stick around the X-axis, with the same button pressed, causes the terrain to rotate around the X-axis (change pitch).

In addition to icon intersection and movement and terrain motion, the flight stick may be employed as a measuring device. Whenever the laser intersects the terrain, a small HUD, in the lower right corner of the workbench, appears and displays the coordinates (lat./lon. and UTM) and the elevation (above sea level) at the point of intersection.

Distances and headings are measured by intersecting the first point of interest (or icon) with the laser and pressing only the left-most button. The second point is then intersected and the left-most button pressed again. A HUD appears along the lower edge of the screen and displays the distance, heading, and elevation change between two points or a series of points. Pressing the right-most button resets the measurements and causes the HUD to disappear.

The flight stick concept was designed to simplify the use of the workbench. Indeed, all the above functions are enabled by movement of the flight stick or combinations of trigger pulls and button pushes involving only one trigger and two buttons. However, as the functionality of the application increases, so will the difficulty of providing simple and intuitive interactions. We are currently planning to perform evaluation testing of interface methods. The results of these evaluations will drive future interface development efforts.

Figure 19.6 An example of the high resolution obtainable with clip texture mapping techniques. (*See color plate on page 335.*)

Networking

An ongoing challenge for VR is to integrate VR with networking to facilitate remote collaboration in problems ranging from manufacturing through modeling and simulation. This issue can be subdivided into two classes. Some applications require complex interactions among a limited number of participants, while others, such as military simulations, require servicing thousands of players. Large, multiuser virtual environments must keep each entity aware of other's actions. This places considerable demands on the workstation I/O, network bandwidth, and the underlying architecture. One approach to this challenge, developed at the Naval Postgraduate School, is NPSNET (Macedonia et al. 1994). NPSNET is a large-scale software package designed for networking that is capable of simulating articulated humans and ground and air vehicles in the DIS networked virtual environment of 250–300 players. NPSNET is the first 3-D virtual environment to make effective use of the multicast backbone of Internet in order to avoid direct connections between all sites. It also makes extensive use of dead-reckoning to predict object position and reduce visual latency in low-bandwidth situations. The software architecture logically partitions a virtual environment by associating spatial, temporal, and functional classes with network multicast groups.

We have begun an investigation into networked Workbenches, jointly with the Graphics, Visualization, and Usability Center at the Georgia Institute of Technology. For the Workbench, the problem is not dealing with a large number of users. Rather, the Workbench emphasizes fine-grained interactions. The detailed interaction raises many interesting issues in human perception, interaction, and collaboration. Questions of how to partition the usable workspace, of what operations you can and cannot perform (remotely) on my Workbench, of how two remote users can share a common object, and similar issues will be the topic of future investigations. We have recently completed a first demonstration of networked Workbenches. The workbenches are connected by an ATM network, and each viewer sees correct perspective. Issues of how joint collaboration should be performed are under investigation.

Conclusions

The Virtual Reality Responsive Workbench is fundamentally different from previous VR systems in that it emphasizes fine-grained interaction rather than navigation through immersed space. Only four years old, the Responsive Workbench is rapidly being accepted as a major VR paradigm. It has transitioned from a research tool, to commercial development of the hardware, to implemented, utilized systems such as the situational awareness application discussed in this article. A number of research and development issues are first being examined. These include graphical representations on the workbench, interface issues, topics in perception and evaluation, and effective systems for networking the Workbench. Hardware improvements are also important; particularly, we look forward to the time when the projector can be replaced by flat panel displays. We anticipate much activity in Workbench development over the next five years by many organizations. We see Workbenches moving out of the lab and into command centers, engineering design centers, medical training centers, and to other end users with similar requirements.

Acknowledgments

The authors thank others who have contributed to these projects: Rob King, Brad Colbert, and Chris Scannell, as well as graduate student summer interns Greg Newton (Georgia Tech), Jim Van Verth (Univ. of North Carolina), and Ken Gordon (SUNY Stony Brook). This work was sponsored by the U.S. Office of Naval Research, the Marine Corps Warfighting Laboratory, and the Defense Advanced Projects Research Agency.

References

C. Cruz-Neira, D. Sandin, and T. DeFanti, "Surround-Screen, Projection-based Virtual Reality: The Design and Implementation of the CAVE," *Computer Graphics, Proc. Siggraph 93,* July 1993, pp. 135–142.

N.I. Durlach, and A.S. Mavor, eds., *Virtual Reality: Scientific and Technological Challenges,* National Academy Press, 1995.

R.A. Earnshaw, and L.J. Rosenblum, "Virtual Reality, Visualization, and Their Application," *Proc. Interface 95,* 1995.

W. Kreuger and B. Froehlich, "The Responsive Workbench," *IEEE Computer Graphics and Applications,* Vol. 14, No. 3, May 1994, pp. 12–15.

W. Kreuger et al., "The Responsive Workbench: A Virtual Work Environment," *IEEE Computer,* Vol. 28, No. 7, July 1995, pp. 42–48.

M.R. Macedonia et al., "NPSNET: A Network Software Architecture for Large-Scale Virtual Environments," *Presence,* Vol. 3, No. 4, Fall 1994.

U. Obeysekare et al., "Virtual Workbench—A Nonimmersive Virtual Environment for Visualizing and Interacting with 3-D Objects for Scientific Visualization," *Proc. Visualization 96,* Oct. 1996, pp. 345–350.

L. Rosenblum, S. Bryson, and S. Feiner, "Virtual Reality Unbound," *IEEE Computer Graphics and Applications,* Vol. 15, No. 5, Sept. 1995, pp. 19–21.

L.J. Rosenblum and R.A. Cross, "Challenges in Virtual Reality," to appear in *Visualization and Modelling,* Academic Press 1997.

L.J. Rosenblum et al., "Shipboard VR: From Damage Control to Design," *IEEE Computer Graphics and Applications,* Vol. 15, No. 6, Nov. 1996.

L.J. Rosenblum et al. "Situational Awareness Using the VR Responsive Workbench," *IEEE Computer Graphics and Applications,* Vol. 16, No. 4, July 1997.

Chapter 20

Inner Space: The Final Frontier

David Leevers

Multimedia Communications, BICC Group

Introduction

Once upon a time the computer was clearly part of the external world, hidden behind layers of data preparation clerks and programmers. Even the recent desktop interface was reassuringly external, explicitly placed at arms length. However, the networked virtual environments that have been described at the *From Desktop to Web-top: Virtual Environments on the Internet, WWW, and Networks* (1997) conference are not just animating the vastness of space beyond the screen, they are also punching out at us in true 3-D movie fashion and starting to colonize the "inner space" of our private mental models.

The catalyst for this impertinent invasion is a hope that shared visualizations can be used to support new forms of collaboration with other people. This paper draws on a number of communications projects that are helping to define a new "enhanced reality" that goes way beyond shared visualization to a seamless integration of the real world, telecommunications, and virtual reality. A conceptual framework known as the *cycle of cognition* is being proposed as a way of disentangling the many modes of communications with individuals, groups, and data, all of which may be local, remote, or virtual.

Because VR applications play with the boundary between real and imagined spaces they are helping to give new insights into how thinking is dominated by the physics of the real world on one hand and by trust in colleagues on the other. Such an understanding is becoming essential as new networked services are helping to reconcile the complete range of fundamental human needs within an ever more complex global society. This society is considerably more than just an "Information Society." It is perhaps the Holy Grail of the Me Generation—a self-actualizing society or "i-society" for short.

Ubiquitous Telecommunications

My own introduction to virtual reality was not a need to simulate a real environment but the desire to support day to day activities in manufacturing and construction. Since the workface is inherently three dimensional and teams can be distributed across a wide area, this was seen as an ideal application for collaborative virtual environments.

The range of social and practical activities in manufacturing and construction is similar to that in the rest of our lives. This has encouraged speculation on how society as a whole will change as local and physical reality is augmented with telepresent and virtually present people and objects.

Most forms of Western leisure involve copious consumption of material resources in the form of possessions and travel. Any attempt to replicate such a lifestyle throughout the planet is unsustainable and would be destined to plunge the world into a new dark age of conflict and destruction. However, by applying emerging understanding of the evolution of humans as social animals, we may be able to construct a new "enhanced reality" in which compatible wishes are fulfilled in the real world and conflicting ones are transferred to the virtual. The route to this enhanced reality is still obscured by the fog function of our personal z-axis buffers. Many sci-fi writers have well-developed private visions, but as the recent 2001 retrospect industry has shown, even the most competent of those private visions can be hopelessly inaccurate. HAL can tell us that, or can he?

We have attempted to build this optimistic scenario by studying the lifestyle of those who already have unlimited resources, the emperors and lottery winners among us, and then by introducing virtual surrogates for what can never be made available to all. Already the compact disc has replaced the emperor's private orchestra, the movie and TV have almost completely taken over from the theatre. Agents are being programmed to provide the obsequious but stimulating conversation of a skilled courtier, and the barest minimum of white goods can replace most slave tasks.

This new world of virtually unlimited resources (or unlimited virtual resources!) will be accepted only if it is satisfying and fulfilling. The enhanced real space must stimulate a new "inner space," a cosmology of the mind that will be far richer and more rewarding than the barrenness of outer space. It will be a space in which we can enjoy fantasy lives while remaining firmly connected to the responsibilities of the physical reality that we share with other minds. The new society is effectively a multiplayer mixed reality game in which everyone can win, and everyone can be virtually famous, infamous (or whatever else they want to be) not for just 15 minutes but for the whole of their lives.

Spatial Metaphors

Millions of years of natural selection have fine-tuned the brains of our prehuman ancestors for competence in the spatial and gravitational world. However, humans also share language, ideals, ethics, and logic. How could such an overwhelmingly powerful thinking capability have emerged over an incredibly small number of generations?

Current thinking in some anthropological circles is that an existing part of the brain was diverted to distinctively human activities. It is pretty obvious which part of the brain was left with little to do as protohumans descended from the three-dimensional complexity of the forest canopy to the flat two-dimensional savannah. Much of our 3-D spatial competence may have been co-opted to handle the increased complexity of social interactions that became possible on the ground.

The archaeological evidence is found not under our feet but in our words. Not only is every language replete with spatial metaphors for social and abstract concepts but, more persuasively, the metaphors are much the same in all languages. Up and down, near and far, left and right, inside and outside, on top or underneath are typical of any conversation about abstract concepts. We appear to have diverted much of our 3-D capability into a richly verbal social life while retaining immediate visual awareness for navigating across the 2-D surface of savannah or city.

The Proscenium Arch

Given that what we are actually aware of is a visual panorama with a minimal amount of depth information, awareness of the surroundings is better described as 2.5-D than 3-D. In a sense, the blind are the only ones that can "see" in 3-D. Because we cannot see things in the round we put much effort into trying to suppress the third dimension. We prefer to stand with our back to a wall and everything is pushed to the periphery of a room. Books and pictures hang on the walls; they are not placed in the center of the room, and most abstract creative activities are recorded on a 2-D paper surface.

This trend to flatten the real world is illustrated in the evolution from the theatre-in-the-round of traditional tribal celebrations to the rectangular stage or screen of the most highly developed forms of narrative entertainment. The 2.5-D stage with its simplified layers of scenery has a more consistent dramatic impact on every member of the audience. The fact that the rectangle of entertainment can occupy a small fraction of the field of view shows that it is not difficult to suspend disbelief, or immerse the mind, as long as the performance is good enough. The experience of physical immersion does seem to be so overpowering that it is best confined to brief dramatic moments—the theme park ride, the actor who enters by bursting out of the audience, or the thunderous opening to *Star Wars*.

Where Is the Virtual Mind?

Perhaps our heads did remain in the clouds when we descended to the ground, the new clouds of fuzzy spatial metaphors replacing the old clouds of the rain forest canopy! Given that our 3-D mental space may be fully occupied with the subject of the discussion, whether it be physical or abstract, 3-D VR representations of remote or virtual colleagues may in fact be confusing rather than supportive.

Strangely enough, the mind-body problem becomes a major issue when trying to build useful shared virtual environments. Although we no longer try to place that mind in the heart or another even more unsuitable organ, we are still not quite sure how the mind relates to the brain. Perhaps we can take advantage of this uncertainty by using more abstract representations of remote and virtual colleagues and thus remove an annoying problem with the real world—the fact that it is not possible for two people to stand in the same place and share the same spatial point of view, an important step towards agreement (i.e., sharing the same conceptual point of view).

It may not be necessary to represent the immediate social group as VR avatars or live video within the shared enhanced reality. Since, in another sense, the network is already starting to liberate the spirit, perhaps the other minds should be represented as "free spirits," only loosely associated with the shared virtual environment. Thus, the telepresent friends watching a virtual football match need not be represented realistically, whereas the virtual players must remain firmly bounded by the physics of gravity, space, time—and the rules of football.

The Cycle of Cognition

Over the last decade the Multimedia Communications Group of BICC has been trying to define the architecture and benefits of what is now called a persistent multimedia communications environment. This work has been partially funded by the UK CSCW project

VirtuOsi and a series of EC RACE and ACTS projects, DIMUN, BRICC, CICC, RESOLV, and MICC, that have all been aimed at the manufacturing and construction sectors.

The conceptual framework for this work is known as the *cycle of cognition.* By including the most probable path from one type of communications to another the cycle helps to define requirements not only for communicating but also for changing from one mode of communications to another. The stages in the cycle—home, map, landscape, room, table, and theatre—cover an idealized working day (Figure 20.1). However, the metaphoric power of the cycle comes from its ability to support tele- and virtual communications and to cover a very wide range of time scales.

The cycle of cognition is effectively an ecology of information and communications services. It presents a dynamic view of such services and indicates how squeezing them in one place is bound to lead to bulges of confusion or bandwidth at others.

The Cycle of Cognition

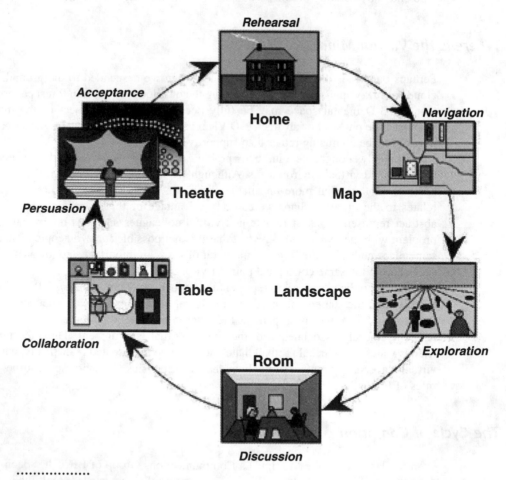

Figure 20.1   The cycle of cognition. (*See color plate on page 336.*)

Home

The home is that private place where we sleep on what we have recently learned and sub-consciously adjust our existing mental model of the world to accommodate new concepts. We rehearse the implications of new ideas, perhaps in dreams or in play, and start to for-mulate what needs to be done next. This metaphoric home is more of a private study or bedroom than family residence because it is private territory, not shared with others.

Map

When we venture out from home we are not immediately ready to confront other people. The map refers to passive information: the morning paper, a book, Yellow Pages, a real map, or a reference book that provide a stable starting point for the tasks of the day. It is not neces-sarily up-to-date, but that is an assurance of its stability. Using the map is a reassuringly familiar mental warming-up exercise that we have probably been through many times before.

Landscape

As the objective becomes clearer it becomes necessary to track down the most up-to-date information. We scavenge in an open landscape made up of people and the documents with which they are associated. Brief conversations are confined to fact finding, and so it is acceptable for others to overhear them. The landscape can be real, such as an open plan office, or virtual, say, a VR visualization of relevant people and documents.

Room

Contentious or unresolved issues need a more committed discussion than is possible in the public landscape. The walls of a real room manifest a complete security and privacy bar-rier that encourages occupants to share information. Similarly, the walls of a shared virtual room displayed on each participant's screen are a reminder of effective firewalls until the group have established mutual trust and understanding. Then they feel confident enough to take their eyes off each other and look down at the shared table.

Table

Most meetings take place around a table, and there is always the problem of letting every-one get an adequate view of documents placed on the table. One advantage of a networked virtual table, usually known as a shared window, is that it can display identical material to all participants. However, each person still has a private area; the part of his/her computer screen that is not shared with the others, or a real-life notepad or filofax.

In our implementation, the virtual table occupies the lower part of the computer screen. Across the top are video windows of the other participants together with shrunken copies of their own screens. This is an example of being able to break the rules of Cartesian space when representing minds rather than things. All the other minds appear to be in ideal positions on the other side of the table, an impossibility in real life.

The video windows indicate the extent to which the others are participating. They maintain rapport and encourage convergence to an agreed conclusion. The shrunken screens serve the same function as glancing across a real table to see what the others are doing. As in real life, these miniatures are not clear enough to read what is being written.

Theatre

The result of a meeting is of use only if it is accepted by the relevant audience: customers, colleagues, or students. A representative of the meeting-room group takes on a theatrical role to broadcast the conclusions. In such a performance the narrative flow is usually decided in advance, but the emotional emphasis can depend on the mood of the audience. It is important to note that the audience is prepared to accept the new story because they know that agreement has been reached around the table and, perhaps more significantly, because each member senses that the rest of the audience is also accepting the message.

Any effective performance changes the way members of the audience think (i.e., how they will react to related situations in the future). The performance can be said to have killed the previous personality and given birth to a slightly different one. This is one reason for the rituals associated with becoming a member of an audience, checking reviews before buying a ticket, studying the mood of others in the foyer, and keeping an eye on them during the performance.

Back Home

Whereas the facts collected in the landscape are used immediately and often subsequently forgotten, the wisdom received during a performance does not trigger immediate action. It is usually necessary to retreat to the home where any new constructs can be absorbed into long-term memory, ready to guide reactions when the time is ripe.

When the potential new construct is physical rather than mental, the home refers to that part of the real world that is temporary personal territory—the location of activities such as bricklaying or manufacturing that are carried out after receiving instructions in the theatre of the foreman's instructions.

A Fractal Cycle

The cycle of cognition echoes our daily cycle: waking at home, setting off using a map, actively browsing the business landscape in the morning, negotiating over the midday meal and collaborating in the hazy glow of the afternoon, then taking a seat in the theatre as the sun sets to surrender the mind to the persuasive powers of the actors, and finally staggering home with new ideas teeming inside our head—ideas that will have been absorbed into and have slightly altered our "inner space" by the time we wake up the next morning.

Similarly, the cycle reflects our journey through life: emerging from the home of the womb, spending a few months in a map of sensations until learning from the landscape of older people becomes possible, then spending adolescence in a lively meeting room of developing personalities, subsequently settling down to more focused and shared activities around the more formal table of career and family, and finally gaining the respect of the community and taking the stage to pass cultural memes to the next generation.

The cycle has fractal properties in that cycles can be nested within other cycles almost indefinitely, from the formulating of a single suggestion and presenting it on the whiteboard during a meeting to the complete lifecycle of a civilization.

The Global Virtual Factory

Our first attempt to evaluate the usefulness of the cycle of cognition is a prototype Intranet known as the Global Virtual Factory. The business objective is to make it as easy to work with people in remote factories as with colleagues on the same shop floor. In this way a distributed global organization can gain the economies of scale that are currently confined to massive centralized plants. The collaboration tools include:

• *Home pages:* These personal Web pages would eventually supersede the company telephone directory. They include video and screen glances that are refreshed once a minute in order to provide the modest degree of activity awareness that is comparable with walking around an open plan office.

• *Nearest neighbors:* Personality awareness is conveyed by asking everyone to nominate their six most important contacts in the organization. This proved to be an effective and acceptable way of capturing the informal social structure of the company as opposed to its formal organization chart.

• *Photo panorama:* A panoramic photo taken from where a person works provides a literal way of seeing things from the other persons' point of view, an essential first step towards reaching an agreement.

• *Shared window:* This is one of the oldest PC-based collaboration tools, allowing all participants in a distributed meeting to see the subject from the same point of view without having to crowd around the same physical document.

• *Video open plan office:* A row of video windows below the metaphoric landscape separates the collaborating minds from the 3-D subject (Figure 20.2).

• *Egocentric meeting:* The video glances of all those who might contribute to a meeting are placed at the top of the screen, as if the shared window is the meeting table and everyone else is on the other side of it—an egocentric seating plan that can only be enjoyed

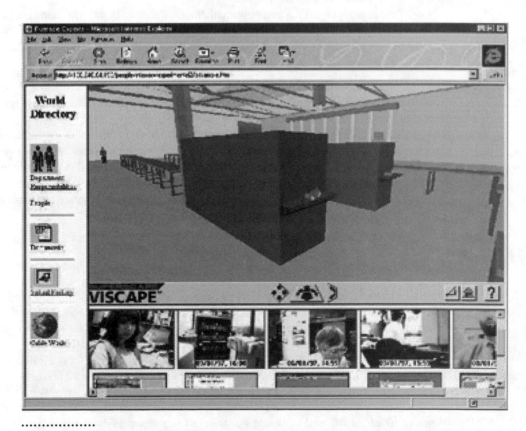

Figure 20.2 The reference factory landscape. (*See color plate on page 336.*)

by one member in a real meeting. In addition, people who are needed for only part of the meeting can continue with their other activities.

• *Reference factory:* People in different factories do not always have the same view of what a factory is. By building a virtual reality reference factory we have been able to reinforce the important common factors between a group of factories that look very different in real life.

• *Photo walkaround:* Unique features of a particular factory can be displayed by selecting from the saturation photo set for the factory of interest. By zooming and panning between views the user gets the impression of moving around a remote factory without having to travel halfway around the world to visit it.

This repertoire of tools has been selected in the light of workplace studies that have analyzed the nature of collaboration in both construction and manufacturing. Three distinct approaches have been used:

• *Distributed cognition:* This refers to identifying how teams work together to solve problems, contributing to an understanding that is somehow spread across the brains of several people.

• *Organizational memory:* Earlier ethnographic studies of the introduction of networked information to such work groups has indicated that it is important to distinguish four types of information: (1) in peoples' heads, (2) in documents, (3) in structured data, (4) in the real world, and ensure that they are all adequately supported. Only when people feel that they can choose freely from all the available sources will they trust the more rigid but also more accurate structured data.

• *Situation theory:* This is a new mathematical theory of information that provides a way of structuring the context of any form of communication. Context is not an issue when everyone is in the same room but is vital if misunderstandings are to be avoided during distributed meetings.

Most of the collaboration tools are emerging as part of the Internet and Web consensus. A few difficulties remain over bandwidth, response times, and privacy, but there is nothing that is inherently either impossible or expensive.

Initial results are very promising, in spite of the immature state of the network and applications. By focusing on the new forms of social interaction that take place across the network we are addressing the organizational issues that are central to the effectiveness of comprehensive data repositories. The communications tools that place information at our fingertips and the IT tools that make sure the information is of value are beginning to complement each other in a very encouraging way.

Enhanced Reality

Initially immersive audio-visual collaborative virtual environments appeared to hold great promise as a way of supporting social interaction for both business and leisure. Unfortunately, most activities remain obstinately attached to the reality of our immediate surroundings—high quality touch, smell, taste, and kinesthetic sensing. We need to enhance reality, not simulate it. Virtual reality technology may be condemned to a rather fruitless asymptotic path towards mere reality—perhaps Moore's law should be updated to state that the difference between the virtual and the real halves every two years! A way out of this Xeno paradox is to leap-frog reality, to ignore the difficult (but challenging) problems of touch, smell, and taste and go straight for a networked enhanced reality.

In the last few decades the silicon technologies of electron and photon grew to dominate our lives. They are now maturing, and it is becoming possible to regain control, to move on from the silicon products and start thinking about the processes that they can support. The user surface of enhanced reality will be made up of many different components, including more comprehensive forms of the 3-D reference factory previously mentioned. The technology itself will have become a new infrastructure, as invisible and ubiquitous as traditional urban services.

The new enhancements to our local realities promise to be just what the human animal needs, a way of mediating both our social nature and our egocentric urges. The new network infrastructure will support negotiation tools that reconcile each person's own need for self-fulfillment with the needs of others. If the conflict is over the location of an activity, then real presence is supported by telepresence. Only one person can occupy a Formula 1 car, but the radio camera allows 1,000 million people to become telepassengers. If the conflict is over physical resources, then virtual resources are introduced. The arcade game replaces the Formula 1 car for the smaller number of enthusiasts who would rather occupy the driver's seat than admire their hero's skills.

The i-Society

The global middle class is starting to use these technologies to achieve a degree of self-fulfillment that was impossible for all but those at the top of a traditional community. Because the tools required for this cosmic sleight of hand are inherently inexpensive there is no reason why the whole world cannot eventually rise to a common plateau of real presence, telepresence, and virtual presence.

Components of this new i-society include

- i-nformation, the i-nfrastructure of cooperation and competition
- i-nnovation, within a sustainable framework
- i-nteraction, the vital spark that brings all of us to life

Not a moment too soon, the Web is becoming the nervous system of the planet, albeit a dry silicon one rather than a wet carbon one. Perhaps this is the time to update the Gaia hypothesis. The new CyberGaia is a global, web-footed amphibian who is equally at home in the world of bits as in the world of atoms, and whose own webbed feat is to provide the physical and the information infrastructure for a sustainable, balanced, and fair society.

References

"Augmented Reality through Wearable Computing,"
   http://www.white.media.mit.edu/tilde/testarne/TR397/main-tr397.html
"The Cycle of Cognition," http://www.hhdc.bicc.com/people/dleevers/papers/cycleof.htm
"The EC Projects," http://www.hhdc.bicc.com/cicc/
"Evolutionary Psychology: Cosmides and Toomby,"
   http://www.psych.ucsb.edu/research/cep/primer.htm
"The Four Types of Information," http://www.hhdc.bicc.com/cicc/pif/pifsample.htm
K. Devlin, "The New Cosmology of the Mind—Goodbye Descartes,"
   http://www.maa.org/devlin/devlin_archives.html (1997).
"Spatial Metaphors in Language," http://metaphor.uoregon.edu/metaphor.html

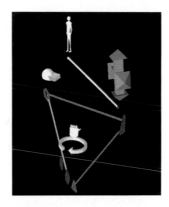

Figure 1.6 Immersive representation of the virtual treadmill navigation metaphor.

Figure 2.1 Aerial view of the University of Essex campus.

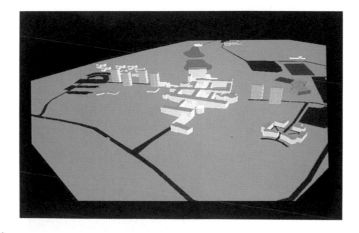

Figure 2.2 View of the University of Essex VRML campus model.

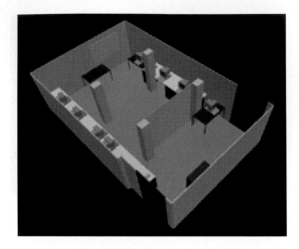

Figure 2.3 View of the VASE laboratory model.

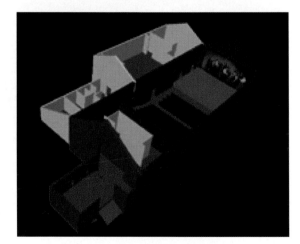

Figure 2.4 Model of the proposed Wivenhoe Theatre: Roof and first floor removed.

Figure 2.5 Model of the proposed Wivenhoe Theatre: View of the stage from a seat.

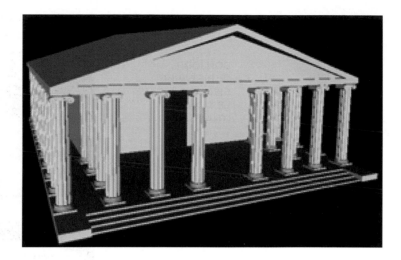

....
Figure 2.7 Roman temple model built using column-generating script.

Community Network is an experimental shared space created at BT Labs, using Cosmo Worlds on Silicon Graphics workstations for use with the Sony Community Place browser on Windows95. Shared spaces are implemented using Sony's Community Place server. The shared space is intended to avoid representing any recognizable locations, but to embody a Suffolk flavor in its design.

Utterances and Movements of the User	*Utterances of the Agent*	*Snapshots Presenting the Movements of the Agent*
	BONJOUR CHRISTOPHE BIENVENUE DANS LE MONDE ITHAQUES.	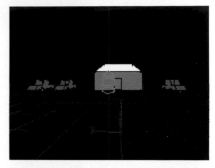
BONJOUR JE VOUDRAIS ALLER DEVANT LA MAISON.		
	C'EST PARTI.	
	J'AI FINI.	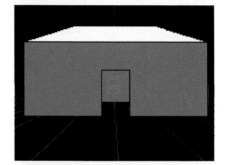
EST-CE QUE TU PEUX REGARDER À DROITE?		
	ET VOILÀ.	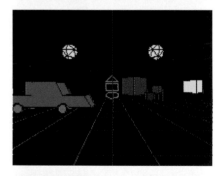
J'AIMERAIS BIEN MONTER SUR LA VOITURE.		
	OUI.	
	J'AI FINI.	
RETOURNE DEVANT LA VOITURE.		
	D'ACCORD.	

Utterances and Movements of the User	***Utterances of the Agent***	***Snapshots Presenting the Movements of the Agent***

J'AI FINI.

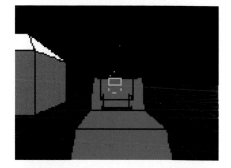

NON DEVANT LA MAISON.

BIEN.

ÇA Y EST.

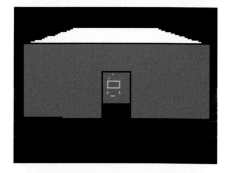

REGARDE À DROITE.

EST-CE QUE TU PEUX TE RENDRE À CÔTÉ DU GRAND CUBE VIOLET NON JAUNE?

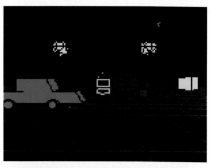

OUI.

ÇA Y EST.

JE VOUDRAIS MAINTENANT QUE TU REVIENNES VERS LA MAISON S'IL TE PLAÎT EN PASSANT SOUS LE DRAPEAU.

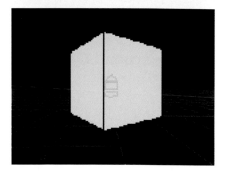

BIEN.

(continued)

Utterances and Movements of the User	*Utterances of the Agent*	*Snapshots Presenting the Movements of the Agent*

	J'AI FINI.	
DIRIGE TOI ENSUITE VERS UNE TABLE.		
	BIEN.	
EST-CE QUE TU VEUX BIEN TE PLACER DERRIÈRE LA CHAISE?		
	OUI.	
	VOILÀ.	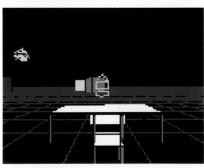
JE VOUDRAIS QUE TU LA CONTOURNES.		
	D'ACCORD.	

Utterances and Movements of the User	*Utterances of the Agent*	*Snapshots Presenting the Movements of the Agent*
	ÇA Y EST.	
EST-CE QUE TU PEUX LA PRENDRE?		
	JE NE SAIS PAS PRENDRE.	

MAIS PEUX TU PRENDRE DE LA HAUTEUR.		
	OUI.	
STOP.		
REDESCENDS DEVANT LA MAISON.		
	IL Y EN A PLUSIEURS.	
CELLE CI. *(L'utilisateur clique on la maison de gauche)*		
	OK.	
	VOILÀ.	
REGARDE À DROITE.		
VA DERRIÈRE UNE VOITURE.		
	OK.	

(continued)

Utterances and Movements of the User	*Utterances of the Agent*	*Snapshots Presenting the Movements of the Agent*
AU REVOIR.		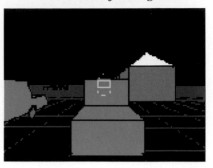
	AU REVOIR CHRISTOPHE.	

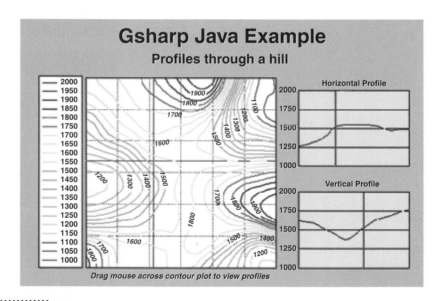

············

Figure 5.2   Example of an "intelligent" visualization client produced with AVS' GSHARP Web Edition. The underlying 2-D graphics were produced with AVS' GSHARP on the server using Java 2-D graphics in an applet, while the profile Java code was transferred to be executed on the Visualization client side.

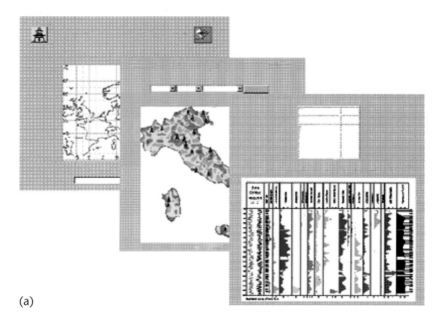

(a)

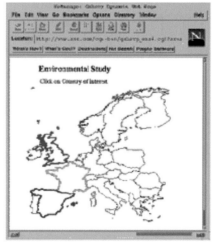

(b)

....

Figure 5.3   (a) Image maps (countries) generated by Gsharp's Web Edition. (b) Information drill-down generated automatically with AVS' Gsharp Web Edition. The user drills down one step at a time. 1. Select country. 2. Select Oil Well 3. Study Well log.

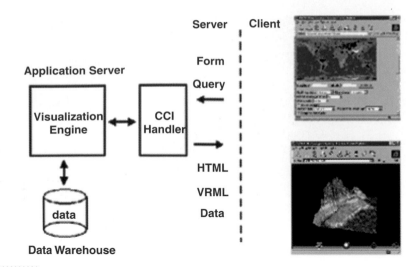

Server   **Client**

Application Server

Visualization Engine ⟷ CCI Handler

Form

Query

HTML

VRML

Data

data

Data Warehouse

Figure 5.4   A diagram of a dynamic Web VRML application.

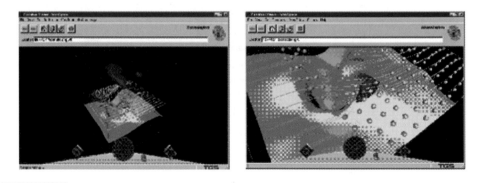

Figure 5.5   Example of a guided tour in the WebSpace Browser.

Figure 5.6   Interactive 3-D information visualization on the Web using the WWWAnchor
Node:"http://www.tel.com/call.htm"
   description
   "AREA CODE: 617; Day of Month: 29; Avg Call; 67; NUM ....."

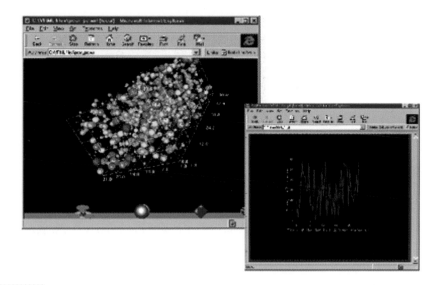

....

Figure 5.7 Example of "Visual Data Mining" on the Web with a large volume of telecommunication data stored in an Oracle database. Dynamic queries are formulated through direct manipulation of graphical widgets, such as icons, buttons, and sliders. The result of the query is a 3-D glyph display, representing selected customers in a market search. A 3-D VRML browser allows the user to walk through the data space in real time. The user can view attached abstract information by selecting objects, "bubbles" of special interest. The scene also highlights the utility of 3-D hyperlink facility in VRML, since every glyph is hyperlinked to some metadata.

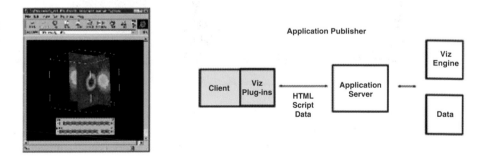

....

Figure 5.8 The Application Plug-in scenario. The mapping of data into geometry and rendering is performed at the Client side. The user can interactively manipulate the data. A script language can be transferred together with the data to set up the appropriate visualization method. This special visualization Plug-in performs "data slicing" through a volume dataset.

Explore 3-D Virtual Worlds!

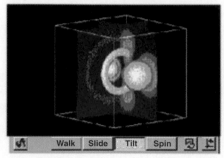

....

Figure 5.9 VRML browser embedded in a Word document as an ActiveX component. The VRML file was produced with AVS/Express advanced 3-D visualization software.

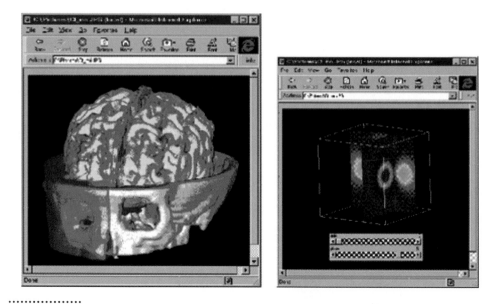

....

Figure 5.10 Example of two Application Plug-ins prototyped by AVS/Express to perform Volume Rendering and Data Slicing at the Client side. Compared to general purpose VRML and http, the Application Plug-ins allow more sophisticated interaction between the client application (the Web browser plus plug-in) and the visualization server. It supports direct manipulation of both data and the visualization parameters. For example, dynamic specification of data/color mapping is accomplished using specialized interaction tools.

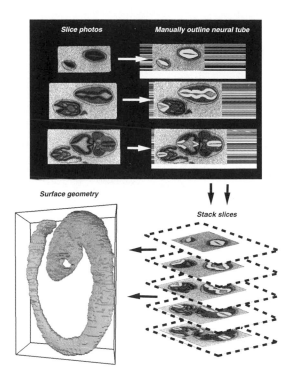

. . . .
Figure 6.3 Stages in Conceptualization of the mouse embryo.

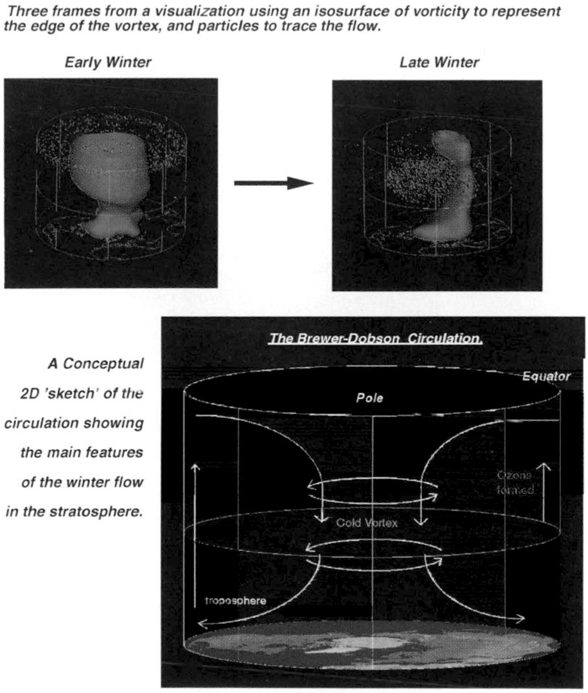

. . . .
Figure 6.4 Visualization of Stratospheric Circulation, and a Conceptual 'sketch' of the circulation.

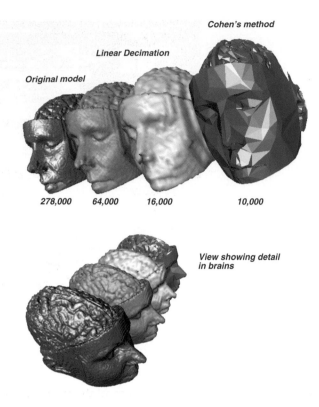

Figure 6.6   Decimation of the MRI surface data.

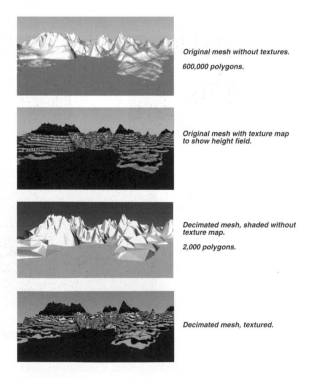

Figure 6.7   Mesh decimation of height fields :-The Isle of Skye.

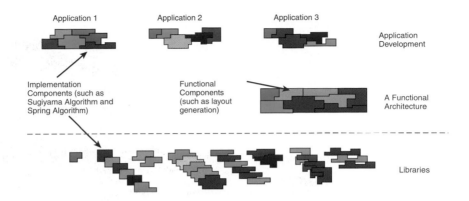

....
Figure 7.3   An object-oriented framework for information visualization.

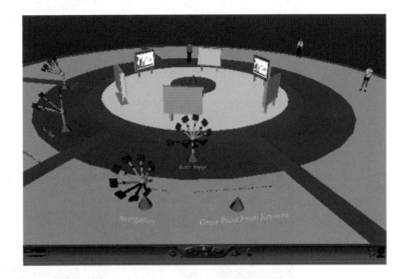

....
Figure 8.1   Growing Information Trees.

....
Figure 8.2   Identity World (from *The Mirror*).

Figure 8.3 An Ecologically responsive environment.

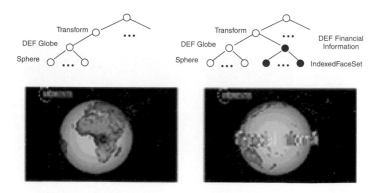

Figure 9.2 The MBone tools in use on a workstation.

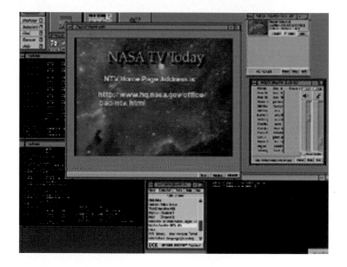

Figure 10.3 Example of an animated presentation of the service currently selected by the user.

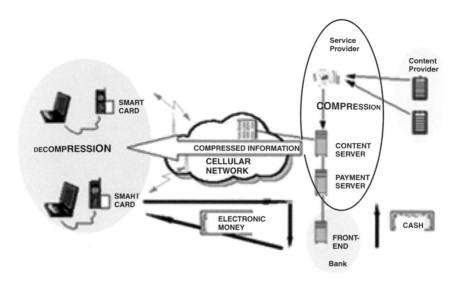

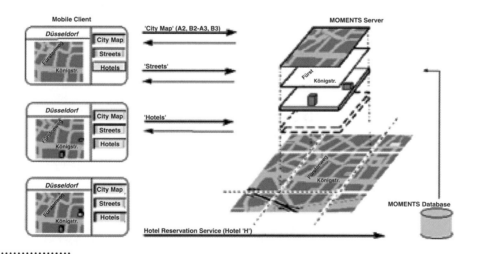

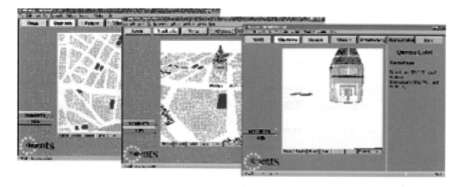

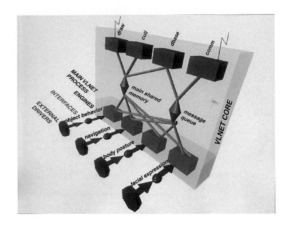

Figure 11.1 Client architecture of the VLNET system and its interface with external processes.

Figure 11.3 Video texturing of the face.

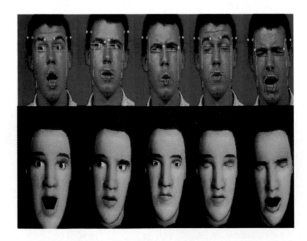

Figure 11.4 Model-based coding of the face: original and synthetic face.

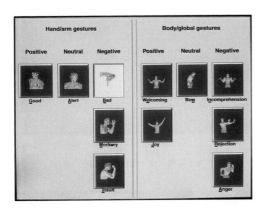
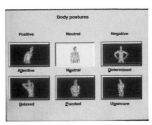

Figure 11.6   Basic set of gestures and postures.

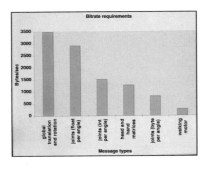
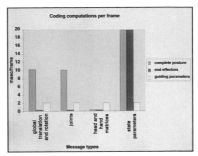
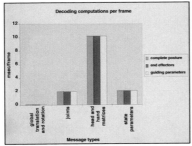
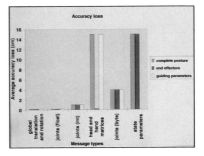

Figure 11.7   Networking body data. (*a*) Bitrate requirements for each message type. (*b*) Coding computations for each body posture. (*c*) Decoding computations for each body posture. (*d*) Accuracy loss with respect to original input data.

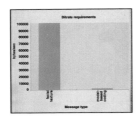
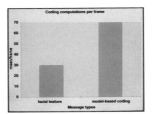
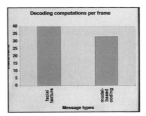

Figure 11.8   Networking face data. (*a*) Bitrate requirements for each message type. (*b*) Coding computations for face data. (*c*) Decoding computations for face data.

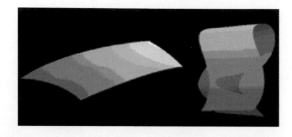

Figure 12.4 Self-collision only occurs in curved surfaces.

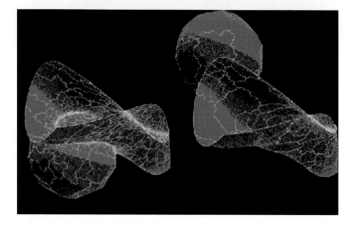

Figure 12.5 Hierarchical detection at work for collisions and self-collisions. The considered hierarchy domains are displayed for the current geometrical configuration.

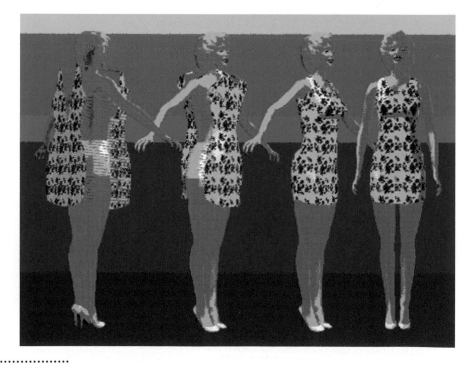

Figure 12.7 Dressing a virtual actor.

Figure 12.8   Cloth animation on a virtual actor.

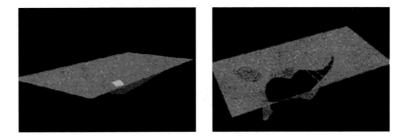

Figure 12.9   Interactive cloth manipulation: Interaction, cutting and seaming.

Figure 13.1   Boolean combination of warped and blended cylinders and blended spheres.

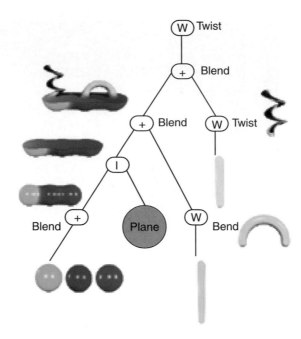

Figure 13.2 The Blob Tree for Figure 13.3.

....
Figure 13.3 Implicit surface model including boolean operations and warping.

....
Figure 13.5 Stamingo designed with SoftEd.

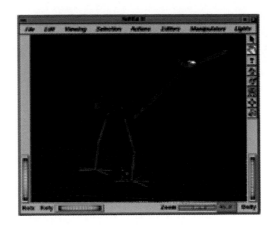

Figure 13.6 3-D view of the Stamingo.

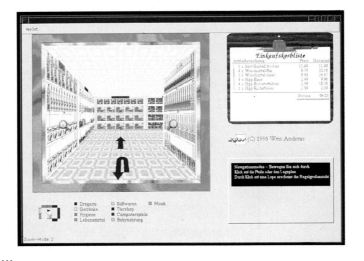

Figure 14.1 Automatically generated, Java-based supermarket with 3-D appearance (Bauer 1996).

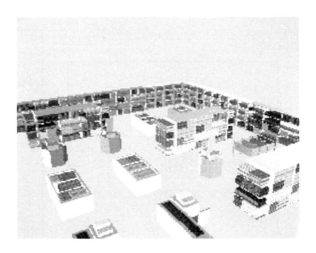

Figure 14.2 Manually produced prototype of a real 3-D supermarket.

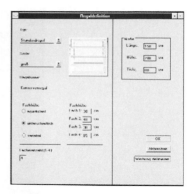

Figure 14.3 User interfaces for the definition of product textures (left) and shelves (right).

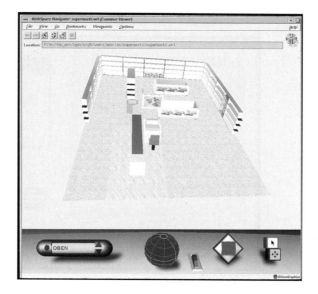

Figure 14.5 A first result of the VRML 1.0 generator. (The shelves are only partially filled for performance reasons.)

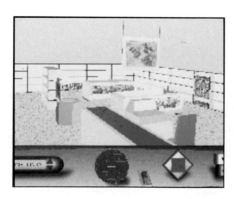
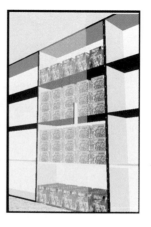

Figure 14.7 Two different views of the supermarket. Please note the open glass door of the fridge in the picture on the right.

····················
Figure 15.3 The mini-pipe in use.

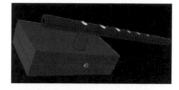

····················
Figure 15.4 VRML 2.0 model of mini-pipe.

····················
Figure 15.5 VRML 2.0 model of mini-pipe.

····················
Figure 15.6 VRML 2.0 model of pp3 9V battery.

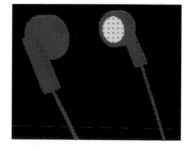

····················
Figure 15.7 VRML 2.0 model of earphones.

Figure 15.8 Close-up of finger pad on chanter.

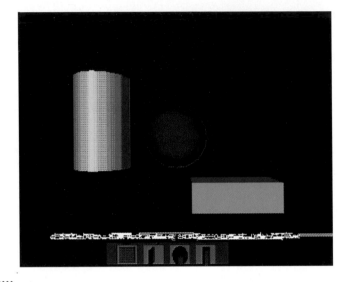

Figure 15.13 Object manipulation interface, mouse over first icon.

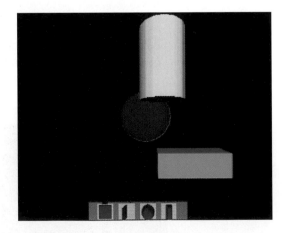

Figure 15.14 Object manipulation interface, cylinder moved.

Figure 16.1　The flow of a circulation folder in a ministry.

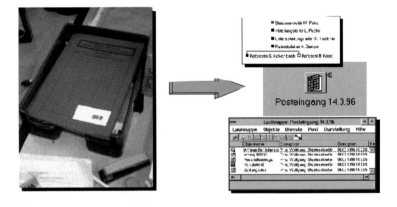

Figure 16.2　Real and electronic circulation folders.

Figure 16.4　An initial virtual POLITeam office.

Figure 17.3 View of doorway showing dimension display.

Figure 17.5 General view of model showing A-frames.

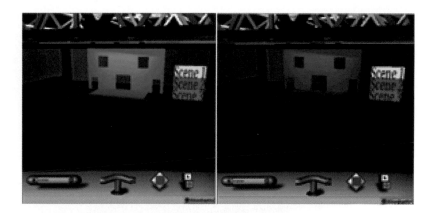

Figure 17.6 Different lighting moods.

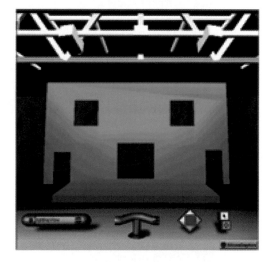

Figure 17.7 View of stage area with three lights.

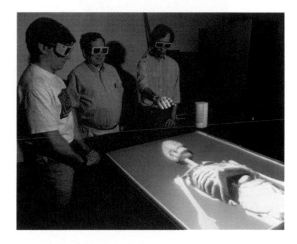

Figure 19.2 A medical application of the Workbench. Both gesture recognition and speech recognition are used to interact with the virtual patient.

Figure 19.3   NRL's VR lab personnel study battlespace terrain for an ONR-sponsored Command and Control research project.

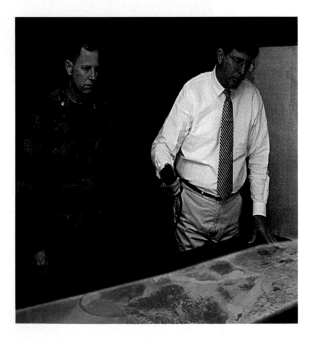

Figure 19.4   Special Purpose Marine Air Ground Task Force (Experimental) Commanding Officer Col. Tom O'Leary receives training on the Workbench at Enhanced Combat Operations Center in preparation for the Hunter Warrior exercise for Sea Dragon.

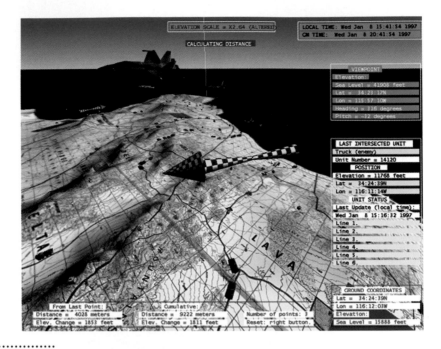

....
Figure 19.5   Icons are overlaid atop the terrain along with textural information. The image seen appears in 3-D on the Workbench.

....
Figure 19.6   An example of the high resolution obtainable with clip texture mapping techniques.

The Cycle of Cognition

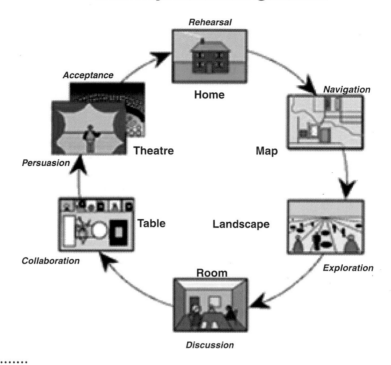

Figure 20.1 The cycle of cognition.

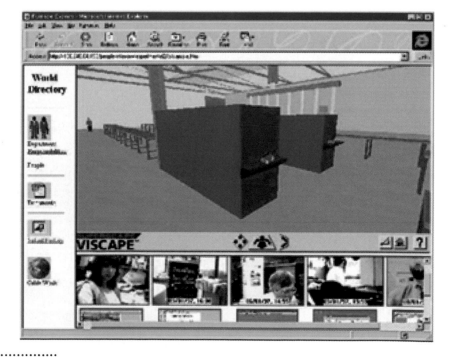

Figure 20.2 The reference factory landscape.

Index

COMPUTER SOCIETY

Press Activities Board

Learning Resources
Centre